PREHISTORIC LIFE

PREHISTORIC LIFE

By

Dr. P.R. Yadav

Lecturer

Department of Zoology

D.A.V. College

Muzaffarnagar (U.P.)

(India)

DISCOVERY PUBLISHING HOUSE

NEW DELHI-110002

First Published-2004

ISBN 81-7141-778-7

Published by

DISCOVERY PUBLISHING HOUSE
4831/24, Ansari Road, Prahlad Street,
Darya Ganj, New Delhi-110002 (India)
Phone: 23279245 • Fax: 91-11-23253475
E-mail:dphtemp@indiatimes.com

Printed at:

Tarun Offset Printers, Delhi-53

Preface

The present title "Prehistoric Life" deals with geological periods, plaeontological localities, plants, invertebrates, fishes, amphibians, reptiles, birds, mammals, the rise of man and morphological topics. The Prehistoric Life describes the earliest forms of plants and animals, most of which have become extinct, and traces the evolutionary development of the creatures whose descendants have survived until the present day.

The present title has been carefully organized and clearly written for the undergraduate students however, it would also supplement the students of postgraduate classes. It has been written in a simple language and in a straight forward manner. Our approach throughout is critical and evaluative, neither old nor new theories are taken trust but are examined rigorously and often critical, the students should be aware of what material is sound and what untested or controversial. Those principles which survive such analysis have put together in a new and current synthesis of the "state of the art."

In the preparation of such a title I have obviously had a great deal of help from a great many people. I would like to take this opportunity to thank many friends and colleagues with whom I have thrashed out the ideas. All of them are most warmly acknowledged.

I have freely consulted a large number of foreign books and journals in the preparation of the present title and also to make it comprehensive and uptodate and to that extent it is not claimed to be our original work.

Any worthwhile criticism and suggestions for improvement would be thankfully acknowledged.

The author expresses his gratitute to Mr. Wasan and staff of M/s Discovery Publishing House for their whole hearted co-operation in the publication of this book.

Author

Contents

1

UNIVERSE

How Did the Universe begin? How will it end? Did it, in fact, have a beginning and will it have an end? What does it contain? What is our place – both in space and in time – in the Universe? These are questions humans have been asking in one way or another for millenia. They have asked them because the answers offer a perspective on how we – humans and the rest of Earth-bound life – fit into the larger scheme of things. They give meaning and content to our existence.

The answers found over the ages are expressed in art, myths, and religions. They have become part of our cultural heritage. In this and the chapters to come, we, too, shall seek answers to the above questions. We shall do this by turning to modern science, which is the product of our own culture. In the present chapter, we shall turn to *cosmology, the study of the large-scale structure and evolution of the Universe. Large-scale structure* means that this chapter is concerned with the cosmic distribution of matter, in which clusters of galaxies measuring many millions of light-years across are the basic units. Details on a finer scale, such as individual galaxies, stars, planets, and life are left for other branches of science and will be discussed in later chapters. *Evolution of the Universe* means that this chapter will try to explain how the distribution of the matter changes with time. In particular, we shall seek answers in this chapter to questions about the origin of the Universe and about its eventual fate.

In our study we shall find that our galaxy does not occupy any special place in the Universe – it is just one billions of galaxies.

The natural laws that operate in our part of the Universe and that have made life on Earth possible are, to the best of our knowledge, the same everywhere, and they probably have been so almost since the beginning. Our bodies are made of the same kind of matter and are subject to the same forces as the stars, the galaxies, and the Universe at large. In short, we believe that the laws of nature are universal and that they apply equally to the very near and very far, the small and large animate and inanimate.

Structure of the Universe

To begin the study of cosmology, this section describes the large-scale distribution of matter in the Universe as it is seen today. How the matter originated and evolved will be discussed in later sections. Let us start by taking an imaginary journey from Earth to the distant galaxies and see what we find along the way. We will pretend that we travel in a spaceship that moves at the speed of light. Travelling at that speed allows us to tell how far we have come just by looking at a clock. Every second of time we will cover a distance of one light-second, which equals 3000,000 km or seven and a half trips around the Earth. Every year we will cover a distance of one light-year (abbreviated LY), which equals 9.5×10^{12} km or 32,000 round trips from the Earth to the Sun. This imaginary journey will last billions of years of time and it will take us across billions of light-years of space.

Milky Way Galaxy

Let us begin the journey. Within four minutes of lift-off from Earth we are already crossing the orbit of Mars and ten minutes later we are passing through the Asteroid Belt, between Mars and Jupiter. Soon the giant outer planets–Jupiter, Saturn, Uranus, and Neptune–briefly fill our viewing window. And in just a little over five hours we have left all of the sun's nine planets behind. During the next few weeks and months we sweep through the outer fringes of the Solar System, coming upon an occasional comet and other debris left over from the Sun's birth. All along, the Sun has been growing fainter and is now just one of many bright stars in the sky.

It will be four years before we encounter the first stars beyond the Sun. They are a triple star system–Proxima Centauri and Alpha Centauri A and B. Alpha Centauri A resembles the Sun in size and temperature, while the other two are somewhat smaller and cooler. Every few years we pass another star. They brighten up as we approach them and then fade in the distance. After just 50 years we lose sight of our Sun. It has become too faint to be seen with the unaided eye.

The centuries and millenia of our journey tick away. We are passing thousands upon thousands of stars. Many are alone, others are double and triple star systems, and still others are grouped in clusters. Some of the stars are very young and are still surrounded by the clouds of gas and dust from which they were born. Others are old and nearing the end of their active lives. Occasionally we pass a star that pulsates or one that explodes, spewing gas in all directions and momentarily blinding us with its brilliance. Eventually the concentration of stars lessens and we break out of the obscuring layer of gas and dust associated with them. Looking back we see an enormous number of stars, star clusters, and clouds, arranged in a huge but very flat disk-like distribution. At its center the disk thickens into a very prominent and bright spherical bulge, like the hub of a giant wheel. From the bulge a distinct spiral pattern of luminous, dark matter winds outward to the edge of the disk. The luminous matter consists of clouds of gas brightly lit by massive stars, while the interspersed dark patches are cold, light-absorbing interstellar dust.

What we are seeing is the Milky Way galaxy –our home in the vastness of intergalactic space. It is a *spiral galaxy* like many others that we shall encounter on our journey. It measures about 100,000 light-years across and contains close to 150 billion stars. Apart from its central bulge or nucleus, its most distinctive feature is the pattern of spiral arms. In fact, our journey began from an inconspicuous spot at the inner edge of one of them, roughly 30,000 light-years from the center of the Galaxy.

Local Group

As we leave our galaxy, we still pass an occasional star or star cluster, whose motions carry them far above the Galaxy's disk. But most of the Milky Way lies behind us. We are entering intergalactic space.

Ahead of us lies the Universe filled with many billions of galaxies. Each is like an island universe, separated from the others by many light-years of emptiness. Many of the galaxies are spirals. Others have spherical or elliptical structures and are called *elliptical galaxies*. Still others look highly asymmetric and distorted; they are known as *irregular galaxies*. The galaxies are not distributed randomly through space, but are grouped together in small and large clusters. The small clusters contain up to a few dozen galaxies, while the large clusters have thousands of them.

The Milky Way belongs to a small cluster of galaxies, known as the Local Group, that contains some thirty members. The two nearest members are the Large and Small Magellanic Clouds, and it is in their direction that we are now steering our ship. Both clouds are composed of young blue stars, bright gaseous nebulae, and dark dust lanes, the same matter that dominates the spiral arms in our galaxy. However, that matter is not distributed in well-defined spiral arms. Powerful forces seem to distort it into rather disorganized shapes. In fact, the Small Magellanic Cloud appears to have been torn into two misshapen components, perhaps by a collision with the Large Magellanic Cloud. Furthermore, an extended stream of hydrogen gas trails behind the clouds, as if it had been pulled out by the forces of such a collision and by tidal interactions with the Milky Way. Because of the clouds' disorganized shapes, astronomers have classified the Magellanic Clouds as irregular galaxies. From Earth they are visible only from the Southern

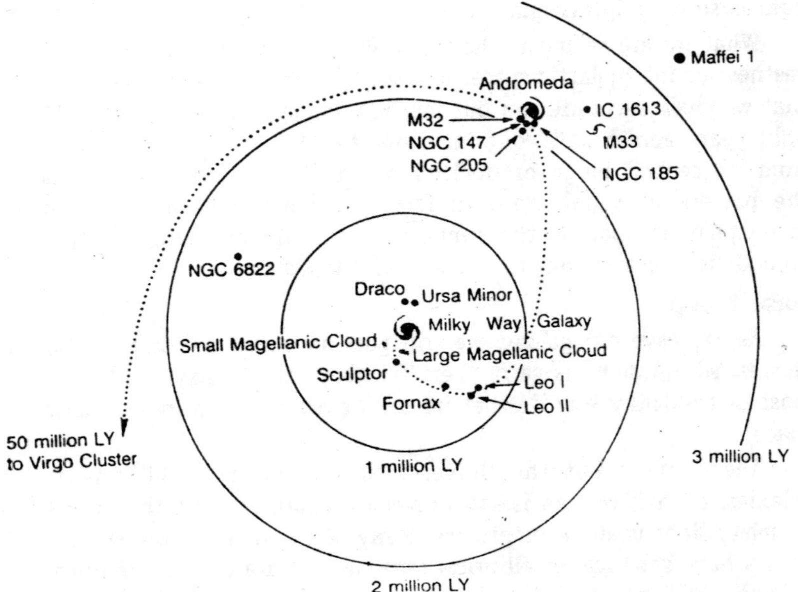

Fig. 1.1. Flight path of the cosmic journey (dotted line). The diagram illustrates the layout of the Local group of glaxies in two dimensions.

Hemisphere and Europeans did not become aware of them until they began their worldwide explorations in the sixteenth century. The two clouds were named in honor of the Portuguese sea captain Ferdinand Magellan (1480-1521), who led the first circumnavigation of the Earth.

Beyond the Magellanic Clouds and after roughly 750 thousand years we pass two dwarf elliptical galaxies – Leo I and Leo II. They impress us with their near perfect symmetry and uniformity. Gas and dust seem to be absent in them, and their light comes entirely from faint, old stars. As our journey continues, we will find that by far the largest number of galaxies in the Universe are dwarf ellipticals.

We are now leaving the Milky Way's immediate neighbourhood and are changing course toward the most spectacular galaxy in the Local Group, the great spiralgalaxy in Andromeda. We pass this galaxy in about 2 million years. From Earth it appears as a small and faint haze of light when viewed with the unaided eye. However, from close up we are able to see many details that remind us of the Milky Way: an almost spherical and very bright nucleus, a great disk of stars, gas, and dust, and an impressive pattern of spiral arms. Like the Milky Way, it also has two nearby neighbours, the small ellilptical galaxies M32 and NGC 205.

Virgo Cluster

After we have inspected at close range the Andromeda galaxy, M32, NGC 205, and several other galaxies in the same vicinity, we change course once again and leave the Local Group. Our next destination is the nearest large cluster of galaxies, the Virgo cluster. This cluster is still some 50 million light-years distant, and along the way we come upon several other smaller clusters similar to the Local Group. Each of them contains one or two large galaxies about which other smaller ones are gathered.

One of the first galaxies we pass is NGC 3115, a highly flattened elliptical. It closely resembles a spiral galaxy, except that gas and dust and, therefore, arms are absent. After roughly 16 million years we fly past a very unusual galaxy, Centaurus A. It has the speherical shape of an elliptical galaxy, but it also contains a disk of gas and dust. From Earth we see the disk edge-on, making the dark, light-absorbing dust contrast sharply against the bright stellar component. A number of violent explosions have occurred in the recent past in the nucleus of Centaurus A, ejecting jets of gas at

Fig. 1.2. Centaurus A (NGC 5128). This unusual galaxy lies at a distance of about 16 million LY, and is the nearest of the giant radio galaxies.

up to 5% of the speed of light. A pair of these jets, tossed out in opposite directions roughly 30 million years ago, are powerful sources of radio emission. Centaurus A's nucleus is also a strong and variable emitter of X rays, indicating that violent events are still taking place today. Astronomers suspect that the source of the radiation and of the explosions is a gigantic *black hole* at the very center of the galaxy, which periodically tears up passing stars and draws their matter into its bottomless gravitational well.

After about 45 million years the frequency of galaxies increases as we approach the Virgo cluster. This cluster contains thousands of members. They are of all sizes and types-spirals, ellipticals, irregulars-and they are rather randomly scattered about. Near the cluster's center lie several giant elliptical galaxies. They form the gravitational hub of the Virgo cluster, about which the other galaxies trace their orbits.

The largest of the elliptical galaxies near the center of the Virgo cluster is M 87. It is an impressive and remarkably symmetric galaxy, containing close to one hundred times as many stars as the Milky Way. This makes it the most massive galaxy known. The radiation emitted from the main body of M 87 is so intense that few details can be discerned. However, toward the outer fringes of the galaxy, the light gradually fades revealing a great number of star clusters. Each contains hundreds of thousands to millions of stars and resembles the globular star clusters in our galaxy (more about them in the following chapter). Short exposure photographs of the nucleus of M 87 show a jet-like structure, with a faint indication of a second one protruding in the opposite direction. The jets must have resulted from an explosion of enormous magnitude. Is the source of this violence a black hole, like the one believed to form the center of Centaurus A? Many astronomers are leaning toward this interpretation. They base their opinion on evidence that within less than 300 Ly of M 87's very center exists a mass equal to at least 5 billion suns. Only a black hole can contain so much mass in such a small volume and also provide the energy to power the observed events.

It takes us well over 10 million years to thread our way through the heart of the Virgo cluster. Besides the kinds of galaxies already described, we pass some that are colliding or have recently (that means during the past billion or so years) undergone collisions. Many of these galaxies are grotesquely distorted by the gravitational forces they exert on each other. Often they are connected by long, thin bridges of gas and dust that they have torn off each other. Others are surrounded by gaseous and dusty rings. Some spiral galaxies are stripped entirely of their gas and dust by these interactions and have become what are known as *SO galaxies*. The Virgo cluster is a world of great fascination and violence.

As we travel on, the concentration of galaxies eventually begins to thin again and, as earlier, we encounter only now and then a small cluster of galaxies. About 100 million years after leaving the Milky Way their frequency diminishes still further and we find ourselves in mostly empty space. We have left our local cosmic neighbourhood.

Superclusters of Galaxies

Looking back we see that we have come from a huge *supercluster of galaxies*, measuring more than 100 million light-years across. Its center is dominated by the Virgo cluster with its rich concentration of galaxies. Surrounding it like a halo is a sparse distribution of smaller clusters, one of which is the Local Group. These small clusters are weakly tied to the Virgo cluster by gravity and circle in giant orbits around it. The whole system contains millions of galaxies of all sizes and types.

The Virgo supercluster is not alone in the Universe. As our journey continues, we encounter such superclusters of galaxies again and again. The Coma cluster and Cluster CI 0024 + 1654 are just two examples. No matter how far we travel or in which direction, the pattern remains similar. On the average, we cross another supercluster every 100 to 200 million years, although occasionally we see stretched-out aggregates of clusters of galaxies that are two or three times that size. The superclusters of galaxies form the basic structural units of the Universe, and between them exist great voids nearly as vast as the superclusters themselves.

Let us end our journey at this point and summarize what we have discovered: The matter of the Universe is concentrated in large superclusters of galaxies, each measuring, on the average, 100 to 200 million light-years across and containing millions of galaxies. There are differences in the exact number of galaxies, in the individual appearance of these galaxies, and in their spatial arrangement. However, if we disregard such fine-scale details, we find that the superclusters are very much alike. Furthermore, they are rather evenly distributed through space, as the map of nearly a million galaxies seen in the northern sky illustrates. We therefore conclude that on the largest scale the Universe is homogeneous and isotropic (*homogeneous* means that the Universe has the same structure and composition everywhere; *isotropic* means that it looks the same in all directions).

Fourth Dimension

Our imaginary journey to the distant galaxies has left unanswered some compelling questions, such as "where is the center of the Universe?", How far is its edge?" and "where exactly are we located?" We did not find answers to such questions because there aren't any. Such questions are perfectly valid when asked in the context of the three-dimensional Earth environment. However, our local experiences do not necessarily apply to the large-scale structure of the Universe, and frequently they lead us completely astray. In the study of the Universe as a whole, a fourth dimension

– time – must be included along with the three spatial dimensions. The above questions then lose their meaning, for it turns out that in four-dimensional space-time the Universe has neither a center nor an edge. It only has a beginning in time and, possibly, an end.

Expansion of the Universe

Up to 500 years ago, most humans believed in a geocentric Universe, with Earth at the center and the Sun, the planets, and the stars circling around it. The Polish astronomer Nicolaus Copernicus (1473-1543) revolutionized that point of view by placing the Sun at the center of the Universe and making Earth one of the planets. In the following centuries, astronomers gradually realized that the Sun is just one star among many millions in the Milky Way and that it is far from the center of the Galaxy. The prevailing view was that the Milky Way galaxy comprised the entire Universe. However, by the eighteenth century a few scientists and philosophers, such as William Herschel (1738-1822) in England and Immanuel Kant (1742-1804) in Germany, dared to consider that the Universe may extend beyond these limits. They suggested that the many faint nebulae visible through telescopes might be separate galactic systems.

Only in the early part of the twentieth century, with the development of modern reflecting telescopes with large light-gathering mirrors, were astronomers able to estimate accurately the distances to those nebulae. The new telescopes allowed detailed study of individual stars in many nebulae. From the faintness of their light, astronomers calculated that they were at distances of hundreds of thousands to many millions of light-years, which places them far beyond the Milky Way. This proved beyond the shadow of a doubt that these nebulae were separate galaxies and that the Universe was much bigger than anyone had imagined. Despite some initial dispute among astronomers concerning this conclusion, by the mid-1920s it was generally accepted and modern cosmology was born.

Hubble Constant

Among the pioneers who participated in these discoveries was the great American astronomer Edwin P. Hubble (1889-1953). After the true nature of the galaxies had been established, he turned his attention to their motions. He was motivated to do this by a puzzling report by Vesto M. Slipher (1875-1969) of the Lowell Observatory in Arizona that many of the faint nebulae are moving away from us with velocities of hundreds and thousands of kilometers per second (a velocity of 1000 km/sec equals 3.6 million km/hr or 2.2 million mi/hr). That seemed peculiar, for the stars in the Milky Way have velocities much smaller than that and some move away while others move towards us. In contrast, the velocities of nearly all galaxies, with the exception of a few nearby ones, are always directed away from us. Slipher had made his observations during the 1910s, before anyone was certain that these nebulae were really galaxies far beyond the Milky Way. Hence, he did not know what to make of their unusual velocities. However, by the mid-1920s Hubble did know that the galaxies were separate galactic systems and he began a systematic study of the relation between their velocities and their distances.

Working with the new 2.5-meter (100-inch) telescope at the Mount Wilson Observatory in southern California, Hubble and his colleague Milton L. Humason (1891-1972) let the faint light of distant galaxies pass through a spectrograph, an instrument that splits electromagnetic radiation into its spectral components (some

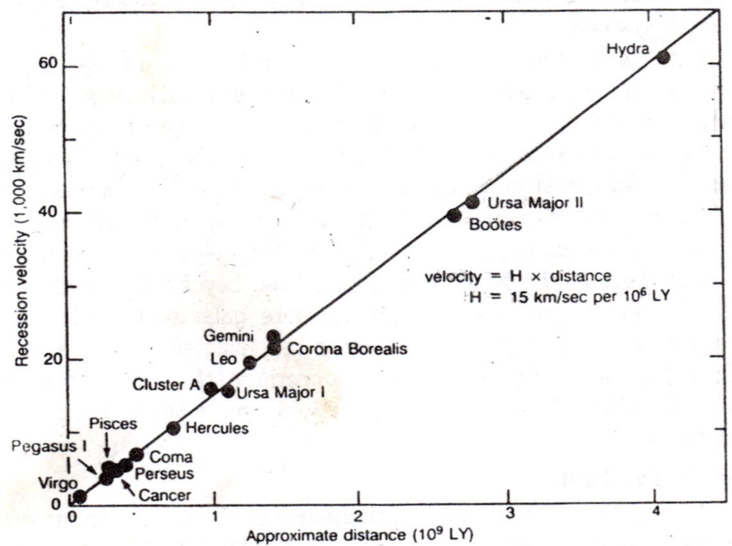

Fig. 1.3. Hubble velocity–distance relation for 15 clusters of galaxies. The alignment of the data along a nearly straight line implies that recession velocity varies regularly with distance. For instance, galaxies at 1 billion LY have a recession velocity of 15,000 km/sec, those at 2 billion LY have a recession velocity of 30,000 km/sec, and so forth.

spectrographs use prisms to accomplish this). In almost every case they found that the light was shifted toward the colour red. This meant that the galaxies are receding from us and confirmed Slipher's earlier result. Furthermore, Hubble and Humason noticed that these recession velocities vary regularly with distances: Galaxies that have two, three, or ten times greater recession velocities than nearby ones are approximately two, three, or ten times farther away. A plot of recession velocity versus distance clearly shows that, on the average, for every increase in distance by 1 million light-years there exists a corresponding increase in velocity by 15 km/sec. Today astronomers refer to the rate of increase of recession velocity with distance as the *Hubble constant* or, more briefly, as H.

Big Bang

The relation between recession velocities of galaxies and their distances has powerful implications for the origin and evolution of the Universe. It suggests that the entire Universe is in a state of expansion. This in turn implies that in the past the Universe was much mor compact than it is today and that, possibly, at some long-ago time it came into existence in one gigantic explosion. The explosion sent matter flying in all directions. Matter that initially acquired large velocities relative to us is consequently very distant from us today, and matter that acquired a small velocity is comparatively still quite close. The situation may be compared with the explosion of a grenade. The energy of the explosion breaks the grenade into many fragments, some of which acquire high velocities and others low velocities. As the fragments fly apart, the ones with high velocities fly farther than those with smaller velocities. This analogy should, however, not be taken too literally, for there exists a definite center in the case of the explosion of the grenade. There is no such center in the case of the Universe.

Observers in other galaxies (if, indeed, there are any) see the expansion of the Universe as we do, except from their vantage point it is our galaxy that is moving away from them. And the more distant we are from them, the faster they see the Milky Way receding. Hence, they share our impression that the Universe was born in a gigantic explosion – the Big Bang.

Age of the Universe : First Estimate

The relation between the velocities and distances of galaxies allows us to estimate the age of the Universe by the following reasoning. From the Hubble constant and also from the graph, we know that, on the average, galaxies at a distance of 100 million light-years move away from us with a velocity of 1500 km/sec, those at 1 billion light-years with 1500 km/sec, and so forth. If, for example, we assume that galaxies at 100 million light-years have been receding from us with a velocity of 1500 km/sec since the Big Bank, we can compute how long it took them to reach that distance:

$$\text{time} = \frac{\text{distance}}{\text{velocity}} = \frac{\text{100 million LY}}{\text{1500 km / sec}} = \text{20 billion years}$$

Thus, we may conclude that the Universe was born approximately 20 billion years ago.

A later section will show that this simple arithmetic leads to overestimating the age of the Universe. Gravity is continuously slowing down the expansion of the Universe; hence, the value of the Hubble constant - that is, the rate of the Universe's expansion - must have been greater in the past than it is today. Consequently, the Universe probably reached its present size in less than 20 billion years.)

Cosmological Principle

Remember that during our imaginary journey through space we discovered that on the average the Universe is *homogeneous* and *isotropic*. This observation (which in reality, of course, was discovered by observations through Earth-based telescopes) plays such a fundamental role in the construction of theories about the origin and evolution of the Universe that it is referred to as the *Cosmological Principle*.

Does the expansion of the Universe not invalidate the Cosmological Principle? Will there not be changes with time that will destroy the homogeneity and isotropy? There will be changes with time, but because the Universe expands uniformly in all directions, those changes will be the same everywhere. A galaxy that today is twice as far away as another will still be that much farther away in a million or billion years because its recession velocity is also twice as great. Therefore, even though the Universe expands and ages, it remains homogeneous and isotropic.

To observe the Universe's homogeneity and isotropy we must compare different parts of it at the same time. When we look at galaxies through telescopes we don't do that because we see them not as they are today but as they were when the light that we pick up with our telescopes left them. The most distant galaxies visible through present-day telescopes are approximately 12 billion light-years away. This means that the light we receive from them was emitted 12 billion years ago and has been in transit every since. It gives us information about those galaxies at that distant time. According to the Cosmological Principle, however, the conditions in our local cosmic neighbourhood 12 billion years ago were, on the average, the same as those that we observe by looking at galaxies 12 billion light-years away.

The delay in transmitting information across intergalactic space is due to the finite speed of light. A similar delay in the transmission of electromagnetic radiation occurs here on Earth, except that it is negligible because the distances are so small. Light or telephone signals take less than 2/100 of a second to travel from the Atlantic to the Pacific coast of the United States. This interval is too short to affect communication. However, when the astronauts landed on the Moon, noticeable pauses existed between the transmission of radio messages from Earth and the reception of answers. They were due to the 2.5 second travel time of radio waves from Earth to the Moon and back.

A message from the nearest star would take approximately four years to reach Earth and from the Andromeda galaxy it would take 2 million years. We do not expect that things would greatly change on Alpha or Proxima Centauri in just four years or in the Andromeda galaxy in 2 million years. However, with the distant galaxies this is not the case. In the course of several billion years and longer, significant evolutionary changes can take place in the Universe. Thus, observing distant objects gives astronomers an important tool for probing the Universe's past.

Cosmic Background Radiation

If we extrapolate the present structure and expansion of the Universe back in time, we find that the galaxies were closer together. Eight billion years ago they were approximately twice as close as they are today and 10 billion years ago they were three times as close. At the time of the Big Bang–well before the galaxies were formed–the matter must have been compressed to exceedingly high densities and heated to enormous temperatures. If the Universe started from such extreme conditions, some remnant of the early, intense heat might still be around today in the form of electromagnetic radiation and perhaps it can be detected.

This was the kind of reasoning that in the 1940s lead George Gamow (1904-1968), a Russian physicist who had emigrated to the United States, to calculate what the cosmic radiation left over from the Big Bang should be like today. He assumed that the radiation, if it existed at all, should be as much part of the Universe as is matter. Hence, it should participate in the expansion of the Universe and all its wavelengths should have become greatly stretched, just as the distances between galaxies became stretched. From the expansion rate of the Universe and its present density of matter, Gamow estimated that the present average wavelength of the cosmic radiation should be about 0.06 centimeters. This places it into the microwave range and gives it a temperature of about 5 degrees Kelvin (for a discussion of temperature scales). The radiation should be very faint, fill the entire Universe, and come to us equally from all directions (very much like white light comes equally from all directions during a whiteout in a snowstorm or heavy fog). Unfortunately, at the time no one took Gamow's bold predictions seriously, and no efforts were made to search for cosmic radiation.

The story continues in 1964 with two scientists of the Bell Telephone Laboratories, Arno Penzias and Robert Wilson. They were observing radio emission from our galaxy using a highly sensitive 20-foot radio antenna built for communication with the *Echo* and *Telstar* satellites. To their dismay, their instrument picked up low intensity microwave noise that resembled the static heard on a radio with poor reception. The noise was very weak and isotropic. At first they suspected that it was produced by faulty equipment, but careful checking showed that this was not the case.

When Penzias and Wilson discussed their problem with other scientists, they learned that Robert Dicke and James Peebles, two theoreticians at Princeton University, might have an explanation for their radio noise. Dicke and Peebles had taken up Gamow's old idea of cosmic background radiation and repeated his calculations. The two teams got together and upon comparing the observations with the theoretical predictions they realized that the microwave noise could only be the relic radiation from the Big Bang. Since then the *cosmic background radiation*, as it is now known, has been observed by radio astronomers around the world at wavelengths from fractions of a centimeter to 50 centimeters. It peaks near one-tenth of a centimeter, which is in the microwave range and corresponds to a temperature of about 3 degrees Kelvin, just a bit lower than Gamow's ingenious prediction 20 years earlier.

In measuring the cosmic background radiation, astronomers have paid much attention to its isotropy. Does it come equally from all directions as the Cosmological Principle predicts? The answer is no. In 1977 several researchers reported that radiation coming approximately from the direction of the Virgo Cluster is slightly blueshifted and radiation coming from the opposite direction is slightly redshifted. Because the two directions are exactly opposite each other and the blue- and redshifts are nearly equal in magnitude, only one conclusion can be drawn: The cosmic background radiation is isotropic, but Earth is traveling relative to the radiation. The speed of travel is about 390 km/sec, which is the combined result of the motion of the Sun within the Milky Way and the motion of the Milky Way through the Universe. If we subtract from this value the Sun's orbital velocity around the center of the Milky Way, we find that our galaxy is falling at approximately 600 km/sec towards the Virgo cluster, the gravitational center of our local supercluster of galaxies.

This result, and the otherwise near-isotropic nature of the cosmic background radiation, suggests that this radiation defines an absolute reference frame for the Universe, relative to which the motions of the galaxies and all other material objects can be measured. The frame is an expanding one, because the radiation is expanding right along with the expansion of the Universe. Any motion relative to the frame is a motion relative to the expansion of the Universe.

It is commonly believed that the superclusters of galaxies have no appreciable velocity relative to the frame defined by the cosmic background radiation. However, individual galaxies move relative to the frame, because each of them has an appreciable velocity within the supercluster to which it belongs.

The discovery of the cosmic background radiation must be ranked among the great achievement of twentieth century science. In 1978 the Swedish Academy recognized this and awarded the Nobel Prize to Penzias and Wilson, commenting that "their discovery made it possible to obtain information about cosmic processes that took place a very long time ago, at the time of the creation of the universe."

Primordial Abundance of the Elements

Since the discovery that the Universe is expanding, the Big Bang theory has had a number of rivals, such as the steady-state theory. However, none of the other theories is able to explain both the expansion of the Universe and the existence of the cosmic background radiation. Nor are they able to explain an additional discovery astronomers made during the past two decades: the discovery of the primordial abundance of the elements.

The primordial abundance of the elements refers to the composition matter acquired when it was created by the Big Bang. Astronomers have estimated this abundance by examining the spectrum of light emitted by the oldest stars in our and other galaxies. They found that out of every 100 atoms, approximately 93 are ^{1}H and seven are ^{4}He. By mass, this amounts to approximately 76% ^{1}H and 24% ^{4}He. (^{4}He has four times the mass of ^{1}H, hence the difference in the percentage values by number and by mass.) Elements heavier than helium are present in only trace amounts. As explained in the final section of this chapter, this ratio of helium to hydrogen in the oldest stars is consistent with the Big Bang model of creation. The Universe started out explosively from a very hot and dense state and quickly cooled as it expanded. The hot and dense conditions lasted long enough for some hydrogen to fuse into helium, but not so long as to allow the production of significant amounts of the heavier elements. They were made much later in the interiors of massive stars.

Cosmological Theory

So far we have been largely concerned with observational information about the Universe: the distribution of galaxies, the expansion of the Universe, the cosmic background radiation, and the primordial abundance of the elements. These observations led to the conclusion that the Universe was created some 20 billion years ago by a gigantic explosion - the Big Bang - and has been expanding ever since. We now face the obvious question: Will the Universe go on expanding forever or will it eventually fall back onto itself? To answer this question, observational information must be merged with theoretical predictions.

The expansion of the Universe may be compared to the flight path of a rocket launched from the surface of a planet. Whether or not the rocket is able to escape the gravitational pull of the planet depends on the planet's mass and radius and on the rocket's velocity at take-off. In the case of the Earth, the rocket's velocity must exceed 11 km/sec (about 40,000 km/hr or 25,000 mi/hr) to escape. If it is less than that, the Earth's gravity is sufficient to reverse the rocket's motion. The rocket will rise to maximum height and then fall back to Earth and crash.

Likewise, the eventual fate of the Universe depends on how much matter it contains, on its present size, and on how fast the matter is expanding. The more matter there is, the greater are the gravitational forces holding the Universe together, and the more likely it is that the Universe is finite and will eventually collapse back on itself. The greater the present size of the Universe, the further are the galaxies and the rest of the matter separated from each other and the weaker is the gravitational attraction. This increases the likelihood that the Universe is infinite and will expand forever. Finally, the greater the velocity of expansion, the harder it is for gravity to slow the expansion to zero and to reverse it. Again, this contributes toward making the Universe infinite and ever-expanding.

The task of this section is to determine the relative contribution of each of these three factors - cosmic mass, size, and expansion velocity - to the evolution of the Universe and then to use them to decide on theoretical grounds what the density of the Universe will be. Before going on, it is important to understand two differences that exist between the launching of a rocket and the expansion of the Universe.

First, in the case of the rocket, there are two separate entities - rocket and Earth. The rocket is flying away from Earth. In the case of the Universe, there is just one entity - the Universe. It is not individual galaxies that fly away from the Universe. It is the entire Universe and all it contains that fly apart.

Second, when calculating the flight path of a rocket it is safe to assume that space is described by the three-dimensional geometry of Euclid (third century B.C.), with time as a separate parameter. Furthermore, the physics of the flight path may be assumed to be the classical one developed by Isaac Newton in the seventeenth century. Neither of these assumptions is appropriate when dealing with the Universe at large. Here space and time must be linked into the four-dimensional geometry of space-time and the physics must be based on the General Theory of Relativity. This theory, developed early in the present century by the German-American physcist Albert Einstein (1879-1955), is considered one of the great intellectual accomplishments of our time.

Einistein's General Theory of Relativity

In Einstein's theory, massive objects warp the geometry of space–time and this warping gives rise to gravitational forces. The greater the warping the greater are the gravitational forces. The Earth and Sun warp space-time only slightly because they are not very massive and they are rather extended. In their vicinity classical physics is a good approximation. However, in the case of the large-scale structure of the Universe or near highly compact objects like neutron stars and black holes, the curvature of space–time is appreciable. Classical physics breaks down and general relativity applies.

General relativity predicts a number of effects that classical physics cannot explain. It predicts that light travels through the Universe and past stars not in straight lines, but along curved

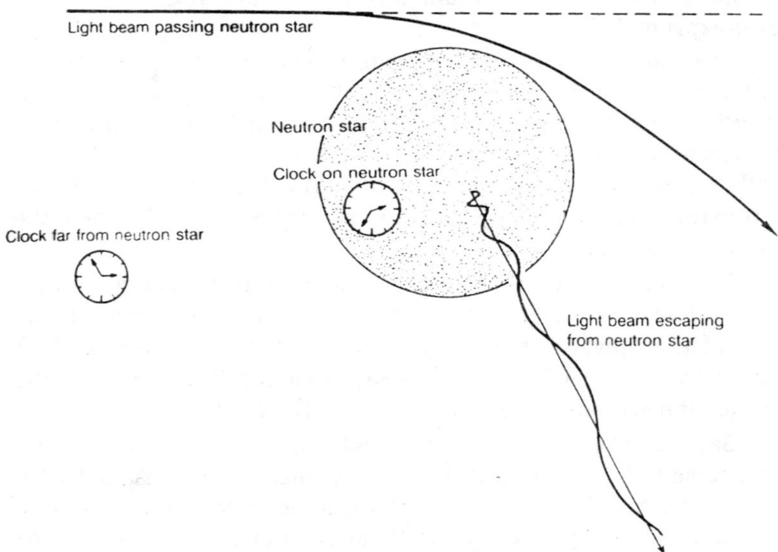

Fig. 1.4. Illustration of three of the predictions of the General Theory of Reltivity: 1. Light travels past a star not in a straight line, but along a curved path. 2. Clocks tick more slowly on a star than in free space. 3. The wavelengths of light emitted from a star become stretched (called the gravitational redshift). According to the General Theory of Relativity, all three effects are caused by the warping of space-time in the vicinity of the star.

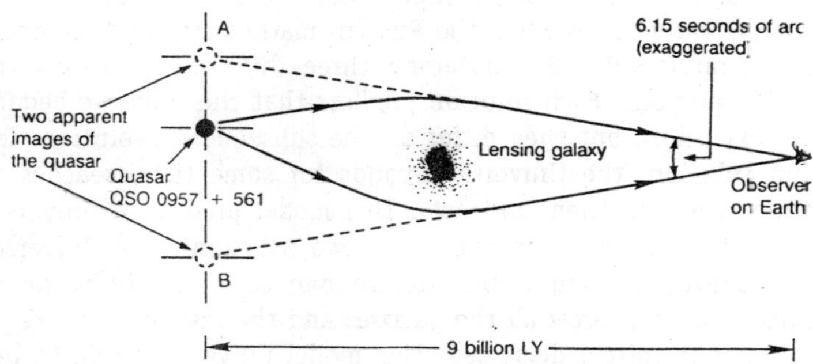

Fig. 1.5. One of the most dramatic confirmations of Einstein's General Theory of Relativity, the gravitational lens effect. The light from a quasar (QSO 0957 + 561) was found to be bent by the gravitational field of a massive intervening galaxy so as to produce two images.

paths; that time does not tick away at equal rates everywhere, but ticks more slowly where gravity is strong; and that light emitted from a star or galaxy loses energy and its wavelengths are stretched to longer values (which means that the wavelengths are shifted toward the red part of the electromagnetic spectrum, an effect that is called the *gravitational redshift*). The greater the warping of space-time, the greater are these effects. Since Einstein first published this theory in 1915, each of these predictions has been observed and verified as occurring in nature.

Perhaps the most dramatic confirmation of the predictions of general relativity has been the discovery in 1979 of the gravitational lens effect. The light from two distant quasars (objects at great distances, characterized by luminous, point-like sources of light) QSO 0957 + 561 A and B, was found to originate from a single quasar and to be bent by the mass of an intervening galaxy so as to produce two separate images 6.15 seconds of arc apart. That the two images (A and B) are, in fact, of the same source is demonstrated by their identical redshifts (velocity of recession = 200,000 km/sec, corresponding to a distance of about 9 billion LY) and identical spectral features. Five other examples of gravitational lensing of distant quasars have been discovered since 1979, and many more are suspected to exist.

Friedmann's Three Models of the Universe

After developing his General Theory of Relativity, Einstein and others applied it to the study of the large-scale structure and evolution of the Universe. The first major success in these efforts came in the early 1920s when the Russian mathematician Alexander A. Friedmann (1888-1925) obtained three different solutions to Einstein's equations. Each solution predicts that the Universe began with an explosion, but they differ on the subsequent evolution. In the first solution, the Universe expands for some time, reaches a maximum size, and then contracts. This model predicts a Universe that is *finite and closed.* In the other two solutions, the Universe expands forever, making it infinite. In one case, the Universe is still expanding even after all the galaxies and the rest of its content have become infinitely dispersed. This model universe is said to be *open*. In the other case, the Universe is just barely able to disperse to infinity, its expansion velocity slowing to zero in the process. This model universe lies at the border between the other two and is called the *flat universe*.

Einstein developed his General Theory of Relativity in order to describe in a consistent manner physical phenomena occurring near stars and other gravitating bodies. It is remarkable that this theory, derived from local phenomena, allowed Friedmann to make predictions about the evolution of the entire Universe, including the prediction of the cosmic expansion several years before Hubble confirmed it by observation. This result attests to the broad applicability of the theory, as well as to the genius of its inventor.

The theoretical predictions of Friedmann are most clearly illustrated by a graph of the scale of the Universe versus time. All three models start from a very compact state – the Big Bang origin of the Universe. For some time thereafter they evolve nearly identically, and their graphs show this. However, as the rate of expansion is slowed by the decelerating effect of gravity, differences appear. The graph of the closed universe reaches a maximum height and then descends. Eventually, the graph dips all the way down as this universe collapses into the compact state of its birth. There may occur a bounce from which the closed universe re-emerges for yet another cycle in an ongoing evolution, though this is not a prediction that follows from a general relativity. The other two models of the Universe expand to infinity, as shown by their open-ended graphs. The graph of the open universe remains steeper and lies above those of the other two. This indicates that its expansion velocity is the largest of the three models and that it is still expanding even after it has become *infinitely* dispersed. The graph

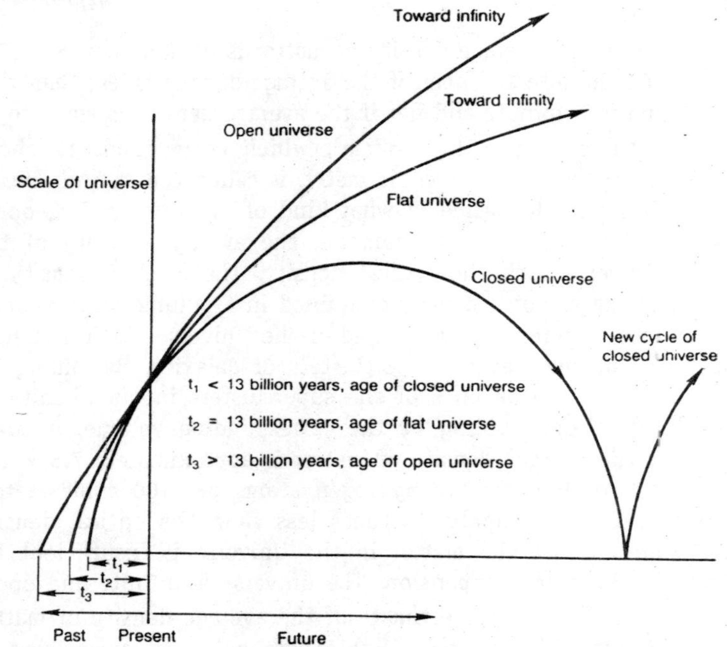

Fig. 1.6. Evolution of Friedmann's three models of the Universe. The distances between clusters of galaxies (labeled Scale of Universe) remain finite in the closed universe, but they become infinite in the flat and open universe. The ages indicated are based on H = 15 km/sec per million light-years.

of the flat universe becomes less and less steep until, after an infinite time, it becomes horizontal while simultaneously reaching infinite height. This means that the flat universe slows more and more until the expansion rate approaches zero while simultaneously dispersing to infinity.

Now we must decide which of Friedmann's three models corresponds to the actual Universe. As stated earlier, this decision needs to be based on observational data and on theory. The observational data pertinent to the problem are the present average density of matter in the Universe and the present rate of cosmic expansion. The theory is Einstein's General Theory of Relativity. It will tell us whether or not the cosmic expansion rate is sufficient to overcome the gravitational pull of the matter.

The present rate of expansion of the Universe is given by the Hubble constant – 15 km/sec per million light-years. For this rate of expansion Einstein's theory predicts that the Universe is finite and closed if the average density of matter is greater than 5×10^{-30} g/cm^3; it is infinite and open if the average density is less than that value; and it is infinite and flat if the average density is equal to it.

The value of 5×10^{-30} g/cm^3, which corresponds to about three hydrogen atoms per cubic meter, is called the *critical density* of the Universe. To decide in what kind of universe we live–open, flat, or closed–we need to compare the average density of the actual Universe with the critical density. The average density of matter is the amount of mass contained in a volume large enough to embrace a representative sample of the Universe. Such a volume must extend over several superclusters of galaxies. By adding up the masses of the galaxies of the superclusters in our vicinity of the Universe and dividing by the corresponding volume, it turns out that the average density of matter is approximately 7.5 × 10-32 g/cm3, or four to five hydrogen atoms per 100 cubic meters of space. This is roughly 70 times less than the critical density and suggests that the matter in the Universe is insufficient for reversing the cosmic expansion. The universe is infinite and open.

Could it be that our estimate of the average density of matter in the Universe is too low? Might there not exist somewhere in the vastness of space matter not yet discovered that is sufficient to close the Universe? For example, there are many more ways for matter to be invisible, or at least hard to see, than there are for it to be visible. Visible matter is either in the form of luminous stars or gas clouds illuminated by stars, but dark or hard-to-see matter may be in such different forms as rocks, planet-like objects, faint burnt-out stars, black holes, dark gaseous halos surrounding galaxies, intergalactic clouds of neutral hydrogen that never condensed into stars or galaxies, or even neutrinos (if they possess mass).

Although dark matter cannot be seen directly its presence can be deduced from its gravitational effect on visible matter. This is done by measuring the velocities of stars and gas clouds within galaxies and of galaxies within galaxy clusters. The velocities tell how much mass (luminous and dark) is present to keep the fast moving objects bound by gravity and prevent them from flying off.

Measurements of stellar motions within galaxies indicate that, depending on the galaxy type, their true masses are anywhere from a factor of 3 to 80 times greater than the masses of the visible matter, with a factor of 6 to 7 being most common. If we include this invisible matter of the galaxies in our estimate, we obtain very approximately a value of 5×10^{-31} g/cm^3 for the average density of the Universe. This value falls still short by a factor of 10 of the critical density and again suggests that the Universe is infinite and open.

Measurements of the velocities of double galaxies around each other and of galaxies within clusters of galaxies indicate that additional matter exists in the space between the galaxies, not accounted for by the galaxies themselves. Without that matter, the clusters would have become dispersed long ago. This is the *missing mass*, whose existence was first suspected nearly 50 years ago by Fritz Zwicky (1898–1974) of the California Institute of Technology. Of course, the mass is not really missing, it is just not visible. As suggested above, this dark mass might be in the form of black holes or intergalactic gas clouds. Some of it could be in the form of small faint galaxies or globular star clusters that have escaped detection over intergalactic distances. Another contribution to the missing mass could arise from the hypercharge force that has recently been suggested by Ephraim Fischbach and his colleagues as a fifth fundamental force of nature. The missing mass might also exist in the form of neutrions, those elusive particles that, according to current theories, have a mass of about 1/100,000 the mass of an electron. There may, in fact, be so many neutrinos, that, despite their tiny mass, they add up to ten times the ordinary mass found in galaxies. If this is correct, the average density of matter in the Universe would be very close to the critical density. That is a very interesting result. It means that the Universe is very nearly flat. However, at present the estimate of the Universe's average density is too uncertain to allow us to be more specific. We do not know whether the Universe is exactly flat, slightly open, or slightly closed.

Age of the Universe: Revised Estimate

The models derived by Friedmann allow us to estimate the age of the Universe more accurately than we did earlier. The estimate of 20 billion years was based on the assumption that the Universe has always been expanding at its current rate of 15 km/sec per million light-years. The steepness of the graphs, however, that this rate was much greater in the past. Hence, it took the Universe less time to reach its present scale than we estimated. If the Universe is flat, its age turns out (assuming H = 15 km/sec per million light-years) to be two-thirds of our earlier result - 13 billion years. If it is open, its age is somewhat greater. And if it is closed, its age is a little less. Because it seems that the Universe is very nearly flat, 13 billion years is the estimate for the age of the Universe assumed in this text.

Geometrical Interpretation of the Model Universes

Much can be learned by considering the geometrical characteristics of Friedmann's three models. To do this we must draw diagrams of their space-time structures. However, it is not possible to draw diagrams of four-dimensional structures. What we can do is "cut through" one of the space dimensions and then draw the remaining three–one time and two spatial dimensions. Presumably, physical events happening along the space dimension deleted by cutting are, on the average, like the events happening along the two space dimensions that remain. Hence, what we learn from three-dimensional space-time diagrams should be correct in principle, if not in detail.

Flat Universe

The simplest geometry is that of the flat universe: it is the Euclidean geometry with which we are familiar from daily life. Cutting through one space dimension results in three-dimensional space-time. The two remaining spatial dimensions are a flat surface that extends to infinity in all directions, and time, which runs perpendicularly to that surface.

Let us focus on a finite square that lies on that flat surface obtained by cutting and see how it and its contents evolve. Dots on the square represent clusters of galaxies. As time goes on, the square moves along the time axis and expands; the dots move farther and farther apart; and after an infinite length of time, all the dots become dispersed to infinity. These changes in the size and internal structure of the square simulate the aging and expansion of the flat universe. If we go back in time, we find that the square started out from a highly compact state, in which all the dots–the galaxy clusters–were crunched together and the Universe was born.

We selected the square quite arbitrarily. There exist infinitely many other squares exactly like it, one next to the other and filing the space of the flat universe. No matter how far we go in any direction, we never come to an end or to an edge. We will

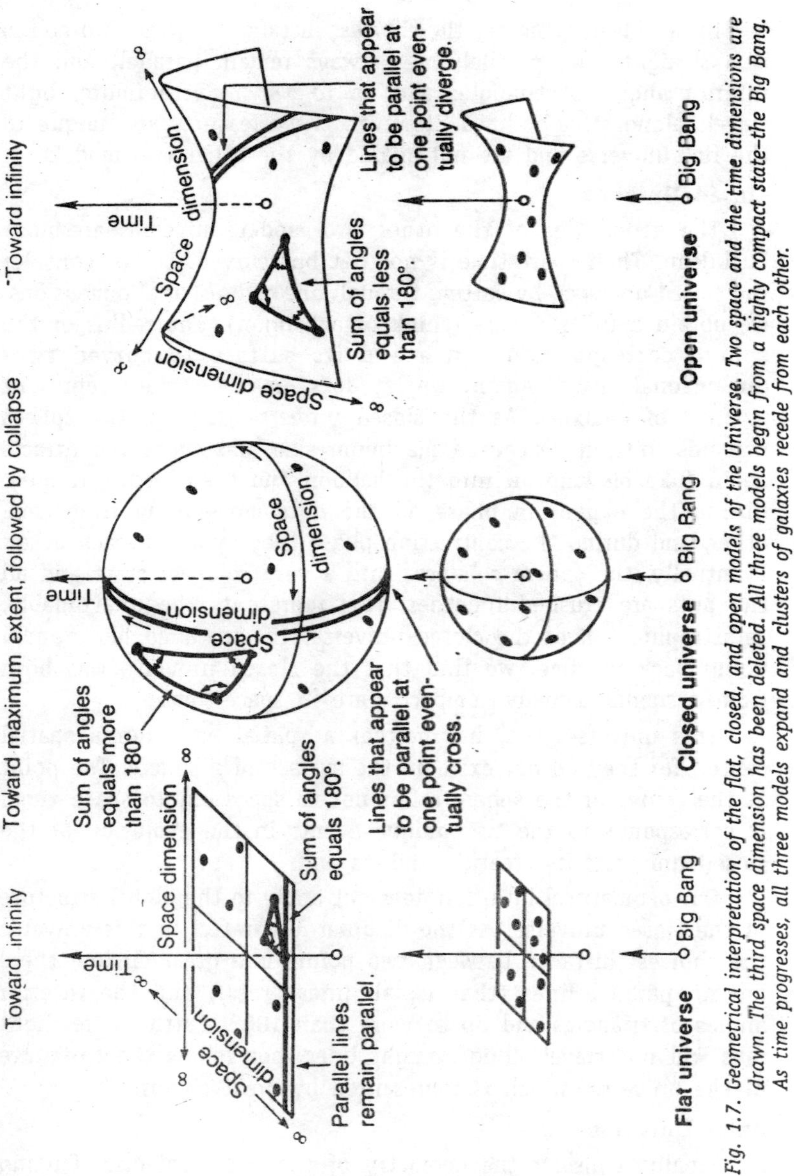

Fig. 1.7. Geometrical interpretation of the flat, closed, and open models of the Universe. Two space and the time dimensions are drawn. The third space dimension has been deleted. All three models begin from a highly compact state–the Big Bang. As time progresses, all three models expand and clusters of galaxies recede from each other.

also never find a center, for that would mean that one square distinguishes itself from all the others. The only edge that we ever find in the flat universe is an edge in time–namely the beginning.

In Euclidean geometry the shortest distance between two points is a straight line, parallel lines always remain parallel, and the interior angles of triangles add up to 180°. Furthermore, light travels along straight lines. However, these features are unique to the flat universe and are not shared by the other two models.

Closed Universe

The geometries of the other two model universes are non-Euclidean. Their space-time is not flat but curved. Let us consider the closed universe. By cutting through one of its spatial dimensions, we obtain a finite sphere (think of a balloon). The radius of the sphere corresponds to time and its surface to curved two-dimensional space. Again, we let dots on the surface represent clusters of galaxies. As the closed universe evolves, the sphere expands until it reaches a maximum size and then it contracts again (like blowing air into the balloon and then letting it out). During the expansion phase all the dots move away from each other, and during the contraction phase they approach each other. Eventually, the sphere collapses into a very compact state and all the dots are crushed together. This represents the unavoidable, catastrophic end of the closed universe, the so-called big crunch. Going back in time, we find that the closed universe was born from a similar, equally compact state in space-time.

This universe, too, has neither a spatial edge nor a spatial center, for they do not exist on the surface of a sphere. The point at the center of the sphere refers not to space but to space–time. It corresponds to the two unique events in the evolution of the closed universe: its creation and its end.

Our geometrical intuition does not apply to the global structure of the closed universe. As the diagram illustrates, in this universe the shortest distance between two points is a great circle,* there are no parallel lines (that is, all lines cross), and the interior angles of triangles add up to more than 180°. Furthermore, light rays will not travel along straight lines, but follow the curvature of the universe, which is represented by great circles.

Open Universe

Finally, consider the geometry of the open universe. Cutting through one of the space dimensions yields a saddle-shaped surface that extends to infinity in all directions. As time marches on, the saddle expands, its curvature lessens, and the dots–representing galaxy clusters–move apart. After an infinite length of time, all the dots will be separated from each other by infinite distances. Going back in time, the section of the saddle shown in the diagram becomes smaller and the dots move closer together. During the Big Bang, they were compressed together into the same highly compact state that we encountered at the beginning of the other model universes.

As in the other universes, neither an edge nor a center exists in the space dimensions of the open universe. No matter how far we go in any direction, we will find galaxy clusters. An edge exists only in space-time, namely at the beginning.

Again, our geometrical intuition fails in the open universe. The shortest distance between two points is a curve (although it is not a great circle), there are no parallel lines (that is, all lines cross or diverge), the interior angles of triangles add up to less than 180°, and beams of light follow curved paths.

In principle, we should be able to tell in which kind of universe we live by observing the geometry of our cosmic environment. Do the interior angles of triangles add up to 180°, or are they greater or smaller? Are there parallel lines, or do all lines cross or diverge? What is the shortest distance between two points? Here on Earth–by drawing lines on paper or surveying the countryside–Euclidean geometry holds as far as we can tell. However, that does not help us decide in which kind of universe we live. Earth extends over such a small portion of the Universe that even if space were curved it could not be distinguished locally from flat space. This is analogous to looking at a microscopic section of the surface of a large balloon or saddle. It, too, cannot be distinguished from a section of a flat surface. Likewise, the section of the Universe that we can observe with precision with present-day telescopes is too small to permit us to discover the nature of the cosmic geometry. It could be flat, spherical, or saddle-shaped. All we can say is that if space is curved, it is not curved very much. Again, as when we compared the average density of the Universe with the critical density, we have failed to discover the Universe's structure and its ultimate fate.

Birth of the Universe

The currently available evidence suggests that the Universe came into existence in one gigantic explosion some 13 billion years ago, as discussed in the preceding sections. During the earliest phase of this explosion, the temperatures and densities were so extreme that neither space and time, nor matter and energy nor the laws of physics were as we know them today. In order to get some idea of what the conditions may have been like then, we need to start from the known conditions of today and extrapolate backward in time. The best presently available theories concerning the behaviour of matter at high energies along with the general theory of relativity shall be our guides.

Such an extrapolation is daring, to say the least. It may even be grossly misleading. Physicists have learned from hard experience that extrapolating beyond solid observational and experimental evidence has generally led them astray. However, at present there are no observational or experimental data of the first instant of creation, and possibly, there never will be. Extrapolation is the only means of such an investigation. Besides, to do so is an exciting intellectual adventure.

Our extrapolation will tell us that the initial extreme conditions passed in rapid succession through a number of transitions, during which the Universe acquired the characteristics that still distinguish it today. A single *superforce*, which ruled in the beginning, split into the four forces we are accustomed to. Matter was spontaneously created and acquired the enormous expansion velocities that we still observe. There occurred a brief burst of primordial nucleosynthesis, during which hydrogen (H) was converted into helium four (^{4}He). One by one, the lightest of the elementary particles-gravitons, neutrinos, and photons-decoupled from the rest of the particles and began to evolve independently. This entire sequence of events, with the exception of photon decoupling, was over in about 5 minutes. The photons decoupled approximately at age 500,000 years and subsequently evolved into today's 3 K cosmic background radiation. From then on, matter (that is, hydrogen and helium) evolved on an independent course as well and began to condense into the lumpy distributions-clusters of galaxies, galaxies, stars, and planets (around at least one star)-that we find in the Universe today.

A more detailed discussion of these events is best divided into two parts; the first 10^{-6} second and the evolution from age 10^{-6} second forward. Only during the first microsecond were conditions so severe that any treatment of them is unqualifiedly speculative and, no doubt, will be subject to major revisions in the future. In contrast, by age 10^{-6} second the content of the Universe and the physical laws governing it are believed to have been the same as they are today. Consequently, discussion of that period is much less in question, though uncertainties do remain.

Theoretical Background

Before discussing the Universe's birth, a few remarks are necessary concerning the theories upon which this description is based. One theory has already been introduced–the General Theory of Relativity–though to be fully applicable to the Universe's first instant of existence Einstein's formulation of relativity needs to be modified. In particular, Einstein's theory needs to be replaced by a quantum theory of gravitation. Unfortunately, no one has yet succeeded in working out such a theory in detail, and this constitutes the greatest source of uncertainty in the description below.

Another theory that we need goes by the poetic name of *quantum chromodynamics* or QCD. According to this theory, protons and neutrons are not really elementary, but are composed of more fundamental particles called *quarks*.

A third theoretical component of the description is the *grand unified theories* or GUTs (plural, because there are many grand unified theories, none of which is at present universally accepted). With GUTs, physicists seek to find relationships between the forces of nature and unify them into one. For example, the theories predict, that at temperatures above approximately 10^{15} K two of the standard forces–the weak and the electromagnetic forces–become unified into the *electro-weak force*. At the same time, electrons and neutrinos lose their identity and behave in identical ways. They are then referred to collectively as *leptons* and they interact by the electro-weak force. Above approximately 10^{27} K, the electro-weak and nuclear forces become unified into the *grand unified force*. Finally, above approximately 10^{32} K, the grand unified force and gravitation are expected to become unified into a single *superforce*.

The concept of a theory that unifies the forces, and simultaneously, describes the fundamental nature of the elementary particles is not new. For example, after completing his General Theory of Relativity, Einstein attempted to unify gravity with the other forces. However, it was not until the late 1960s that real progress in this regard was made. The first to succeed in unifying the electromagnetic and weak forces were Steven Weinberg, Sheldon Lee Glashow (both of the United States) and Abdus Salam (of

Pakistan), and in 1979 they were awarded the Nobel Prize in physics for this work. Then in the 1970s, Howard Georgi (of the United States), Glashow, and others began working out theories (namely GUTs) unifying the electro-weak force with the nuclear force, an effort that continues today. No one has yet succeeded in unifying the grand unified force with gravity, in part because there is no quantum theory of gravitation. Therefore, the earlier reference of the existence of a single superforce above approximately 10^{32} K needs to be accepted with particular caution.

The application of GUTs to the early Universe described below was first made in 1980 by Alan H. Guth of the Massachusetts Institute of Technology and has been developed further by him and others since then. Guth called his theory the *inflationary model* of the Big Bang because it predicts an enormously fast initial expansion of the Universe, during which most of the matter that is still with us today came into existence.

First Microsecond

In the discussion of the birth of the Universe that begins here, the focus is on the sequence of transitions mentioned earlier. These transitions, in a step-by-step fashion, are believed to have marked the evolution from the initial extreme conditions to the more familiar ones of today.

Age 10^{-43} second

Age 10^{-43} second is the earliest point in time about which physics has anything to say. Neither matter and energy nor the laws of physics as we know them today existed then. Space-time was so severely curved that black holes were continually forming and bursting. They were microscopic in size, with diameters 10^{20} times smaller than a proton, and they lasted on the average about 10^{-43} second. Many black holes may have come into existence and then burst; we don't know. Along with the creation and destruction of black holes, space and time were also continually being broken and reconnected. Thus space and time were not continuous. They were "quantized" analogous to the way today's photons and other elementary particles are quantized. Under these conditions even the concept of causality, namely such words as *earlier* and *later*, had little meaning.

At that early instant of time, the temperature was approximately 10^{32} K and the force that governed the black holes was the superforce introduced earlier. This force constituted a unificatin of our present-day gravitational, weak, electromagnetic, and nuclear forces. Sometime between ages 10^{-43} and 10^{-35} second, this era of black hole creation and destruction ended. The black holes that existed then were expanding, the temperature was dropping, and the superforce split into gravity and the grand unified force.

Stop and consider. Space-time was continually being broken and reconnected between ages 10^{-43} and 10^{-35} second. Does reference to these times make much sense when time itself was discontinuous and causality had little meaning? There is at present no clear answer. There may well have existed an infinite stretch of time before 10^{-43} second. Or that point in time may have been preceded by earlier cycles of oscillating universes. Or any number of conditions that, with our limited understanding, we cannot even conceive of may have existed then. We just don't know.

Age 10^{-35} second

By age 10^{-35} second, the temperature had fallen to about 10^{27} K, and the grand unified force separated into the nuclear and electro-weak forces. Within each of the black holes, this transition was accompanied by the spontaneous and rapid creation of quarks and leptons, as well as antiquarks and antileptons. We may compare this creation to the spontaneous formation of raindrops in the present Earth atmosphere, when water vapour cools below the condensation point, though the energies involved were far more extreme. According to GUTs, the quarks, leptons, and their antiparticles came into existence because the vacuum of that time contained energy (unlike today, when a vacuum contains no energy and, hence, cannot produce matter). In the language of physics, the creation of the particles constituted a phase transition from pure energy into matter.

The rapid creation of quarks, leptons, and their antiparticles led to enormous pressures, which forced the black holes to expand at superfast rates. This was the expansion that Alan Guth called *inflation*. Even though each of the black holes, along with the disconnected regions of space-time associated with them – called *domains* – expanded extremely quickly, they moved away from each other still more rapidly. Consequently, they never became connected and, in the course of time, got more and more out of touch. We do not know what became of any of these domains except one. It continued to expand and evolved into the Universe in which we live, as described in this book. The other domains may have become annihilated. Or they may still exist as separate universes, forever inaccessible to us. We shall never know.

Age 10^{-12} second. By about age 10^{-12} second, the temperature had dropped to 10^{15} K and the last of the unified forces divided. The electro-weak force separated into the electromagnetic and weak forces. Thus, all four of the forces that we experience today had become established.

Quarks, leptons, and their antiparticles were now no longer created spontaneously, and the epoch of inflationary expansion ended. Inflation had increased the size of the Universe by roughly a factor of 10^{25} from its embryonic state at age 10^{-35} second. Furthermore, the distribution of the particles throughout the Universe was nearly perfectly homogeneous, a condition still observed today on the largest scale.

Even though inflation had come to an end, the Universe continued to expand simply because of the momentum it had acquired during inflation. However, the expansion was not unopposed. The global attractive force of gravity –that is, the force of gravity exerted by all matter and energy in the Universe – began to slow the expansion. As a result the large-scale evolution of the Universe followed–and still follows today–one of the three curves derived by Alexander Friedmann. As already noted, we do not know which of the three curves it was, and GUTs give us no answer either. We only know that the cosmic expansion rate at that early time must have been extremely delicately balanced against the global gravitational attraction to ensure that the Universe would coast to its present size. Any deviation from this finely tuned balance would have become greatly magnified with time. If the expansion rate had been only a fraction greater, the Universe would be totally dispersed today. Had it been slightly smaller, the Universe would long ago have recollapsed into its initial, highly condensed state. Put differently, our Universe is nearly perfectly flat, as recognized in the previous section.

With the separation of the electro-weak force into the weak and electromagnetic forces, the leptons and antileptons no longer remained single species of particles. They separated into electrons and positrons (electron antiparticles), and neutrinos and antineutrinos. The electrons and positrons interacted by the electromagnetic force, while the neutrinos and antineutrinos interacted by the weak force. No corresponding change occurred in the case of quarks and antiquarks.

In addition to these particles and antiparticles, which are characterized by having mass, there also existed massless particles, namely gravitons and photons. The gravitons had been created during the Universe's rapid inflationary accleration, and they continued to be created now, though at an ever diminishing rate.

The particles that filled the Universe at age 10^{-12} second were in a chaotic state. There was no structure or order. The particles were flying around at high speed in all directions, continually colliding and interacting violently with each other. Upon colliding, the particles and their antiparticles annihilated each other, creating high-energy photons, which, in turn, quickly decayed and changed back into matter and antimatter particle.

The only particles that did not participate in these collisions were the gravitons. They passed right through the other particles without interacting directly. They were decoupled from the rest of the matter and the photons. This happened because they interacted only by the gravitational force, which is the weakest of the four forces. However, despite being decoupled, the gravitons still felt the global gravitational attraction of the rest of the matter and energy in the Universe. Consequently they traveled along curved paths as dictated by the geometry of space-time and then, presumably, they still do so today. Should we ever succeed in detecting them, they would constitute the most ancient direct evidence of the birth of the Universe.

Age 10^{-6} second

At age 10^{-6} second the temperature dropped below 10^{-13} K and the quarks and antiquarks combined into protons and antiprotons as well as neutrons and antineutrons. Simultaneously, the protons and neutrons collided with the antiprotons and antineutrons and became annihilated, producing a huge number of photons. At the existing temperature, the photons were no longer energetic enough to create either protons and antiprotons, neutrons and antineutrons, or quarks and antiquarks. Thus, the back-and-forth reactions between heavy particle-antiparticle pairs and photons came to an end. The heavy particle and their antiparticles annihilated each other without being recreated by the photons. (This was not true of the electron-positron pairs. They continued their back-and-forth reactions with the photons for about another 15 seconds)

Not all of the protons and neutrons were annihilated, however. At age 10^{-35} second, slightly more quarks had been created than antiquarks–approximately one unmatched quark for every billion quark-antiquark pairs. These unmatched quarks combined into protons and neutrons that did not find antiprotons and antineutrons with which to annihilate. Hence, they survived. These excess protons and neutrons (together with excess electrons) constituted the matter that eventually evolved into the stars, galaxies, and clusters of galaxies that mark the Universe today. Had matter and antimatter particles been created in exactly equal numbers, they would have completely destroyed each other and left the Universe in a rather impoverished state, with photons, neutrinos, and gravitons as its only content.

To summarize, the Universe at age 10^{-6} second was in a state of rapid expansion. It contained protons, neutrons, electrons and positrons (both in nearly equal numbers), neutrinos and antineutrinos, photons, and gravitons. The forces governing the interactions among the particles were the nuclear, electromagnetic, weak, and gravitational forces.

All the particles were in continuous chaotic motion, incessantly colliding and interacting with each other. The only exception was the gravitons, which were decoupled from the rest. The electrons and positrons were still continually annihilating each other and creating photons. The photons, in turn, quickly decayed and changed back into electrons and positrons.

None of the collisions produced nuclei of helium or heavier elements, nor did protons and electrons combine to form atoms. At the high prevailing temperatures, the collisions were still too violent, by many orders of magnitude, for that. Had nuclei of heavy elements or atoms formed, they would have been instantaneously broken up into their component particles by further collisions.

After the First Microsecond

Space and time, matter and energy, and the four forces all came into existence during our Universe's first microsecond of existence. Let us now examine what happened next.

Age 1 second

By age 1 second, the temperature had dropped to about 10^{10} K. As well as causing this drop in temperature, the cosmic expansion had reduced the neutrinos' energies and changed their interaction properties such that they ceased to feel the presence of the other particles. Thus, neutrinos and antineutrinos became the second species of particle to decouple from the rest. Like the gravitons, they still were part of the Universe and subject to its global gravitational attraction. Presumably, these primordial neutrinos are still racing through the Universe today, though they have not yet been detected. Should they be detected in the future, they will be the second oldest evidence from the Big Bang. In fact, there may be so many of them that their combined mass closes the Universe, as noted earlier.

Age 15 seconds

As the Universe aged and expanded further, its radiation and matter continued to cool. By age 15 seconds, the temperature had fallen below 3 billion degrees Kelvin and the photons no longer had the energy to produce electron-positron pairs. The electrons and positrons experienced the same fate that had befallen the protons and neutrons earlier. They annihilated each other without being recreated by the photons. The only exception was the small excess number of electrons with which the Universe was born. Thus, the last vestige of primordial antimatter (with the exception of antineutrinos) disappeared. The Universe now consisted of equal number of electrons and protons, of neutrons, neutrinos, antineutrinos, photons, and gravitons. The neutrinos, antineutrinos, and gravitons were decoupled from the rest.

Age 1 minute

The time between ages 1 and roughly 5 minutes was the epoch of primordial nucleosynthesis. During this interval, the temperature dropped from about 1.3 billion to 600 million degrees Kelvin (which brought it into the range of 40 to 80 times the central temperature of the Sun). Collisions between the particles became much less violent than they had been earlier. Consequently, collisions between protons and neutrons produced stable nuclei that were not torn apart again by further collisions. In particular, protons, and neutrons fused to form nuclei of hydrogen-2 ($^2H^+$ also called deuteron), helium-3 ($^3He^{++}$) and helium-4 ($^4He^{++}$).

$$p^+ + n \rightarrow {}^2H^+ + \gamma$$

$$p^+ + p^+ \rightarrow {}^2H^+ + e^+ + \nu$$

$$p^+ + 2H^+ \rightarrow {}^3He^{++} + \gamma$$

$${}^3He^{++} + {}^3He^{++} \rightarrow {}^4He^{++} + 2p^+ + \gamma$$

(The same nuclear reactions, with the exception of the first one, occur today in the interiors of the Sun and of many other stars. In fact, they are the major source of the stars' radiant energy.)

By the time the Universe was five minutes old, the temperature had dropped below 600 million degrees Kelvin and nucleosynthesis came to an end. All the neutrons had become used up in the nuclear reactions. The matter now consisted of roughly 76% hydrogen (^1H), 24% helium-4 (^4He), and traces of deuterium (^2H) and helium-3 (^3He), all in fully ionized form. (Percentages refer to mass.) Virtually no heavier elements had been manufactured. The primordial abundance of the elements remained unchanged until the birth of the first generations of stars hundreds of millions of years later. Only then did the buildup towards the heavier elements–carbon, nitrogen, oxygen, the metals, and the rest - continue.

With continued expansion the temperature kept falling further - to 10 million degrees, 1 million degrees, and lower. Although these temperatures were below the threshold for nuclear reactions they were easily high enough to keep the matter fully ionized. Hydrogen still existed as free protons and free electrons. As soon as a proton captured an electron to form a neutral hydrogen atom, it was torn apart again into a proton and an electron either by a photon or by collision with some other particle. The same was true of helium. It existed in the form of free helium nuclei and free electrons.

All the particles–except the neutrinos and gravitons - were still colliding with each other incessantly, In particular, the photons were unable to travel very far before striking an electron, a proton, or a helium nucleus. This made the Universe opaque. It resembled thick fog, except that it was much hotter. Because of the constant collisions, the photons and the material particles shared their energies of motion; and because temperature is a measure of this energy photons and matter shared a common temperature.

Age 500,000 years

Up to roughly 500,000 years, the state of the Universe was as just described. By then the temperature had fallen to approximately 3000 K. As a consequence, the energy of the photons had become so greatly reduced that they were less and less able to keep the nuclei and electrons from combining. Soon they were totally unable to do so and matter changed into neutral atomic form. This also changed the appearance of matter from opaque to transparent, and the photons did what the gravitons and neutrinos had done earlier. They ceased to interact with the matter and the Universe became transparent, as it still is today.

The decoupling of the photons from the matter is the last transition included in this description of the birth of the Universe. From here on, photons and matter evolved independently, each according to its own physical laws. The matter evolved independently, each according to its own physical laws. The matter evolved into the structures we see all around us today–galaxy clusters, galaxies, stars, planets, and life. The photons evolved very little. Like the gravitons and neutrinos, the photons remained part of the Universe and continued to feel the global gravitational attraction. However, they merely partook in the cosmic expansion. They did not acquire any structure. Their energies were reduced more and more, their wavelengths became longer and longer, and the temperature characterizing them kept falling. Today we observe them as the 3 K cosmic background radiation, a faded remnant from the birth of the Universe.

One final comment is in order. As noted earlier, the 3 K cosmic background radiation observed today is nearly perfectly isotropic. For example, if we measure the intensity of this radiation from one direction and then from the opposite direction, we find that the two results are identical to roughly one part in 10^4. This is rather surprising. The two opposite locations were approximately 10^7 LY apart when the radiation that we measure today was emitted, at age 500,000 years (the age of photon decoupling). In 500,000 years no signal–not even light, the fastest mode of communication–could have traveled 10^7 LY to "instruct" the two places to emit the same intensity background radiation. Hence, these two locations should not have been causally linked. Yet they emitted radiation of nearly identical intensity.

Apparently, the isotropy had been established much earlier, at age 10^{-35} second, when the Universe was much more compact and, despite its young age, all of its parts were in communication and causally linked. During the subsequent inflationary expansion, when its size increased by 10^{25} orders of magnitude, the various parts of the Universe got causally out of touch with each other. However, they all inherited the same conditions. That is why at age 500,000 years the intensity of the cosmic background radiation was virtually the same everywhere and why it still is so today. It is also the reason why the distribution of matter at that time was nearly the same everywhere and why, on the largest scale, the distribution of clusters of galaxies is still homogeneous and isotropic today.

2

Precambrian Period

Since the last century, the interval of earth history that preceded the Phanerozoic Eon has been known as the Precambrian. Although the term Precambrian has no formal status in the geologic time scale, it has traditionally been employed as if it did. The Precambrian includes nearly 90 percent of geologic time, ranging from 4.6 billion years ago, when the earth formed, to the start of the Cambrian period about 4 billion years later.

Two eons are formally recognized within the Precambrian : the Archean and the Proterozoic. The Archean Eon, which is the subject of this chapter, includes about 45 percent of the earth's history-the interval from about 4.6 to 2.5 billion years ago. The Archean Eon was an important span of time during which, as we will learn in this chapter, the earth underwent enormous physical change and life developed on its surface. Many details of Archean history however, remain unknown or poorly understood.

In fact, much less is known of both the Archean and Proterozoic eons than of the succeeding Phanerozoic Eon. One reason for this discrepancy is that, although the precambrian constitutes such a large span of geologic time, precambrian rocks form less than 20 percent of the total area of rocks exposed at the earth's surface. Erosion has destroyed many precambrian rocks, and metamorphism has altered others to the degree that they can no longer be dated and therefore recognized as precambrian. Still other Precambrian rocks lie buried beneath younger sedimentary and volcanic rocks that were laid down upon the surface of the earth. Furthermore, index fossils are seldom found in Precambrian rocks because primitive organisms without durable skeletons predominated until the end of precambrian time. As a result, stratigraphic correlation of precambrian rocks has been based largely on radiometric dating.

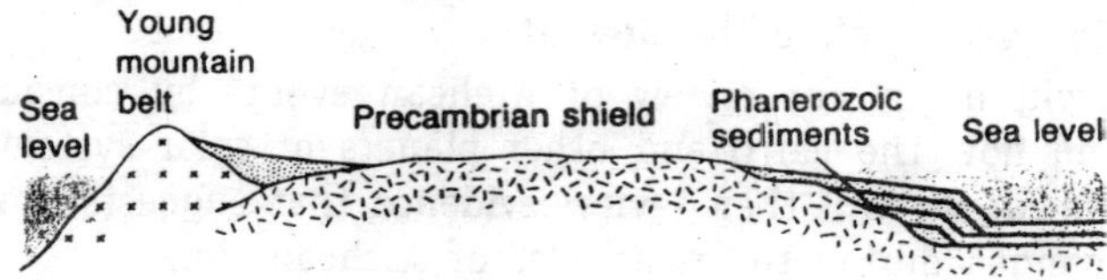

Fig. 2.1. Diagram illustrating how Precambrian shields are often flanked by younger (Phanerozoic) mountain belts and sediments.

Most geologic information about the precambrian has been derived from *cratons,* which are large portions of continents that have not experienced tectonic deformation since precambrian or early Paleozoic time. All the continents of the present world include cratons that consist primarily of precambrian rock. A *Precambrian shield* is a largely precambrian portion of a craton that is exposed at the earth's surface. The largest of these is the vast Canadian Shield, which has recently (geographically speaking) become more fully exposed by the action of pleistocene glaciers. Although shields contain some sedimentary rocks, they consist primarily of crystalline (igneous and metamorphic) rocks. As we will see, mountain belts that formed during precambrian time left traces that can be recognized within precambrian shields, but their elevated topography was destroyed long ago by erosion. Today, the only precambrian rocks that stand at high elevations within mountain ranges are those that have been uplifted by phanerozic orogenies.

The Archean Eon, as previously stated, was a time of major change in the structure of the earth itself, beginning with the events that gave the planet its basic configuration, in which the mantle and crust surround the core Early in Archean time, the earth's crust seems to have differed from what it is today. The earth was much hotter than it is now, and there were apparently no large felsic cratons. Moreover, Plate-tectonic processes may not have operated precisely as they do in the modern world. By the end of Archean time, however, large cratons had begun to form, and plate tectonic processes were modifying these cratons in the same way that they do now. The transition to this modern style of tectonism ushered in the proterozoic interval of precambrian time, which will be discussed in the two chapters that follow. In the absence of useful biostratigraphic data the boundary between the Archean and Proterozoic intervals is defined by its absolute age of 2.5 billion years before the present.

We will begin our review of Archean events by considering when and how the earth and other planets of solar system may have formed. We will then review evidence that suggests how the earth changed during the remainder of Archean time.

Ages of the Universe and the Planet Earth

All seriously considered theories of the origin of our solar system are based on the premise that its planets formed almost simultaneously. Part of the evidence for simultaneous origin is the observation that the planets' orbits around the sun lie in nearly the same plane. Had the planets formed independently, we would expect their orbits to occupy different planes.

Precisely when the planets came into being has been a more difficult issue for scientists to resolve. Astronauts report that the most beautiful object they see from space is the earth, whose surface is partially balnketed by swirling white clouds set against the blue backgroud of extensive oceans. But while the earth's water is aesthetically pleasing and necessary for life, its abundance near the earth's surface makes rapid erosion inevitable. Continuous alteration of the crust by erosion and also by igneous and metamorphic processes make unlikely any discovery of rocks nearly as old as the earth. Thus, geologists have had to look beyond this planet in their efforts to date the origin of the earth. Fortunately, we do have samples of rock that is generally believed to represent the primitive material of the solar system. These samples are *meteorites*—extraterrestrial object that have been captured in the earth's gravitational field and have subsequently crashed into planet.

Some meteorites consist of rocky material and, accordingly are called *stony meteorites*. Others are metallic and have been designated *iron meteorites* even though they contain subordinate amount of elements other than iron. Still others consist of mixtures of rocky and metallic meterial and are thus called *stony-iron meteorites*. Meteorites come in all sizes, from small particles to the small planets known as asteroids; no asteroid, however, has struck the earth during recorded human history. Many meteorites appear to be fragments of larger bodies that have undergone collisions and broken into pieces.

Meteorites have been radiometrically dated by means of several decay systems, including rubidium-strontium, potassium-argon, and uranium-thorium. The fact that the dates thus derives tend to cluster around 4.6 billion years suggests that this is the approximate age of the solar system. After many meteorites had been dated it was gratifying to find that the oldest ages obtained for rocks gathered on the surface of the moon also approximated 4.6 billion years. Ancient rocks can be found on the moon because the lunar surface, unlike that of the earth, has no water and is characterized by only weak tectonic activity.

The universe, however, seems to be much older than the solar system. Stars tend to be clustered into enormous galaxies within the universe, and many of these galaxies have disclike spiral shapes. We see our own galaxy, the Milky Way, edge on as a cloudy band of stars, but if it were viewed from a distant perspective along a line of sight perpendicular to the plane of the spiral arms, it too, would have a spiral shape.

It is now widely believed that the universe is expanding. Evidence of this movement comes from the so-called red shift, a shift toward longer wavelengths in the spectrum of colours that reach us from stars moving away from the earth. This phenomenon is an example of the *Doppler effect,* as is the progressive lowering of a train whistle's pitch, as the locomotive recedes from a person standing near the tracks. Studies of red shifts indicate that nearly all galaxies are moving away from each other at extraordinary

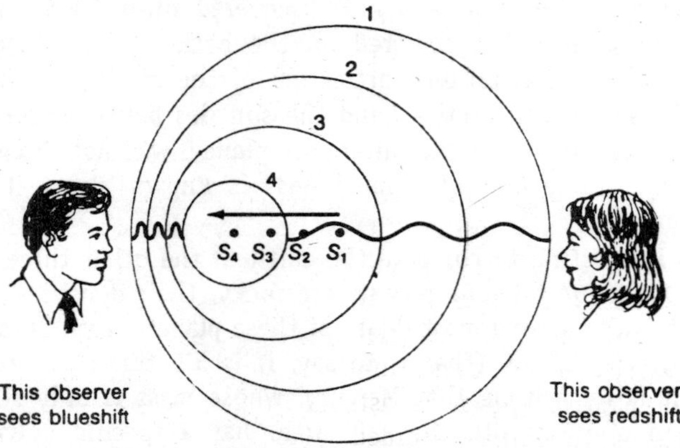

Fig. 2.2. The Droppler effect. The points labelled S_1 to S_4 mark successive positions of the light source.

rates; indeed, calculations have shown that between 10 to 20 billion years ago, all of the galaxies occupied the same spot. This has been taken to indicate the approximate date of the so-called big bang—the primordial explosion of matter from which stars have formed. As we will see, however, the earth and other planets of our solar to system formed long after the big bang took place. In order to understand the origin of the earth, it is important that we know something of our solar system's other planets, which formed at the same time as the earth.

The Planets

The planets of our solar system differ from one another in many respect. The earth, for example, has unique features that allow it to support living creatures ; no other planet can now sustain life as we know it. Like Mercury, Venus, and Mars, the earth is relatively small and close to the sun, All four of these planets are composed of rocky material, but despite their high density, they are so small that they possess little angular momentum (which is the product of mass, velocity, and distance from the center of the sun). The next four planets, which are known as the jovian planets, are jupiter, Saturn, Uranus, and Neptune. These planets consist largely of frozen elements and compounds that exist as gases on the surface of the earth—primarily hydrogen and helium for jupiter and heavier compounds such as ammonia (NH_3) and methane (CH_4) for Uranus and Neptune. Because they have larger masses and lie farther from the sun, jupiter, Saturn, Uranus, and Neptune have more angular momentum than the earth, Mercury, Venus, or Mars. Neptune was not discovered until 1846, shortly after its existence was predicted on the basis of its measurable perturbation, by gravitational attraction of the motion of Uranus. A belt of asteroids in orbit around the sun lies between the inner and outer planets. Pluto the outermost planet, was not discovered until 1930, and even today much less is known about it than other planets.

It is interesting to compare the earth to the other three inner planets. Although all four planets are rocky, their densities differ. The earth is the second most dense of these planets, with an overall specific gravity of 5.5 (that is to say, it is 5.5 times as dense as water). It is remarkable that Mercury, whose mass is only slightly more than onetwentieth that of earth, has a specific gravity of 5.7, making it even denser than the earth. Mercury's small mass

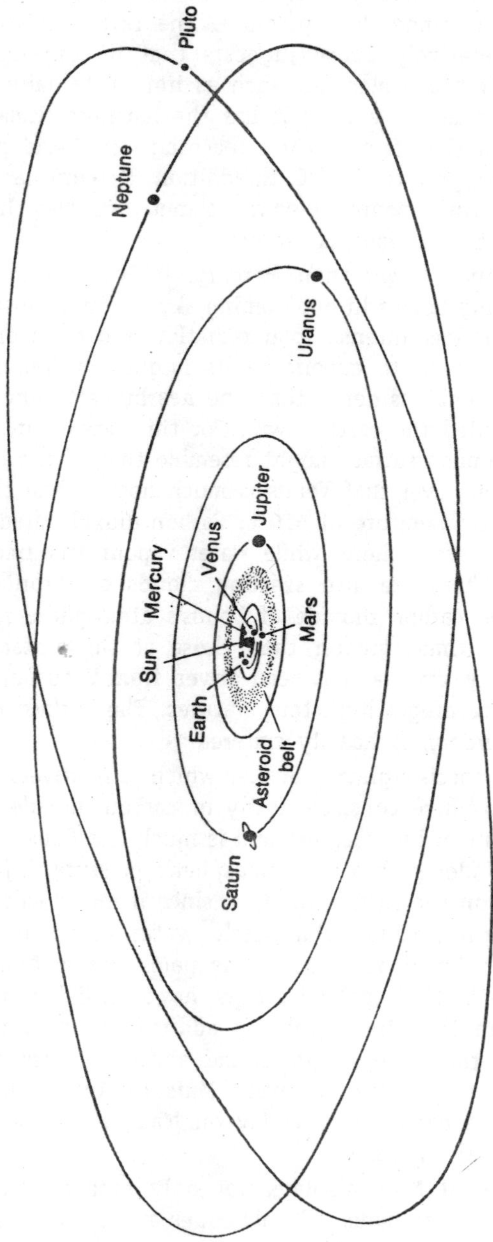

Fig. 2.3. A diagram of the planetary orbits. The fact that the orbits lie in nearly the same plane represents evidence that the planets formed simultaneously from a rotating dust cloud.

indicates that its gravitational force is much weaker, and thus its inner regions are under less pressure. The fact that Mercury is none theless relatively dense suggests that it contains a high concentration of heavy elements such as iron. Externally, Mercury differs greatly from the earth ; it has the harshest climate of all planets, with daytime temperatures reaching 425°C and nighttime temperatures dropping to –175°C. In addition, Mercury has virtually no atmosphere. The topography of the planet resembles the surface of the moon, with myriads of craters.

Venus is much larger than Mercury. It is often one of the brightest heavenly bodies in our evening sky, orbiting close enough to the sun to reflect intense solar radiation and close enough to the earth to allow us to experience its brightness. Venus is only slightly smaller and less dense than the nearby earth and, in fact, is sometimes called the earth's twin. For this reason, it was once believed that Venus's surface might resemble that of the earth. We now recognize however that Venus is much hotter than the earth, often reaching temperature of 450°C. Carbon dioxide predominates in the Venusian atmosphere, while water vapour and free oxygen are rare, and there are also swirling clouds of strongly acidic compounds. The carbon dioxide of Venus's atmosphere is present at pressures 90 times greater than those at the surface of the earth. Apparently, this gas has been driven from Venusian rocks by the planet's exceedingly high temperatures. The surface of Venus, like that of Mercury, is heavily cratered.

Mars has a specific gravity of 3.9, which is relatively low. Like Venus, its atmosphere consists mostly of carbon dioxide, but this atmosphere is under low pressure and is much less dense than that of the earth. Under such a low atmospheric pressure, liquid water cannot survive on the surface of Mars, since if introduced it would quickly evaporate; thus, what little water exists on Mars is primarily locked up on ice caps at the poles, where temperatures are about –125°C. These polar ice caps, however, differ from those of the earth in that they grow by addition of carbon dioxide "snow" during the Martian winter and shrink by loss of carbon dioxide during the Martian summer. Mars rotates at almost the same rate as the earth; thus, a day on Mars is only 41 minutes longer than a day on earth.

The surface of Mars displays not only numerous craters of varying sizes, but enormous volcanic cones as well. It also exhibits channels that look very much like water-formed features on earth, displaying meanders and water appear to be river mouths enlarged in a downstream direction. Although temperatures at the equator approximate what we regard as "room temperature" on earth, the dry conditions that predominate on Mars would appear to be hostile to life as we know it. Still, the presence of abundant frozen water and evidence that it once flowed in a liquid state suggests the possibility that at some time in the past, life may have existed on the so-called red planet. Perhaps the life of Mars will have to be studied by paleontologists rather than by biologists!

Origin of the Solar System and Earth

It is interesting to note that we know more about the origin of distant stars than we do about the origin of our own solar system. This is because our solar system formed long ago, and we have no other examples of young solar systems to observe at close range. Scientists can train their telescopes on multitudes of large stars in various stages of development—but unfortunately, these stars are too far away to allow us to observe how planets may be forming in association with them.

At one time, it was widely believed that planets formed when a star passed close to the sun, tearing away matter that subsequently condensed into planetary bodies. Subsequent calculations have shown that matter removed in this way would either fall back to the sun or fly into outer space. Today, most experts favour some variation on the *solar nebula theory*, which holds that the solar system formed from a cloud of particulate matter called cosmic dust. The details of this theory, however are widely debated.

Galaxies, which are enormous clusters of stars, form by the gravitational collapse of dense clouds of gas (mainly hydrogen) into stars. Our galaxy, which is believed to have originated less than 10 billion years ago, is made up of approximately 250 billion sunlike stars. Even after a galaxy forms by the establishment of some stars, secondary stars are continually born within the spiral arms, where galactic matter is concentrated. Pressure varies in time and space within a spiral arm matter contracts and expands.

Although all stars form in essentially the same manner their histories vary according to their size. The sun, for example, is apparently the remnant of a star that imploded, or collapsed violently, to form heavy elements. After this collapse, a *supernova*—

or an exploding star that casts off matter of low density-was formed. What remained was a dense cloud that condensed as it cooled after the explosion. This dense cloud, or *nebula*, is assumed to have had some rotational motion when it formed and must then have rotated more and more rapidly as it contracted, just as an ice skater automatically spins more rapidly (conserving angular momentum) as he pulls in his arms.

It is possible that the sun formed alone from a nebula and then captured a cloud of cosmic dust that formed the planets. Alternatively, the sun and the planets may have formed from the same dust cloud. What does seem certain, however, is that the planets formed either during or soon after the birth of the sun. In fact, the sun and the planets appear to have originated during an interval of time no longer than about 100 million years. This has been deduced from the excessive concentration in some meteorites of the isotopes xenon 129 and plutonium 244. ("Excessive" in this case means present in an amount greater than is normal for meteorites.) Excessive amounts of these isotopes in a meteorite indicate that some of the short-lived parent isotopes were originally present in the meteorite material and then decayed to xenon 129 and plutonium 224. Thus, the planet or planetlike body of which such a meteorite is a fragment must have formed soon after the elements of the solar system came into being; otherwise, the short-lived parent isotopes of xenon 129 and plutonium 244 could not have been incorporated into the material of meteorites in a sufficiently high concentration to leave excessive amounts of their daughter products. The conclusion is that the sun and the heavy elements that were synthesized in its formation cannot by very much older than the meteorites, the moon and the planets, which have ages of about 4.6 billion years.

Whether it was captured or formed with the sun by the gravitational collapse of galactic matter, however, the dust cloud that became the planets must have been rotating. Today, the angular momentum of the solar system resides less in the sun than in the planets, as evidenced by the fact that the planets are positioned far from the sun and rotate rapidly even though they contain little mass. It is not known precisely why this distribution of angular momentum developed, but it is known why the outer planets formed from volatile (i.e., easily vapourized) elements. It would appear that these elements were expelled from the hot inner region of the nebula to solidify in colder regions far from the sun. More dense materials were left behind, and these formed the inner planets, including the earth. When the rotating dust cloud reached a certain density and rate of rotation, it automatically flattened into a disc, and the material of the disc then segregated into rings, which later condensed into the planets. Each planet began to form by the aggregation of material within one of these rings. The aggregates eventually reached the proportions of asteroids, which subsequently coalesced to form planets.

It would appear that the sun passed through the socalled T Tauri stage that some other stars can be seen to represent today. This stage is an early interval in the history of a star during which matter is emitted at an exceedingly rapid rate. The powerful "solar wind" that characterizes this stage would have driven a large percentage of preexisting hydrogen and helium (light elements) out of the solar system. Obviouosly this did not occur until the outer planets had already trapped the hydrogen and helium that forms so much of their bulk.

After the planets formed, asteroids also remained in orbit around the sun. While some of these asteroids—including those that form the asteroid belt survive to this day, most have been swept up by larger planets. Others have undoubtedly had their motions so severely disturbed by near collisions that they have passed out of the solar system.

The origin of the moon, which circles the earth, is still debated. Was the moon an independent body that was captured by the earth's gravitational field? Was it produced by the separation of a large chunk of the earth? Or was it accreted from matter in orbit around the primitive earth? Lunar rocks obtained in the Apollo space program exhibit an isotopic composition of oxygen that is remarkably similar to that of terrestrial rock. This suggests that the moon formed near the earth from the same portion of the solar nebula. (Meteorites, in contrast, which are presumed to form in a variety of regions, show widely varying isotopic compositions of oxygen). Thus we have some evidence that the moon is not a totally foreign body. If it was captured, it was captured, it was previously a near neighbour.

How the Earth and its Fluids became Concentrically Layered

The ideas described earlier in this chapter explain the manner in which our planet may have come into being, but they do not account for many of the earth do not account for many of the earth's particular features. How, for example, did the concentric structure of the solid earth and the fluids above it develop? How did metal and rock become segregated into core, mantle, and crust? Have there always been continents and ocean basins? How did the oceans and the atmosphere develop? These are questions that we will consider in this section.

Core, Mantle, and Crust

The concentric layers of the earth, from the core to the atmosphere, are arranged according to density, with the least dense layer, the atmosphere, situated on the outside. To understand how this layering may have developed, let us begin with the earth's core, which is thought to be composed primarily of iron. The traditional view has been that the earth formed by *homogeneous accretion*—that is, that the condensed material that formed the planet initially aggregated in a haphazard fashion. When we begin with this assumption, however, our problem is to explain how iron and nickel, which are dense, abundant metals sank to the earth's center to form the primary constituents of the core. The explanation would be simple if the primordial earth were known to have been a molten mass, in which case the high-density materials would automatically have sunk toward the center while less dense components would have floated toward the surface and formed the crust. The mantle, which is of intermediate density would then have remained between crust and core.

Was the earth initially molten? The earth's early thermal history is difficult to evaluate because its temperature of formation is assumed to have been controlled by the rate of accretion of earth materials, about which little is Known. Rapid accretion would concentrate a great deal of heat, but slow accretion would allow components to cool while falling into place. Many experts currently believe that the earth accreted components in a solid state—but even if this was the case, the earth must have developed a liquid interior when it heated to a critical temperature upon reaching a certain size. At this point, gravitational collapse would inevitably have moved heavy elements (primarily nickel and iron) to the center. Such contraction would then have released an enormous amount of energy—enough, some would claim, to raise the temperature of the earth by as much as 1200°C! Under such conditions, the mantle would also have segregated in a liquid state, and radioactive elements (especially isotopes of uranium, thorium, potassium, and rubidium) would then have moved to positions near the surface as a consequence of their tendency to combine chemically with other elements in low-density silicate minerals.

It is not universally accepted that the earth's core formed secondarily by liquefaction and gravitational collapse. Some have suggested that the primordial earth formed instead by *inhomogeneous accretion*, with a naked core developing first as a dense nebular condensate upon which the less dense silicates of the mantle and crust later collected. What is certain, however, is that the earth's core formed early in the planet's history. Remnant magnetism found in the oldest rocks now recognized, which are about 3.8 billion years old, has established a minimum age for the earth's magnetic field—which, as we have seen, is generated in the outer, liquid portion of the core.

Atmosphere

The asteroids that coalesced to form the earth were too small for their gravitational fields to have held gases around them as atmospheres. We can thus conclude that the earth did not inherit its atmosphere from these ancestral bodies : instead, the gases that the primitive earth retained as an atmosphere must have been emitted from within the planet after it formed, many of them while it was in a liquid state so that they could easily escape to the surface. If the core, mantle, and crust became differentiated when the earth first became liquefied, extensive *degassing*—or loss of gases to the earth's surface—would have accompanied this differentiation.

Degassing by way of volcanic emissions has continued to present, albeit at a much lower rate than early in the earth's history. The chemical composition of gases released from modern volcanoes provides a general picture of what the early atmosphere contained: primarily water vapour, hydrogen chloride, carbon monoxide, carbon dioxide, and nitrogen. In the modern world, plant photosynthesis is responsible for most of the oxygen in the atmosphere today. In the absence of plants, little oxygen entered or formed within the early Archean atmosphere and these small quantities were quickly removed by oxidation of iron minerals and other materials at the earth's surface. The early atmosphere would thus have been inhospitable to animal life.

Hydrogen and helium are the only elements of low enough density to have escaped from the earth's gravitational field during the period of rapid degassing. It is therefore strange that more dense volatile elements and compounds, such as carbon nitrogen, water and noble gases (neon, argon, and their chemical relatives), are also very rare in the earth and its atmosphere today in comparison to their abundance elsewhere in the solar system. Clearly, these more dense volatiles must have escaped from planetary condensates before the latter coalesced to form the earth. Although they are rare, carbon, nitrogen, and water are more abundant in the earth than the other volatiles, and it seems evident that they were preferentially retained as a result of being partially locked up in solid compounds. Noble gases rarely combine with other elements.

Oceans

The rapid degassing of the liquid earth released hot clouds of water vapour. Initially, the great heat of the earth would have kept the water in a gaseous state. Only when the planet cooled to the point at which its surface temperatures fell below 100°C (boiling point of water) would water have fallen as rain and remained on the earth as lakes, rivers and oceans.

Like modern rain, the rains that formed the earliest oceans are assumed to have contained few salts. Salts accumulated in early sea water by the reaction of the water and carbon dioxide dissolved within it with natural minerals such as clays and carbonates. Calculations show that sea water should have attained a salinity comparable to that of the modern world early in Archean time. Since that time, salts have been precipitated on the sea floor as evaporite sediments about as rapidly as they have been added to the oceans, and thus the salinity of sea water has not varied greatly. At the same time, the global hydrological cycle has moved water but not salts from the oceans to the atmosphere and back again.

Continental Crust

As stated already, continents consist primarily of thick felsic crust and are surrounded by the thinner mafic crust of ocean basins. An important question is how the two kinds of crust came into being. To understand the answer, it is important to recognize that the earth's interior cooled from its fully molten state primarily by means of convection, which carried hot material to the surface. The primitive crust probably formed rapidly during this brief interval of cooling as magma flowed to the surface at a high rate. Most of the early crust was composed of mafic material that rose from the denser ultramafic mantle. Mafic materials at the earth's surface weather more rapidly than do felsic materials. The result in early Archean time was that erosion and weathering of mafic rocks standing above sea level left a residue of felsic minerals, especially clay. These minerals, having accumulated along the margins of mafic island, were then turned into felsic crystalline rocks by the metamorphism and melting that accompanied continuing igneous activity. It is assumed that quartz was an abundant mineral in many of these felsic rocks.

In as much as mafic minerals are preferentially destroyed and are not replaced by other minerals, the sequence of weathering, erosion, metamorphism, and remelting serves, in effect, as a natural "machine" for the production for granite and other felsic crystalline rocks of the sort that form most continental crust. This machine began to operate as soon as the crust had cooled enough for felsic igneous and metamorphic rocks to form within it. Since that time, the machine has continually increased the volume of continental crust by extraction felsic material from more mafic, material—priamarily mafic material that has moved up from the mantle to form oceanic crust.

Continental crust was probably present earlier than 4 billion years ago, but doubts have been expressed about the validity of the few radiometric ages that have been found to be this great. Of the Archean rocks whose ages are widely accepted, the oldest are metamorphic rocks of various kinds found in a small area near the southern tip of Greenland. Among these are rocks that were metamorphosed about 3.8 billion years ago after they had accumulated as sediments sometime earlier. In fact, metamorphism has undoubtedly altered many bodies of rock that are older than 3.8 billion years and reset their radiometric clocks to give younger ages. Other early Archean rocks have been remelted or destroyed by erosion.

Great Meteorite Shower

Before moving to a general discussion of surviving Archean rocks, we will consider what must have been one of the most remarkable episodes in earth history: the pelting of the earth by large numbers of meteorites and asteroids over a period of several hundred million years. Given the rarity of early Archean rocks, our only evidence that this extraterrestrial shower took place is provided by other planets and particularly, by the moon. These other bodies of the solar system have not undergone the weathering and erosion that constantly alter the surface of the earth.

Earthbound observers have long commented on the moon's pockmarked appearance. The moon's craters, which are known as *maria*, (singular, mare), the Latin word for "seas" were first sketched by Galileo. It is only from manned lunar exploration and from photographs provided by artificial satellites, however, that we have gained detailed knowledge of these enormous craters. The maria that face the earth have an average diameter of approximately 200 kilometers (~ 125 miles) and are distributed in a crescentshaped pattern. Initially, it was not known whether the maria were craters produced by volcanoes or were indeed the impact scars of huge meteorites, but meteoritic origin has now been established by detailed study of the maria and neighbouring lunar terrane. The entire surface of the moon is, in fact scarred with craters, most of which are much smaller than the enormous maria. The lunar highlands surrounding the maria consist of rock fragments that testify to the pulverization of the lunar crust by the impact falling meteorites. The maria facing the earth are floored by immense flows of dark volcanic basalts, which give the maria their dusky appearance. These basalts are themselves scarred by smaller craters, and the relative ages of successive lava flows can be deduced from the density of cratering.

The dating of associated rocks has revealed that most large lunar craters are quite old, ranging in age from slightly less than 4.0 billion years to about 4.6 billion years. Many rocks of the lunar highlands exhibit ages of approximately 4.0 billion years, and it is believed that their radiometric clocks were reset some 600 million years after the moon formed and that the highlands formed with the reset of the moon about 4.6 billion years ago. The clocks may have been rest by the intense meteorite bombardment whose effects are so evident.

It has been estimated that during the early cataclysmic interval of lunar history, the frequency of meteorite impacts was more than 1000 times the present rate. Radiometric dating shows that basalts flooring the lunar maria tend of be slightly younger than the maria themselves; most of these basalts range in age from 3.2 to 3.9 billion years. The relatively modest cratering of the basalts shows that shortly after 4 billion years ago the intensity of meteorite showering subsided to something like its present low level. Not only the large maria but most of the lunar craters have been shown to have formed early in the history of the solar system. We can see that craters of all sizes also formed on planets of the solar system whose surfaces have not since been as heavily altered as that of the earth. These craters formed during a time when the solar system was, in effect, being cleared of many small remaining pieces of solid material that had consolidated from the solar nebula but had not been assimilated into planets or captured as planetary satellites. It seems an inescapable conclusion that the primitive earth was subject to the same kind of meteorite bombardment as was its neighbouring moon. The composition and structure of the earth's crust must have been altered by the material contributed by this enormous shower of meteorites. Unfortunately, however, the rock record that might have provided us with details has been all but destroyed by igneous and metamorphic activity as well as by weathering and erosion.

Archean Rocks

Before the advent of radiometric dating scientists found it difficult to establish even relative ages for precambrian rocks. Early on, however, geologists concluded on the basis of the principle of superposition that a particularly distinctive association of rocks seemed to represent the oldest Precembrian interval, which became known as the Archean.

Most Archean rocks can be divided into two groups: the even-grained, high-grade metamorphic rocks known as *granulites* and the bodies of rock known as *greenstone belts*. Greenstone belts, which are unique to the Archean Eon, are dominated by volcanic rocks and by sedimentary rocks that are derived primarily from volcanics. The rocks of greenstone belts are frequently metamorphosed, and the metamorphic mineral chlorite gives them the greenish colour from which they derive their name. Granulites, many of which have formed since Archenan time as well as during it, result from such severe metamorphism that they teach us little about environments of the Archean. Greenstones are more revealing, and it is upon them that we will focus in this section.

Greenstones and Ancient Sediments

Archean greenstones include or are associated with a limited variety of rocks characterized by a particular structural pattern. The composite features of greenstone belts are without conterpart in the modern world and thus reveal that the earth's crust in Archean time differed from what it is today.

The Canadian Shield and areas to the south include several small areas of Archean terrane and three large areas : the Superior, Slave and wyoming provinces. The largest uninterrupted greenstone belt here is the Abitibi Belt of southern Ontario and Quebec. The Abitibi greenstones resemble those found in other regions of the world in that they occur as podlike bodies of volcanic rock associated with sedimentary deposits. In cross section, a typical greenstone pod is seen to have the configuration of a syncline. Granitic rocks often surround a body of greenstone, and sometimes they also intrude it, proving that they are of younger age.

Typical greenstone basins reveal an interesting pattern: Volcanics at the base of the sequence are more mafic than those at the top. In the Canadian Shield, the upward trend is normally

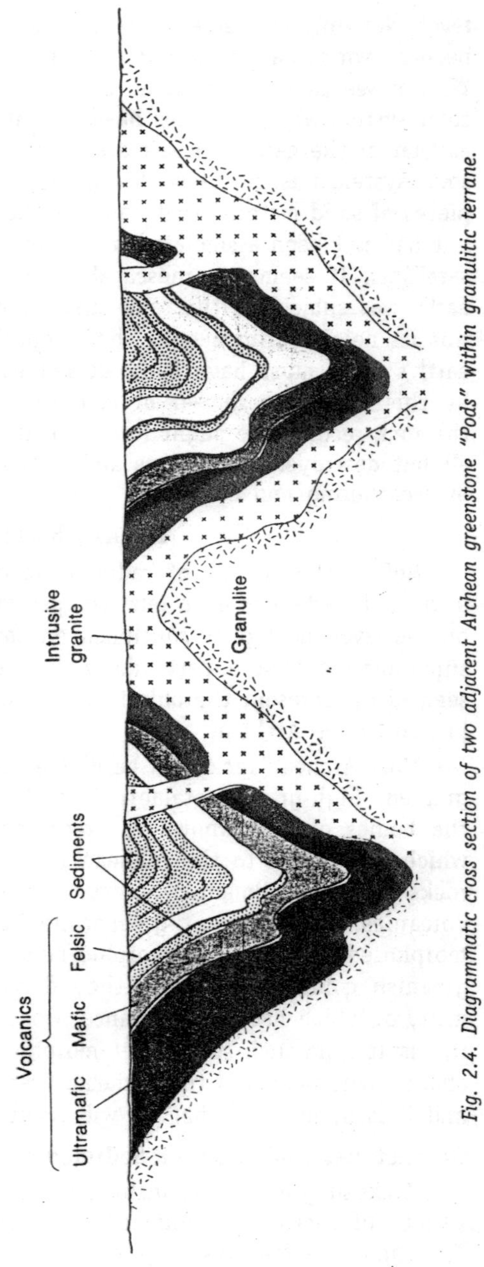

Fig. 2.4. Diagrammatic cross section of two adjacent Archean greenstone "Pods" within granulitic terrane.

from mafic to felsic, but older greenstone belts (those exceeding 3 billion years in age) tend to display a fuller sequence, ranging upward from ultramafic through mafic to felsic. Ancient greenstones that include basal ultramafics are especially well exposed in Western Australia and southern Africa. Illustrates the sequence of development of a major greenstone belt of southern Africa. In all cases, sedimentary rocks are most common at or near the top of the volcanic sequence.

The typical ultramafic-to-felsic sequence may reflect the operation of the "machine" for the production of sialic crust described earlier in this chapter. In this machine, weathering rapidly destroys ultramafic and mafic minerals, leaving a residue of felsic products such as clays. Some of these felsic minerals, when buried deep within the earth, become incorporated by magma.

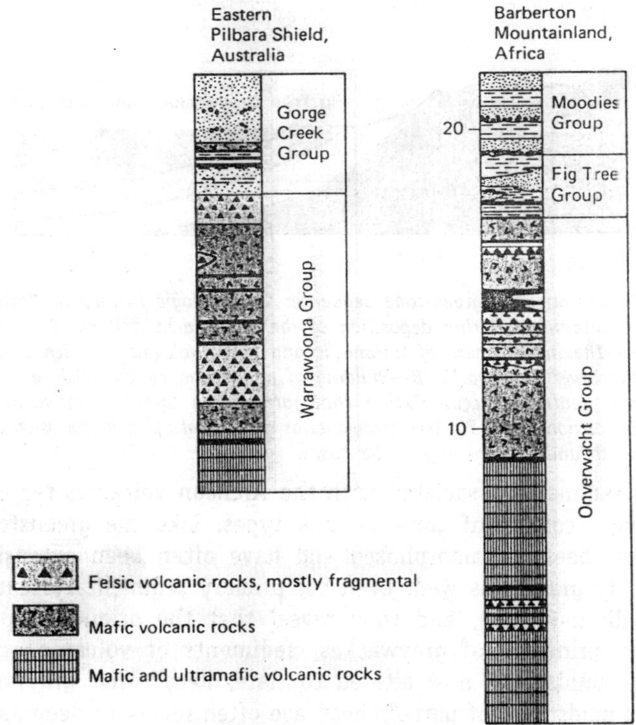

Fig. 2.5. Representative stratigraphic sections through two greenstone sequences. In both of these sequences there is a transition from ultramafic and mafic volcanics to felsic volcanics.

As greenstone belts evolved, their magmas may have incorporated more and more sialic materials as these materials continued to form.

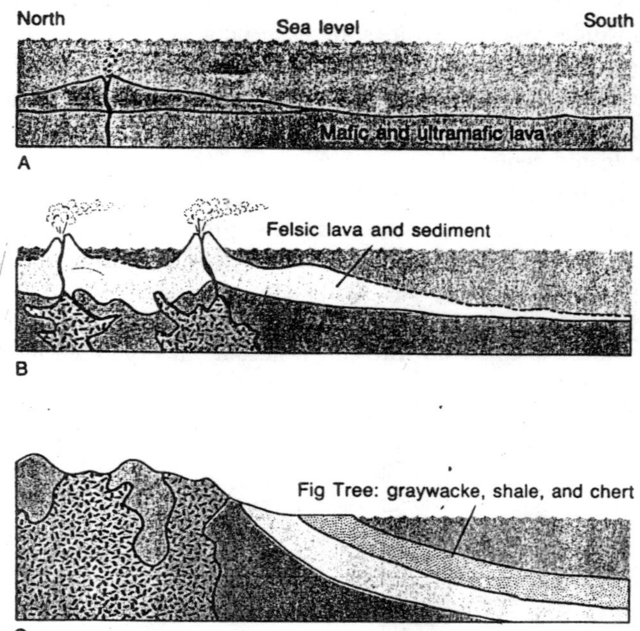

Fig. 2.6. Formation of a greenstone sequence: The geologic history of Barberton Mountainland during deposition of the Onverwacht and Fig Tree Group. A—The initial phase of ultramafic and mafic volcanism, which produced the lower Onverwacht. B—Building of a platform by the felsic extrusions and sediment accumulation that formed the upper Onverwacht. C—Deposition of the Fig Tree sediments as a result of uplift and erosion in the south and a deepening of the basin.

The sediments associated with the Archean volcanics represent a strikingly consistent suite of rock types. Like the greenstones, they have been metamorphosed and have often been extensively invaded by granite as well. Even so, primary sedimentary features can be distinguished, and they reveal that the original deposits consisted primarily of graywackes, sediments of volcanic origin, and dark mudstones now altered to slate. Recall that graywackes and dark mudstones of post-Archean age often represent deep marine environments adjacent to continents. The Archean examples occur both in large belts adjacent to the greenstone synclines and within the synclines themselves. The volcanics characteristically display relict pillow structures, indicating submarine cooling of the original lava. The volcanic belts may have extended along the margins of early landmasses, whereas mudstones and graywackes formed seaward in basins. Certainly the mudstones represent the "background" sedimentation of sedimentary basins, while the graywackes represent turbidite deposits containing material brought in from shallower water.

Other characteristic but less abundant Archean sediments are coarse conglomerates containing large, rounded cobbles that suffered considerable stream or beach abrasion. These conglomerates are not crossbedded like stream gravels, however; they seem instead to have been dumped into place, perhaps by enormous turbidity flows traveling down steep slopes. It is therefore not surprising that the conglomerates tend to occur adjacent to or folded into greenstone belts; they seem to represent nearshore facies. The conglomerates contain pieces of greenstone, but also pieces of granites and other plutonic rocks from unknown source areas. While very few granitic rocks now found in the Archean terrane of the Canadian Shield are older than the sediments, the granitic cobbles in the conglomerates seem to point to the presence of some ancient granitic crust located either between sedimentary basins or on islands within them. The graywackes, like the granitic cobbles, contain abundant quartz, indication derivation from felsic crystalline rocks.

In summary, the Archean rocks seem to indicate the presence of basins of moderate depth flanked by volcanoes that spewed out lava and by small, steep-sided blocks of felsic crust. Under ordinary circumstances, muds were deposited offshore, but turbidity flows periodically dumped conglomerates near the shore and carried finer erosional material (future graywacke) out toward the centers of basins.

Cherts and iron-rich sedimentary rocks known as *banded iron formations* are also found in the Archean sedimentary belts. These structures are occasionally widespread but are seldom of great thickness. Banded iron formations are rare in Phanerozoic rocks but became far more prevalent shortly after the end of Archean time, and thus they will be discussed at some length in the following chapter. It is notable, however, that the banded iron formation at Isua, West Greenland, represents one of the oldest known bodies of rocks on earth. Like other banded ironstones of the Precembrian, it consists of iron-rich layers alternating with quartz layers. The Isua rocks are believed to have originated by chemical precipitation in marine basins, and the quartz with in them is thought to have existed initially as chert precipitated from sea water.

The Rarity of Nonmarine and Continental-shelf Deposits

It is a striking fact that most Archean sediments are of deep-water origin. These include graywackes, mudstones, iron formations, and sediments derived from volcanic activity. In contrast, carbonates and quartz sandstones, which are abundant in younger sedimentary

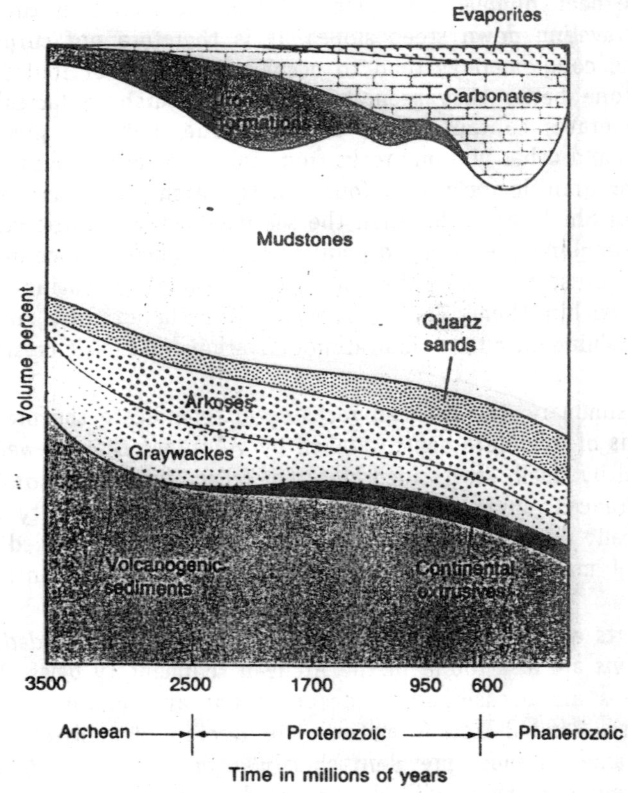

Fig. 2.7. Changes in the relative proportions of rock laid down on cratonic surfaces in the course of geologic time. Notice the increase in evaporites, carbonates, and quartz sands after the Archean.

sequences, are rarely found in Archean sediments. Today, of course, carbonates and clean sands accumulate extensively in shallow marine settings along the continental shelves of large cratons.

While sediments of shallow seas are rare in Archean terranes, terrestrial and freshwater deposits are virtually unknown. Two hypotheses can be proposed to explain this phenomenon: (1) such sediments failed to form extensively during Archean time, suggesting that terrestrial and freshwater environments were rare; or (2) such deposits actually formed in many areas but were later almost entirely destroyed. How can we evaluate these two possibilities? The answer can be found in the relative vulnerability of various kinds of sediments to erosion.

All sedimentary deposits tend to be cycled through the earth's crust; whether they become lithified or not, they are likely to be destroyed following deposition by erosion, igneous activity, or metamorphism. This is especially true of sedimentary evaporites, which are vulnerable to chemical solution and thus tend to be destroyed even more rapidly than sedimentary silicates. Certainly this may partially account for the rarity of evaporites in the Archean record.

Quartz sandstones and carbonate rocks, on the other hand, are chemically and physically resistant rocks, and thus a lack of inherent durability cannot explain their scarcity in Archean terranes. One possible reason for this scarcity is that Archean cratonic deposits, including quartz sands, formed at higher elevations than basin deposits and were thus particularly vulnerable to erosion. We have seen, however, that in more recent times, crustal subsidence has caused shallow marine sediments such as those of the Bahamas and nonmarine sediments such as those of rift valleys to be buried quite deeply in very short spans of time. Why, then, were no similar sequences preserved in Archean time? Preservational bias may partially account for the scarcity of Archean cratonic sedimentary deposits (especially evaporites), but other factors must also have been operating. These appear to have been related to the tectonic framework of the Archean world: There are no widespread continental or continental-shelf deposits in the Archean record simply because there were no large continents during most of Archean time. This condition will be the topic of the following section.

Archean Tectonics

The lack of evidence supporting the existence of large continents, taken together with the floodlike character of many

Archean pillow basalts and the relative abundance of graywackes, has led many workers to postulate that cratonic bodies existed in the Archean Eon only as small steep-sided felsic protocontinents along which conglomerates accumulated. If such protocontinents were present, they must have been separated by numerous marine basins that accumulated lava and volcanic sediments to become greenstone belts. It is not clear whether the sedimentary basins were semi-isolated structures scattered over an ancient crust or bodies whose waters covered most of the globe and were interrupted only by protocontinents of felsic crust or volcanic

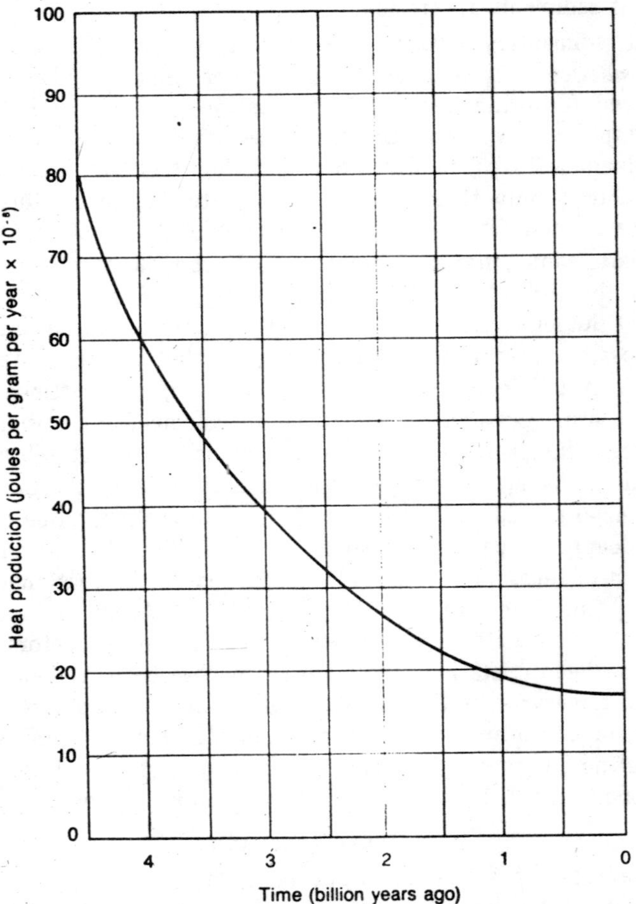

Fig. 2.8. Decline in the earth's rate of heat production through time.

material scattered here and there. The second configuration would account for the unusual sedimentary record of the Archean Eon; if basins covered most of the earth's surface, conglomerates and graywackes would have been derived from the erosion of scattered felsic protocontinents, and the absence of extensive shelf deposits (carbonates and quartz sandstones) would reflect the absence of large continental shelves.

The history of radioactive decay supports the notion of a fragile, mobile Archean crust with many areas of weakness forming basins between or around small protocontinents. Recall that the earth's heat source is constantly diminishing through time as radioactive isotopes decay without renewal. Because these decay rates are constant, geologists can calculate the approximate difference between the earth's rate of heat production today and those of times past. Such estimates show that during the Archean, the rate of heat flow to the earth's surface was much higher than it is today. Near the end of Archean time, the total rate of heat production was perhaps twice as high as it is now, and earlier it was even higher. This substantial rate of heat flow must have resulted in a less stable crust. Under such conditions, lavas may have burst quickly to the surface, flooding basinal areas that we now see as greenstone belts.

Large Cratons Appear

Radiometric dating shows that during early proterozoic time large bodies of magma were intruded into cratonic rocks. It can therefore be concluded, in accordance with the principle of intrusive relationships, that cratons of moderate proportions had come into being very late in Archean time. The hardened magma often forms upright tabular structures, or dikes, that range in age from about 2.1 to 2.5 billion years large mafic dikes of this age, often occuring in swarms associated with other mafic dikes of this age, often occuring in swarms associated with other mafic intrusives, have been identified in North American, Greenland, the Baltic Shield, South America, southern Africa, India, and Australia In many parts of the world, divdence also indicates that major metamorphic episodes occurred slightly earlier, about 2.7 to 2.3 billion years ago. Geologists do not yet understand the origins of this metamorphic interval, but they do know that it resulted in the resetting of many radioactive clocks and in the consolidation of many crustal elements into sizable cratons.

There is also evidence, however, that "cratonization" did not occur simultaneously throughout the world. In most areas, typical Archean greenstone associations formed until approximately 2.5 billion years ago, but in southern Africa a large craton was already present about a half million years earlier. Here, the Barberton Mountainland greenstone sequence, whose age exceeds 3 billion years, is immediately overlain by a substantial body of clastic sediments known as the pongola supergroup, and this is followed by the Witwatersrand sequence, which is famous for the gold deposits that accumulated as detrital material within it. The Pongola Supergroup consists of deposits that are 3 billion years old but nonetheless strikingly similar to intertidal sequences of younger portions of the stratigraphic record. The presence of a broad intertidal belt during Pongola deposition indicates that a sizable land mass existed 3 billion years ago in southern Africa.

The Witwatersrand sequence offers further evidence that cratons did not appear simultaneously. This sequence, whose sediments range in age from about 2.5 to 2.8 billion years, consists of alternating coarse-grained and fine-grained sediments that cover about 40,000 square kilometrrs and have a maximum thickness of nearly 8 kilometers (~5 miles). These deposits accumulated in nonmarine environments on the surface of a craton. The coarsest sediments in the upper division of the Witwatersrand, and it is from these sediments that gold in mined which are widespread conglo merates, are most common. Gold is a very dense metal and thus tends to settle from moving water with much larger particles. During Witwatersrand deposition, bits of gold accumulated with larger silicate pebbles. The gold-bearing gravels apparently formed bars within braided streams that seem to have washed enormous quantities of sedimentary debris from highlands across alluvial fans into one or more lakes. Lake environments were the sites of accumulations of mud and, occasionally, of sand as well. The alternation of coarse and fine sediment throughout the Witwatersrand seems to reflect periodic uplifts in the source area followed by rapid erosion and deposition of gravels.

It is remarkable that the Witwatersrand sequence, like the Pongola Supergroup, has largely escaped metamorphism. Both in thickness and in areal extent, the Witwatersrand is rivaled by no other recognized Archean sequence of shallow-water sedimentary deposits. It could only have formed along the margin of a sizable

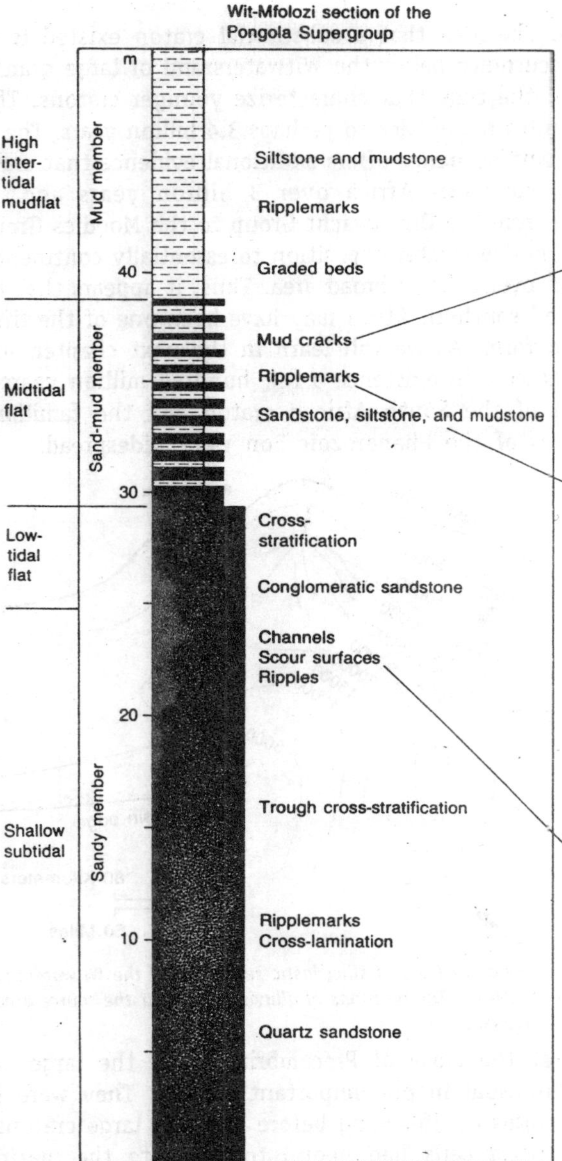

Fig. 2.9. Depositional environments represented within a stratigraphic section of the Pongola Supergroup of southern Africa. The section represents a 3-billion-year-old regressive sequence in which a tidal flat prograded seaward over subtidal environments.

landmass. The idea that a substantial craton existed is supported by the occurrence below the Witwatersrand of large granitic bodies of rock of the type that characterize younger cratons. The ages of these granites range back to perhaps 3.4 billion years. The Barberton Mountainland sequence offers additional evidence that cratonization began in southern Africa over 3 billion years ago. Here the transition from the Onverwacht Group to the Moodies Group records a shift from deep-water deposition to essentially continental, clastic deposition over a fairly broad area. Thus, it appears that the region we now call southern Africa may have been one of the first modern cratons to form. As we will learn in the next chapter, many large continents were in existence a few hundred million years after the formation of the South African craton and the familiar tectonic mechanisms of the Phanerozoic Eon were widespread.

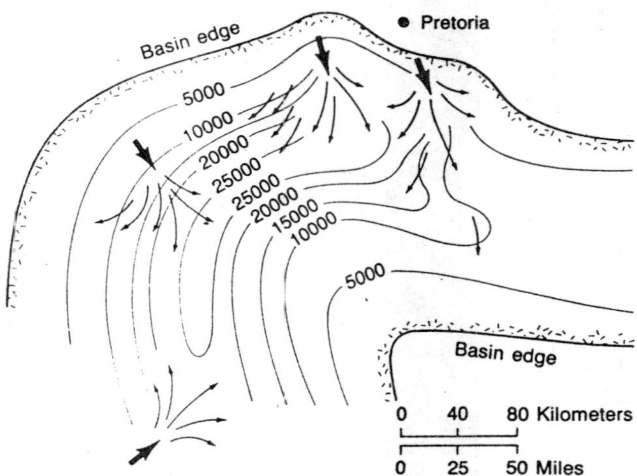

Fig. 2.10. Thickness (in feet) of siliciclastic sediments of the Witwatersrand Basin of South Africa. The positions of alluvial fans near the source area are shown by arrows.

Even at the close of Precambrian time, the large continents remained unusual in one important respect: They were barren of advanced forms of life. Long before the first large cratons existed, however, living cells had begun to populate the marine realm, where they remained at a primitive stage of development for perhaps a billion years or more. These early forms of life are the subject of the following section.

Archean Life

Of all the planets in our solar system, only the earth is well suited to life as we know it. One of the reasons is that its size is right. On a much larger planet, the grantational pull on the atmosphere would be so great that the resulting atmospheric density would exclude sunlight, which is the fundamental source of energy for life. A much smaller planet, on the other hand, would lack sufficient gravitational attraction to retain an atmosphere with life-giving oxygen. In addition, the earth's temperatures are such that most of its free water is liquid, the form that is essential to life. Even Venus, our nearest neighbour closer to the sun, is much too hot to allow water to survive in a liquid state, whereas Mars, our nearest neighbour father from the Sun, has an atmosphere so thin that liquid water would evaporate from the planet's surface almost immediately.

In this section, we will first examine the fossil evidence of early life on earth, and we will then review some of the fundamental chemical steps in the origin and early evoution of life. finally, we will consider possible scenarios for the evolution of the earliest single-celled organisms

Fossil Evidence

We can never hope to possess clear fossil evidence of the origin and early evolution of life on earth, since cells break down easily, and even chemical components that might be indicative of past events can quickly deteriorate beyond recognition. Further complicating this issue is the fact that seemingly dense rocks are in fact riddled with minute cracks and pores into which contaminants can intrude. Thus, while molecules found in Archean cherts might represent primary organic compounds, researchers cannot rule out the possibility that these substances actually seeped into the rock at much later dates. Nevertheless, certain facts do suggest that life is at least as old as the oldest rocks now known. For one thing, graphite, a mineral consisting of pure carbon, is present even in the oldest known sedimenta rocks, the banded iron formations of Isua, Greenland—perhaps reflecting concentration of carbon by organisms. In fact, carbon found in sedimentary rocks of Archean age tends to have nearly the same ratio of isotopes (^{13}C to ^{12}C) that characterizes biological systems today.

More direct evidence of very early cellular life is provided by stromatolites, the internally layered structures shaped like domes, mounds, or pillars. Today, stromatolites form along the margins of warm seas but are scarce in comparison to their former abundance. Recall that within a well-preserved stromatolite, layers of carbonate sediment rich in organic matter alternate with layers of purer carbonate sediment. Stromatolites usually grow side by side in large numbers; some stromatolites, especially those of the geologic past, have attained heights of several meters.

Because structures resembling stromatolites sometime form by deposition of layered sediment in the absence of algae, we cannot always identify fossil stromatolites with certainty. The oldest structures that are thought to be stromatolites occur in the Pilbara Shield of Australia in rocks 3.4 to 3.5 billion years old. Stromatolites are known to occur in slightly younger rock units such as the Bulawayan Group of Rhodesia, for which indirect dating gives an age of approximately 2.8 billion years, and the 3-billion-year-old Pongola Supergroup. They are also found in a few other Archean units, but in general they are poorly represented in comparison to their abundance in Proterozoic sediments. The rarity of Archean stromatolites probably reflects the rarity of Archean shelf deposits. We cannot be certain, however, that the organisms that formed Archean stromatolites were photosynthetic, and therefore we cannot draw definite conclusions concerning the environments in which these stromatolites formed. Certain threadlike bacteria that are closely related to blue-green algae from stromatolite-like structures today in hot springs such as those of yellowstone National park, but most of these bacteria are not photosynthetic; that is to say, they do not synthesize food using light as an Energy source. Thus, we cannot exclude the possibility that similar bacteria formed the stromatolite-like structures found in Archean rocks.

Actual fossils of blue-green algae and bacteria have also been tentatively identified in Archean rocks. Among the oldest cells yet known are filaments from a rock at North pole, Western Australia, where the apparent stromatolites are also found. These filaments which are thought to be approximately 3.5 billion years old, are about the same size as those of modern blue-green algae and display similar transverse partitions. Somewhat younger spheroidal (nearly spherical) structures resembling other types of living bacteria or blue-green algae occur in the Fig Tree Group of southern Africa,. These structures, which are about 3 billion years old, are preserved in what appear to be various stages of cell division.

Bacteria and blue-green algae seem to be the only life forms represented in Archean rocks, and they resemble each other so closely that many experts now advocate changing the name of blue-green algae to blue-green bacteria. Because bacteria and blue-green algae differ markedly from all other groups of cellular life in the modern world, they are placed by themselves in the kingdom *Monera*. Recall that Monera are single-celled organisms characterized by a primitive kind of cell that does not have a nucleus and whose DNA is not clustered into discrete chromosomes. This type of cell, which is known as a *prokaryotic* cell, also lacks certain other internal organlike structures (organelles) that are present in more advanced forms of life. Life forms that do posses chromosomes as well as nuclei and other organelles are known as *eukaryotes*. Of all the cellular forms of life in the world today, only bacteria and blue-green algae are prokaryotes. All other celluler taxa are eukaryotes, and fossils of these are unknown from the Archean and early portion of the Proterozoic record. Thus, with respect to the history of life on earth, the Archean might well be labeled the Age of Prokaryotes.

Chemical Evidence

Although the fossil record will never tell us a great deal about the earliest stages of organic evolution, such information has been derived from other sources. In the early 1950s, for example, researchers found that they could readily produce amino acids in the laboratory by sending electrical sparks through sealed vessels containing ammonia, methane, hydrogen, and steam. These sparks may duplicate what lightning did in the Archean world. Similar laboratory results have since been obtained with simple starting mixtures such as carbon monoxide, nitrogen, and hydrogen. In some cases, ultraviolet light has been substituted for electrical sparks as a source of energy. *Amino acids* are the building blocks of proteins, and proteins are important compounds in living systems. Thus, experiments demonstrating the generation of amino acids under conditions that might well have characterized the primitive earth are highly significant. It has also been found that the heating of amino acids drives off water and links them into chains called *polypeptides*, which resemble proteins but are less complex. This, too, could have occurred in many parts of the primitive earth.

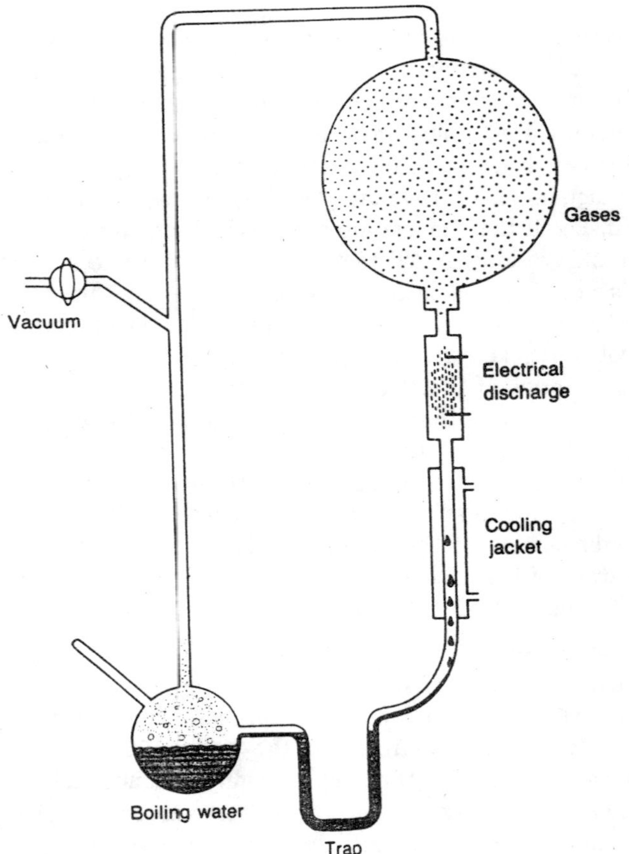

Fig. 2.11. Experimental apparatus of S.L. Miller in which amino acids were produced by circulating ammonia (NH_3) methane (CH_4), water vapour (H_2O), and hydrogen past an electrical discharge. Amino acids accumulated in the trap.

Recent discoveries suggest that the formation of amino acids need not necessarily have begun on earth. Radio-spectroscopy has detected the presence in outer space of many compounds that represent intermediate steps in the formation of amino acids—compunds such as formic acid and methylamine, which can combine to give the amino acid glycine.

Meteorites offer additional evidence of extraterrestrial synthesis of molecules that are precursors of important biological compounds. Stony meteorites containing compounds of carbon are termed *carbonaceous chondrites*. In 1969, fragments of the famous Murchison meteorite of Australia were collected on the same day the meteorite landed, and they yielded numerous organic compounds, including amino acids. A remarkable feature of the murchison and other carbonaceous chondrites is that each of their amino acids tends to be represented by more or less equal proportions of left-handed and right-handed structural configurations. (An amino acid molecule is three dimensional and can exist in either of two mirror-image forms, just as a screw or bolt can have left-handed or right-handed threads.) Amino acids constructed by living systems on earth happens to have a left-handed configuration, which suggests that those of the Murchison meteorite are indeed extraterrestrial. Supporting evidence comes from the presence in these meteorites of organic compounds that are not found on earth and from the occurrence of carbon in a higher isotopic ratio of ^{13}C to ^{12}C than is known in any terrestrial biological system.

While amino acids and proteins are essential features of terrestrial life, they do not account for a basic aspect of living systems: the capacity for self-replication. On earth, this capacity resides within the DNA and RNA of cells. Recall that the genetic messages of chromosomes are chemically encode in DNA; these messages are transcribed by RNA, which is similar in structure to DNA, and are then carried by RNA to regions of the cell where, according to the chemical prescription, proteins are constructed from amino acids. Both RNA and DNA consist of structures called bases, which lie along a chain formed of alternating sugars and phosphate structures. Nucleic acids are not complex chemical structures, and it does not seem at all preposterous to imagine that the first nucleic acid formed in nature by the assembly of sugars, phosphates, and nucleatide bases; but we do not know exactly how this happened or precisely how the codes for protein synthesis to came reside in DNA.

Another problematic step was the origin of cell-like bodies having the ability to build true proteins. Although such bodies have not been created in the laboratory, simpler spherical bodies have been produced. These are structures that have formed from proteinlike compounds in water or salt solutions, especially when the liquid has been cooled. Some of these spherical structures aggregate into chains that superficially resemble those of the living bacteria *Streptococcus*, forms of which cause infections in humans.

A major question that remains, however, is how nucleic acids might have been incorporated into spherical structures composed of proteinlike compounds.

The earliest forms of cellular life that employed nucleic acids to build proteins and to reproduce may have obtained their nutrition in either of two ways: They may have been animal-like (consumers) or plantlike (producers). Consumers obtain food directly from the environment, whereas producers manufacture their own food from simple raw materials that they obtain from the environment. We will consider two scenarios for early Archean evolution-one in which the earliest single-celled organisms are animal-like and the other in which they are plantlike. Modern bacteria live in many different ways, and as very primitive organisms, some of which may have survived from Archean time with little modification, they offer clues about evolution during Archean time.

Animal-like Beginnings?

It has been argued that because the earliest true cells were very simple forms of life they must have been consumers—that is to say, they must have required food and energy from their environment in order to maintain themselves and to reproduce. Like all modern cells, early cells probably employed a compound known as ATP (adenosine triphosphate) as a source of energy. Unlike modern cells, however, early cells may have obtained the ATP that they needed from their environment—by eating it, in effect. In the laboratory ATP is easily produced from simple gases, and it may well have formed inorganically in the Archean world. Even very primitive animal-like cells must have had at least a limited ability to synthesize other essential compounds, although they fed no amino acids, sugars, and other molecules necessary for their existence. It has often been assumed that these organisms lived in a lake or an ocean filled with a sort of natural "soup" of essential molecules.

In time according to this scenario of animal-like beginings, certain cells developed the ability to ferment organic compounds—that is, to break down such compounds into simpler ones and to use the energy liberated in this way to build some of the compounds that they needed. In other words, they became plantlike. Many bacteria employ fermentation today. Most of them ferment sugar or cellulose (the material that forms the cell walls of plants) into products such as ethanol (ethyl alcohol). Energy liberated by fermentation is stored as ATP until it is used to build compounds essential to the operation of the bacterial cell.

Some bacteria that are fermenters also obtain energy by converting compound that contain sulfate into others that contain sulfide—that is to say, they remove oxygen atoms that have been attached to sulfur atoms. These bacteria cannot tolerate oxygen, and most therfore live in the muds of swamps, ponds, or lagoons. The hydrogen sulfide that these bacteria liberate into their muddy environments generates the characteristic "rotten egg" smell of those settings. If, as is widely believed, the earth's atmosphere lacked oxygen during Archean time, such *sulfate-reducing bacteria* may have occupied a much wider renge of environments than they do today.

Bacteria that conduct fermentation or reduce sulfate are said to engage in *chemosynthesis*. The origin of the energetically more efficient process of *photosynthesis* in living organisms represented an important breakthrough that altered the ecosystem profoundly. Photosynthesis is the process by which single-celled algae and multicellular plants employ the green pigment chlorophyll to transform the energy of sunlight into chemical energy. This energy is then used to transform carbon dioxide and water into energy-rich sugar. When it is later released from the sugar, the energy also fuels essential chemical reactions.

Presumably, blue-green algae were not the only Archean Monera that conducted photosynthesis. Forms called *photosynthetic bacteria* probably existed as well. Today these organisms are represented by the purple and green bacteria that inhabit moist area lacking free oxygen, a substance that is poisonous to them. Like blue-green algae, photosynthetic bacteria employ sunlight to produce sugar. Unlike blue-green algae, however, they use hydrogen sulfide rather than water as a raw material, and sulfur rather than free oxygen, is a by-product of their photosynthesis.

The evolution of the blue-green algae, which do release free oxygen, had a profound effect on the global ecosystem. The oxygen that was liberated by these organisms accumulated in the atmosphere to from the reservoir from which humans and other animals were later able to breathe. In the following chapter, we will examine the current controversy over the time of accumulation of the earth's atmosphere.

Plantlike Beginnings: Another Alternative

Recently, some scientists have proposed a new set of ideas about the evolution of primitive cells. This set of ideas conflicts with some of the ideas outlined in the previous section-particularly the notion that the first cells were consumers that fed on molecular soup. The new ideas are based on the discovery during the 1970s of hot springs located along midocean ridges in the deep sea. According to the new scenario, the full series of steps from the origin of complex molecules necessary for life to the origin of cellular organisms took place within similar hot springs early in Archean time.

Especially well studied are modern hot springs along the ridge adjacent to the Galapagos Islands, where Charles Darwin gained insight into the process of organic evolution. Because of the many fractures in the newly formed oceanic crust adjacent to a midocean rif, sea water seeps far down into the crust, where it eventually reaches magma rising up from the mantle along the rift. Under great pressure resulting from the weight of water above, the water is then heated to temperatures that may sometimes reach 1000°C. The hot water subsequently rises along the rift and reaches the surface through a narrow conduit, a roughly cylindrical fracture system called a vent. The water cools as it rises so that it reaches the overlying sea water, which is nearly freezing, at temperatures ranging from 10 to several hundred degrees centigrade.

The water that rises from a midocean vent is rich in various dissolved compounds, some of which are derived from the magma below. These include carbon dioxide (CO_2) ammonia, methane, and hydrogen sulfide (H_2S). Interestingly these are some of the compounds from which many chemosynthetic bacteria derive their energy in the modern oceans. Accordingly, it has been suggested that such bacteria resemble the earliest forms of cellular life and that these forms evolved in the hot waters of deep-sea vents in Archean oceans. By this argument, the earliest cellular organisms would have been plantlike, but they would have been chmosynthetic rather than photosynthetic.

Fig. 2.12. shows how such vents might have been settings for all the major steps in the early evolution of life. High temperatures deep in a vent near the magma could have led to the formation of amino acids from simple compounds such as hydrogen methane, and ammonia. Higher in a vent, under moderate-to-high

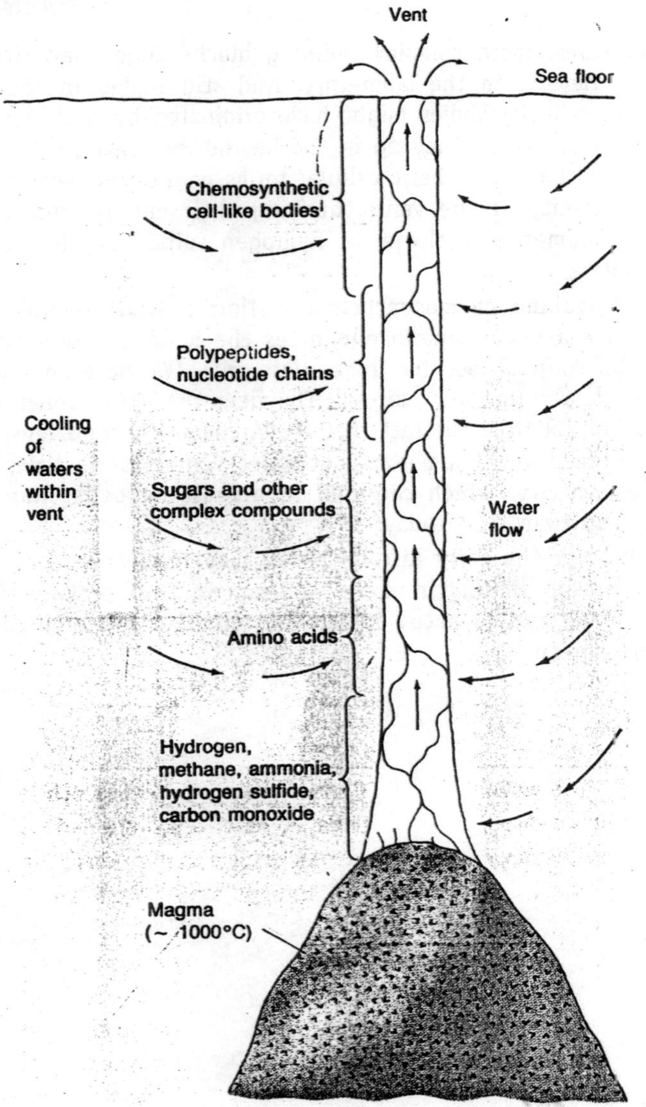

Fig. 2.12. Possible sequence of evolution of early life within a deep-sea vent. Sea water seeps into the vent, which is a roughly cylindrical fracture system above a body of magma. In the vent, the water is heated to high temperatures and then rises. Increasingly complex structures would have originated toward the top of the vent as the rising water cooled. Ultimately, simple chemosynthetic organisms might have evolved near the sea floor, deriving their energy from hydrogen, methane, or other simple compounds formed deep within the earth.

temperatures, more complex building blocks might have formed, just as they do in the laboratory. And still higher in the vent, spherical cell-like bodies might have originated by cooling of the vent waters, just as they do in the laboratory when similar fluids are cooled. Chemosynthetic cellular forms might also have evolved in the vicinity of the vents, deriving their energy from carbon dioxide, ammonia, methane, or hydrogen sulfide supplied by the vent waters.

By invoking chemosyntheic nutrition for early cellular life, the deep-sea–vent scenario eliminates the need for some type of molecular soup as food for the earliest cells. (We have no evidence that such a soup ever exited). The deep-sea–vent scenario also derives appeal from the fact that the Archean sea floor must have been studded with numerous hot-water vents; as we have seen, igneous activity, which centered in greenstone belts, was more widespread than it is today.

Whatever the early history of life may have been, one thing seems certain: Missing from the Archean world of prokaryotic life were animals and advanced animal-like cells that fed upon bacteria and blue-green algae.

3

PROTEROZOIC ERA

The Proterozoic Eon, which succeeded the Archean 2.5 billion years ago, was in many ways more like the subsequent Phanerzoic Eon, in which we live. We have already seen a foreshadowing of this difference between the Proterozoic and Archean eons in the origin of large cratons late in Archean time. The persistence of large cratons throughout the Proterozoic Eon produced an extensive record of deposition in broad, shallow, seas—a pattern that differed substantially from the Archean record of deep-water deposition, which is now largely confined to greenstone belts and adjacent areas. In addition, a larger number of Preterozoic than Archean sedimentary rocks remain unmetamorphosed and therefore accessible for study.

In this chapter, we will learn about the major events that have recorded in the extensive deposits of proterozoic age. These deposits not only document ancient mountain-building events strikingly similar to those of the Appalachians and other younger orogenic belts but also reveal records of major intervals of glaciation, at least one of which seems to have affected most of the world. Also present in Proterozoic rocks is a fossil record of organic evolution that reveals a transition from the simplest kinds of single-celled life at the start of the Proterozoic Eon to more advanced single-celled forms and, finally, to multicellular plants and animals, some of which belonged to modern phyla.

A MODERN STYLE OF OROGENY

The cratons of modern proportions first began to form about 3 billion years ago, late in Archean time. This was also the time when the oldest known sedimentary deposits were laid down over substantial continental areas in southern Africa. Although mountain-building processes resembling those of the phanerozoic world were undoubtedly in operation by this time, it is in rocks about 1 billion years younger that geologists have found the oldest well-displayed remains of a mountain system that is thoroughly modern in character. This is the Wopmay orogen of Canada, which formed along the margin of an early continent that developed between about 1.8 and 2.1 billion years ago, over a large area now approximately 100 kilometers (~60 miles) to the west of Hudson Bay. Today, remarkably well preserved sedimentary rocks of this orogen are exposed along the low-lying surface of the Canadian Shield as a result of continental glaciation that has repeatedly soured the orogenic belt over the past 2 million years.

The present Epworth Basin, which is the site of the Wopmay orogen, lies along the western margin of the geologic region known as the Slave Province and is floored by a ancient fold-and-thrust belt. Although it has long been planed off by erosion, this ancient Zone of deformed rocks bears a striking resemblance to the younger belts described in. In the Wopmay orogen, thrusting was toward the east and igneous intrusions associated with the deformation now lie primarily within the Bear Province to the west. A belt of metamorphism lies between the igneous belt and the fold-and-thrust-belt. To the east, epicontinental sedimentary rocks

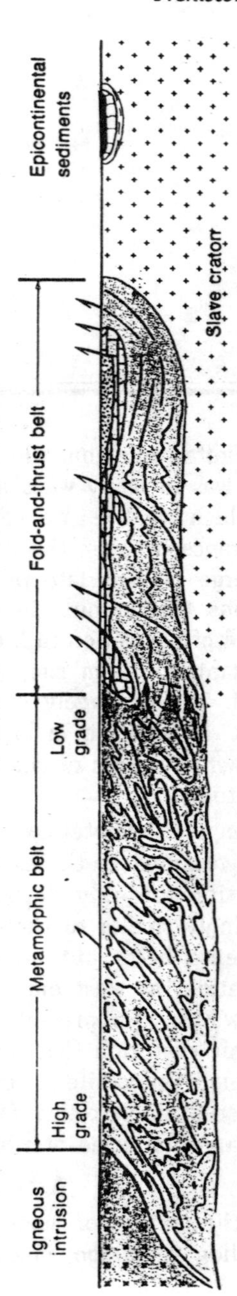

Fig. 3.1. A cross section of the northern part of the Wopmay orogen, depicted with no vertical exaggeration. Compare the three belts as shown here in cross section with their appearance in map view.

continuous with those of the fold-and-thrust belt are flat-lying except where they lie adjacent to local upwarps of basement rock. Like sedimentary deposits of younger fold-and-thrust belts, those of the Wopmay belt show a clear relationship to tectonic history. Near the end of Archean time, before the orogen was formed, most of what we now call the Slave Province existed as a discrete carton. Then, early in the Proterozoic Eon, thin Sedimentary sequences developed on the interior of this carton, and thick shelf deposits accumulated along its western margin. As in younger mountain belts, these were succeeded by flysch and then by molasse deposits. In fact, the thick sequence of deposits in the Wopmay fold-and-thrust belt closely resembles that found in each of the tectonic cycles of the Appalachian orogen. The Wopmay sequence has the following characteristics:

1. The first thick deposit, which formed along the edge of the continental shelf, is a quartz sandstone that prograded toward the basin. This unit resembles the sequence of Lower Cambrian shelf sands at the base of the first Appalachian tectonic cycle. The quartz sandstones of the Wopmay orogen grade westward into deepwater mudstones and turbidites that now lie within the metamorphic belt.
2. Next come rocks in which stromatolites and dolomite predominate. These represent a Proterozoic carbonate platform that resembled the Cambro-Ordovician platform of the first Appalachian tectonic cycle. In the proterozoic platform, sedimentary cycles record repeated progradation of tidal flats across a shallow lagoon. Laminated dolomite representing the lagoonal environment forms the base of each cycle, while at the top of each cycle are oolitic or stromatolitic deposits that must have formed in environments fringing the lagoon on its landward side. It would appear that the shoreline seldom if ever migrated westward as far as the shelf margin, since cycles are less evident here, while at the same time enormous stromatolite mounds are present. These mounds formed a persistent barrier behind which the fine-grained deposits of the lagoon were trapped; thus stromatolites bounded the lagoon on both its landward and seaward margins. The present metamorphic zone consists of a thinner sequence of mudstones that represent deeper environments beyond the shelf edge, together with beds of dolomite breccia that contain blocks as

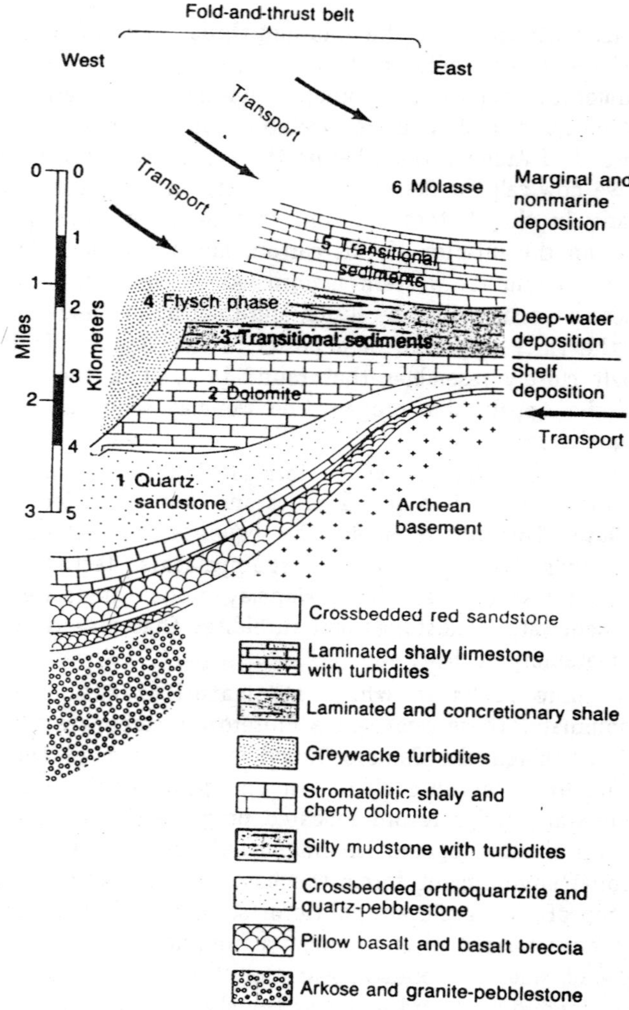

Fig. 3.2. Diagram showing the sequence of development of sediments in the fold-and-thrust belt of the Wopmay orogen; numbers refer to depositional units described in the text. Units 1 and 2 represent marine deposition along a shallow continental shelf. Units 3 and 4 are deep-water deposits, including flysch, that accumulated when the shelf foundered as mountain building began to the west. Unit 5 consists of shallow-water deposits transitional between flysch below and molasse above. Unit 6, the molasse phase of deposition, followed the exclusion of marine waters by a heavy influx of sediment from the west.

large as 50 meters (~ 165 feet) in length. It is obvious that these beds were transported down the steep slope in front of the shelf edge by catastrophic submarine debris flows.

3. The carbonate-platform deposits give way to transitional mudstones, which reflect a downwarping of the platform as a foredeep formed.
4. The mudstones are followed by flysch deposits (turbidites) similar to Upper Ordovician flysch of the first Appalachian tectonic cycle. A westward thickening of the Wopmay flysch and the inclusion within it of particles derived from plutonic rocks to the west indicate that the source area of siliciclastics now lay seaward of the orogenic belt, just as it did in the development of each Appalachian tectonic cycle when the foredeep formed. It is clear that the Wopmay foredeep formed when plutonism and orogenic uplift began in the offshore area to the west.
5. The deep-water turbidites grade upward into beds containing mud cracks and stromatolites, both of which represent shallow-water enviornments and thus point to a shallowing of the foredeep.
6. The influx of sediments eventually pushed marine waters from the Wopmay foredeep, and the wopmay cycle, like younger tectonic cycles, ended with an interval of molasse deposition. The Wopmay molasse consists largely of river deposits in which cross-bedding is conspicuous.

Two kinds of evidence suggest that the Proterozoic Wopmay orogen had the same pattern of formation as a modern orogenic system. First, the parallel igneous, metamorphic, and fold-and-thrust belts resemble similarly arranged belts of younger mountain ranges. Second, within the fold-and-thrust belts, shallow-water shelf deposits are succeeded by flysch deposits that give way to molasse deposits.

Another indication that normal plate-tectonic processes were operating in northern Canada about 2 billion years ago is evidence of continental rifting. East of the Wopmay orogen, slicing into the platform from the northwest and southwest, are deep, narrow troughs containing thick sedimentary sequences. These troughs represent failed rifts—that is, rift systems that ceased to be active before they cut across the entire Slave craton.

It seems likely that the westward-facing edge of the Slave craton along which the Wopmay orogen formed came into being more than 2 billion years ago, when a rift system broke apart a slightly larger continental mass. More or less simultaneously, the two rifts that later failed began to form at right angles to the newly forming shelf edge. Shortly thereafter, about 2 billion years ago, the shelf edge must have foundered when it was rafted up against a subduction zone. Then, as we have seen for younger orogens, igneous activity elevated the crust seaward of the shelf edge, and mountain building began.

Proterozoic Glacial Deposits

The fact that the Slave terrane along the Wopmay orogen behaved like rigid continental crust when it was rifted and deformed indicates that by 2 billion years ago, the earth was markedly cooler than it had been 3 billion years ago, when magmas were pushing up from the mantle to the surface throughout much of what is now the Canadian Shield. That climates in this region were also quite cool approximately 2 billion years ago is shown by evidence that glaciers spread over the land.

Early Proterozoic Glaciation

Just to the north of Lake Huron in southern Canada are some of the most spectacularly exposed of all ancient glacial deposits: those of the Gowganda Formation, which forms part of the Huronian Supergroup. Well-laminated mudstones in this formation almost certainly represent varves that formed in standing water in front of glaciers. These deposits were compared to the strikingly similar glacial varves that formed nearby, where Toronto is now located, just a few thousand years ago. Some of the laminated Gowganda mudstones contain dropstones-pebbles and cobbles that appear to have fallen to the bottom of a sea or lake from ice that melted as it floated out from a glacial front. Whether the mudstones accumulated in an ocean or in a lake is not clear, but it is evident that they alternate with tillites, which seem to represent the periodic encroachment of glaciers into the body of water. Some of the pebbles and cobbles of these tillites are faceted or striated from having slid along at the bases of moving glaciers.

Although the exact age of the Gowganda deposits has not been determined, it is known that the unit is slightly more than 2 billion years old, since it rests on 2.6-billion-years-old crystalline rocks and is intruded by rocks 2.1 billion years in age. Tillites of similar age are found elsewhere in Canada as well as in southern Africa and India. Thus, it would appear that there was an interval

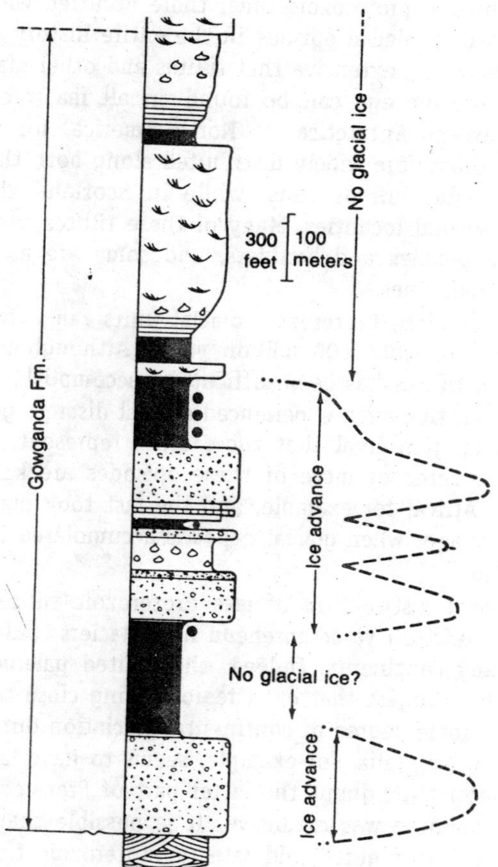

Fig. 3.3. Stratigraphic section of the Gowganda Formation of southern Canada in which the position of tillites is shown. These tillites appear to represent two major glacial advances, the second of which was divided into three pulses. Between these glacial advances, fine-grained, varved sediments with dropstones accumulated (dots).

of extensive continental glaciation not long after the transition from Archean to Proterozoic time.

Late Proterozoic Glacial Interval

Rocks ranging in age from less than 1 billion years to about 2.3 or 2.4 billion years show little evidence of glacial activity, indicating that the early Proterozoic period of glaciation was followed by an interval of about 1.5 billion years in which continental glaciers were rare or absent. Then, 800 or 850 million years ago, in late proterozoic time, there occurred what may have been the greatest glacial episode in the entire history of the earth. This episode was so extensive that tillites and other glacial deposits of late Proterozoic age can be found on all major continents of the world except Antarctica. In North America, for example late Proterozoic tillites are widely distributed along both the Cordilleran and the Appalachian orogens, while in Scotland they are well exposed at several localities. Many of these tillites contain striated and faceted pebbles and boulders, and some are associated with varve-like mudstones.

In general, late Proterozoic glacial units range in age from 1 billion years to about 600 million years. Although precise dating of individual tillites has been difficult to accomplish, it is strongly suspected that the earth experienced several discrete glacial episode during the long interval that these units represent. At least two and possibly three or more of these episodes are known to have occurred in Africa, for example, and the last took place about 600 million years ago, when glacial deposits accumulated in many parts of the world.

The global distribution of late Proterozoic tillites is puzzling in that it is difficult to comprehend how glaciers could have spread over so many continents. Indeed, the limited paleomagnetic data now available suggest that even regions lying close to the equator experienced some degree of continental glaciation during this time. Almost all of Australia, for example, seems to have lain within 30° of the equator throughout the latter part of Proterozoic time, and yet glaciation here was extensive. It is possible that most of the earth was at times quite cold late in Proterozoic time.

Atmospheric Oxygen

In the last chapter, we learned that the earth's primitive atmosphere, which formed by the degassing of the planet, contained little or no free oxygen. It is not known precisely when atmospheric oxygen reached its present level, but it is now widely agreed that a moderate level was reached about 2 billion years ago, early in Proterozoic time. Thus, the presence of very low levels of atmospheric oxygen seems to be another characteristic that distinguishes the Archean world from all subsequent times except the earliest part of the Proterozoic Eon. In this section, we will review the evidence supporting this conclusion, but first we will explain why the atmosphere serves as a reservoir for free oxygen today.

It would appear that the concentration of oxygen in the earth's atmosphere has not varied substantially over the past few hundred million years despite the fact that oxygen is continuously removed from the atmosphere by processes such as oxidation of minerals at the earth's surface and respiration. (Respiration is the process by which organisms employ oxygen to obtain energy from their food; in effect, organisms use oxygen to burn up their food in the same manner that fire oxidizes a flammable material by causing it to combine with oxygen and thus to release energy.) The atmospheric reservoir of oxygen persists because oxygen is continuously returned to it by photosynthesis and, to a lesser extent, by the breakdown of water in the upper atmosphere by the sun's rays.

A number of factors prevent the concentration of atmospheric oxygen from increasing to a significant degree. Were oxygen to build up much beyond its present level, for example, weathering would become more intense and would consequently "soak up" excess oxygen. Further-more, if plants markedly increased their biomass on earth and thus began to liberate more oxygen, animals, and simpler respiring organisms including bacteria, would also increase in biomass because more plants and plant debrls would be available for them to feed on. The resulting increase in the rate of oxygen consumption by respiration would then offset the increase in oxygen production by plants.

With these factors in mind, let us now examine the evidence supporting the assumption that low levels of atmospheric oxygen existed until early Proterozoic time.

Case for Early Anaerobic Evolution

One piece of evidence suggesting that the earth's early atmosphere was anoxic (oxygen-free) lies in the fact that the chemical building blocks of life could not have formed in the presence of atmospheric oxygen. Indeed, chemical reactions that yield amino acids from simpler compounds in the laboratory are inhibited by even smaller amounts of oxygen than these found in the earth's atmosphere.

Furthermore, oxygen has been found to be toxic to the most primitive living bacteria, including purple photosynthetic bacteria and bacteria that obtain energy from fermentation. As we have seen, photosynthetic bacteria as well as bacteria that derive energy from fermentation rather than from respiration are currently restricted to anoxic habitats such as swamps, ponds, and lagoons and are consequently termed *anaerobic*. The fact that the simplest living cellular organisms are anaerobic suggests that the earth's first forms of cellular life had similar metabolisms, and it also suggests that these organisms may have evolved in this way because there was virtually no atmospheric oxygen when they came into being. If this was the case, the subsequent buildup of atmospheric oxygen limited these bacteria to the environments that they now occupy.

Oxidation State of Minerals

Also widely cited as evidence of an anoxic Archean atmosphere is the distribution of uranium and iron minerals in Precambrian rocks. Under the atmospheric conditions of the modern world, the uranium oxide mineral uraninite (UO_2) is readily oxidized further and quickly dissolves from rocks. The iron sulfide mineral pyrite (FeS_2) also disintegrates readily when it is exposed to the modern atmosphere. Today, uraninite and pyrite are seldom found in the sediments of rivers and beaches, and they are similarly rare in ancient siliciclastic sediments younger than about 2 billion years. In contrast, both minerals are relatively abundant in buried nonmarine and shallow marine silicicalstic deposits older than 2 billion years; uraninite, for example, has been found in such rocks on five continents.

Significantly red beds display the opposite pattern: They are never found in terranes older than 2.2 or 2.3 billion years. Hematite, a highly oxidized iron mineral gives red beds their colour. Often, the hematite found in red beds has formed secondarily by oxidation of other iron minerals that accumulated with the sediments. During Phanerozoic time, which has been characterized by the presence of a well-oxygenated atmosphere, this secondary oxidation has often occurred within a few millions or tens of millions of years afters deposition. It would appear that oxidation of this type did not occur early in the earth's history.

Some researchers have argued that red beds cannot be found in Archean sedimentary sequences because the latter formed mainly in deep marine basins, whereas most red beds are of nonmarine origin. The Huronian and Witwatersrand sequences are especially significant in this regard because they formed more than 2 billion years ago on sizable cratons. Although both of these sequences comprise large volumes of nonmarine sediments and would therefore be expected to include some red beds under modern atmospheric conditions both lack red beds but contain substantial quantities of detrital pyrite and uraninite. This condition has been widely cited as evidence that oxygen concentrations remained low until 2 billion years ago or later.

Banded iron formations are other sedimentary rocks whose distribution may reflect the history of the earth's atmosphere. Rocks of this type are rare except in Precambrian terranes. Although banded iron formations are present in Archean terranes and, in fact, are among the oldest known rocks on earth, most accumulated early in the Proterozoic Era between about 2.5 and 1.8 billion years ago. The term *banded iron formation* refers to a bedding configuration in which laminae of chert that is sometimes contaminated by iron alternate with laminae of other minerals that are richer or poorer in iron than the chert. The iron in these formations may take any of a variety of forms including iron oxide, iron carbonate, iron silicate, and iron sultide- but the present chemical form of the iron is not necessarily the form in which it was originally deposited, and in many cases there is evidence that the mineralogy of the iron has altered over time. Banded iron formations account for most of the iron ore that is mined in the world today. One reason for their economic value is that many contain iron in the form of magnetite (Fe_30_4), whose iron-to-oxygen ratio is higher than that of hematite (Fe_20_3).

The origin of banded iron formations is problematic. In some of them, iron minerals accumulated as chemical muds, while in others they were deposited as sand-sized grains, and cross-lamination is some times present in the coarser accumulations. Some of the constituent particles of these formatios, such as those that became cross-laminated were loose grains. It is known that most banded iron deposits are marine, since they often occur in close eonjunction with marine deposits such as offshore turbidites mudstones, and pillow lavas. Some researchers believe that nearby volcanism provided the silica and iron of banded iron formation, while others maintain that most of the original iron mineral in these formations were washed into marine basins form the land.

The fact that the iron in banded iron formations was at least weakly oxidized when these structures first formed implies that some oxygen was present in the environment at this time. In fact, some iron formations are underlain by red beds that are even more highly oxidized. This suggests that atmospheric oxygen existed at a relatively high level between 2.2 and 1.8 billion years ago, attaining greater concentrations in some depositional environments than in others but remaining sufficiently abundant to produce at least weakly oxidized iron minerals in many environments. In any case, the abundance of red beds in rock sequences younger than banded iron formations testifies strongly in favour of a relatively high level of oxygen during the past 1.8 billion years, when banded iron formations have no longer been forming.

Stromatolites and Oxygen

It is widely agreed that photosynthesis caused atmospheric oxygen to build up during Precambrian time. Before such a buildup could occur, however, natural reservoirs known as *oxygen sinks* had to be filled. Essentially, oxygen sinks are chemical elements and compounds that combine readily with oxygen and that are believed to have been present in the crust or atmosphere immediately after the earth formed. Sulfur and iron are two of the most important oxygen sinks. (Note how, on a much smaller scale in the modern world, iron that we extract from naturally occurring compounds rust when exposed to atmospheric oxygen.)

Today photosynthesis liberates oxygen at such a high rate that it could refill the atmosphere within just a few thousand years. Given the probable presence of photosynthetic bacteria and blue-green algae by at least 3.5 billion years ago, how could it have taken nearly 2 billion years for atmospheric oxygen to fill major sinks and then approach a high level? The answer may well be that there were simply not enough stromatolites to liberate oxygen at a high rate until 2.3 or 2.2 billion years ago, when the geologic record reveals that stromatolites became very abundant. The reason for their earlier rarity may be attributable to certain geologic factors. Perhaps the absence of large continents and continental shelves during Archean and early Proterozoic time restricted the growth of stromatolites to the steep margins of small Archean landmasses offering little space or colonization.

It is also likely that during Archean time nutrients were not easily elevated from the deep sea to shallow water for use by stromatolites. Today, upwelling occurs along the margins of continents where currents drag surface waters away from the shore, and nutrient-laden waters from great depths replace these surface waters. During Archean time, however, the absence of large continents may have caused upwelling to be weak and nutrient supplies sparse. Then, as continents grew, more extensive upwelling

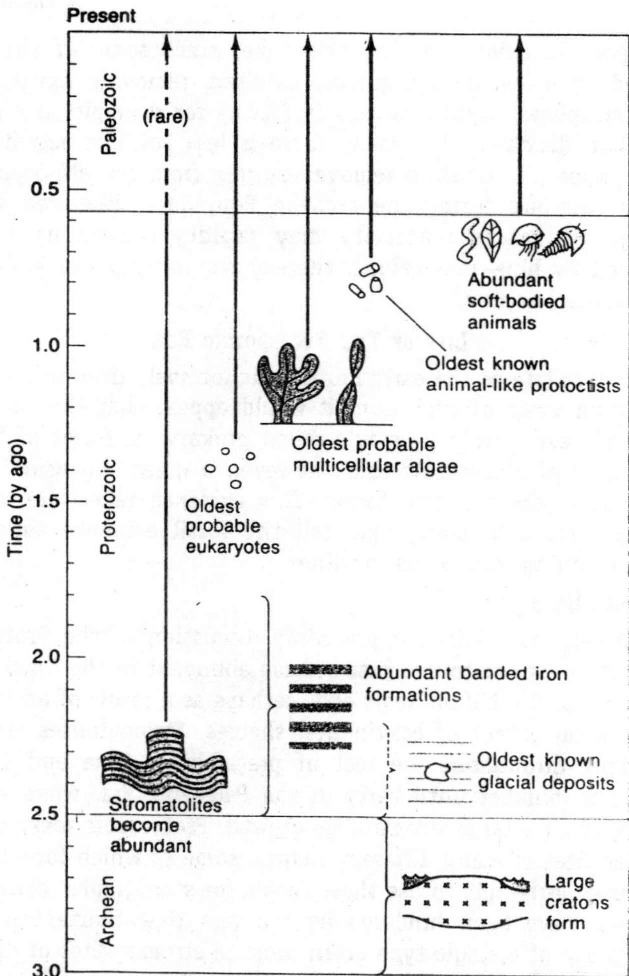

Fig. 3.4. Evolutionary developments during Proterozoic time, with arrows signifying survival of biological forms to the present time. Also shown are developments in the physical environment. All of the occurrences shown in this diagram are based on what we know from the rock record.

may have accelerated the rate of photosynthesis and oxygen production.

Volcanism and Oxygen

The widespread volcanism that seems to have characterized the hot Archean earth may also have impeded the accumulation of atmospoheric oxygen during this time, since some of the gases emitted by volcanoes are quickly oxidised, removing oxygen from the atmosphere. Carbon monoxide ((CO),) for example, is oxidized to carbon dioxide (CO_2). Today there is less volcanic activity, and so this type of oxidation removes oxygen from the atmosphere at a slow rate-but during the archean Eon, gases liberated by the widespread volcanic activity may rapidly soaked up oxygen produced by blue-green algae, thereby suppressing the buildup of atmospheric oxygen.

Life of The Proterozoic Eon

Even as late as the early Proterozoic interval, after an estimated 1.5 billion years of evolution, it would appear that the earth was populated exclusively by single-celled prokaryotic forms of life. In the course of Proterozoic time, however, a great expansion of life issued from these simple forms. This evidence takes the form of stromatolites and microscopic cell the fossil evidence supporting the continuing dominance outlines.

Stromatolites

As emphasized in the preceding discussion of the Proterozoic atmosphere, stromatolites first became abundant in the fossil record about 2.3 or 2.2 billion years ago, perhaps as a result of an increase in the areal extent of continental shelves. Stromatolites remained abundant throughout the rest of precambrian time and did not decline in number until early in the Paleozoic Era, when animals diverstied on a large stromatalite growth. Proterozoic rocks contain stromatolites of many different shapes, some of which form reeflike structures. Attempts to use these fossils for stratigraphic correlation, however, have been hindered by the fact that filamentous blue-green algae of a single type often produce stromatolites of different shapes in different environments. Similar ecologic influences on form in Precambrian time have made it difficult to identify evolutionary trends that might make fossil stromatolites useful for dating purposes.

Fossil Prokaryotic Cells

One of the most interesting fossil discoveries since the middle of this century has been the identification of cell remains preserved in Precambrian rocks. Such remains were first identified in a formation of Ontario and northern Minnesota known as the Gunflint Chert. The Gunflint fossils, which are about 1.9 billion years old, display a wide variety of shapes and appear to represent bacteria or blue-green algae, as do the members of other microscopic fossil assemblages of early Proterozoic age. Analysis of these fossils has revealed not only that prokaryotic forms of life persisted from Archean time into the Proterozoic Eon, but also that many of the forms found in precambrian rocks closely resemble forms that are alive today.

Earliest Eukaryotes

All forms of life except blue-green algae and bacteria are eukaryotes—that is, forms whose cells contain chromosomes, nuclei, and other advanced internal structures. Unfortunately, we do not know exactly when the first eukaryotic cell evolved. One promising approach to solving this problem involves size comparisons of fossilized cells of sphere-shaped algae. While very few living prokaryotic algae with a spherical shape are larger than 20 microns in diameter, spherical algae with diameters exceeding 60 microns are invariably eukaryotic. In the fossil record, microfossil assemblages older than 1.4 billion years are dominated by small cells whose diameters seldom exceed 10 microns, and it is only in rocks younger than 1.4 billion years that cells with diameters larger than 60 microns are abundant. These findings suggest that the eukaryotic condition evolved about 1.4 billion years ago or slightly earlier.

By far the most convincing evidence supporting the presence of eukaryotes during Proterozoic time is provided by fossil cells called *acritarchs*. These near-spherical or many-pointed forms are the dominant group of algal plankton found in the Paleozoic fossil record. Some acritarchs are believed to have represented the resting stages (or cysts) of dinoflagellate cells. Although dinoflagellates constitute one of the most important groups of planktonic algae today and are known to be eukaryotic, we do not know for certain that all acritarchs were dinoflagellates. In fact, some experts believe that Proterozoic acritarchs were not dinoflagellates but instead belonged to the group known as green algae. In any event the large size of many Proterozoic acritarchs, together with the chemical composition of their cell walls, suggests that acritarchs were eukaryotic. It is also significant to note that some Proterozoic acritarchs had cell walls whose complex patterns resembled those that today are restricted to eukaryotes. Interestingly, the fossil record of acritarchs, like that of large cells in general, begins in rocks about 1.4 billion years old.

Despite the antiquity of the acritarch fossil record, acritarchs are rarely found in rocks older than about 850 or 900 million years. In contrast, they are both abundant and diverse in younger rocks. In fact, the fossil record reveals that acritarchs underwent a rapid adaptive radiation between 900 or 850 and 700 million years ago. For this reason, acritarchs are very useful for dating rocks of latest Precambrian age.

It is interesting to note that acritarchs suffered a mass extinction at the time of worldwide glaciation about 600 million years ago. Only a few simple spheroidal types survived this crisis, which may have resulted from worldwide cooling.

Early Evolution of Eukaryotes

Because the fossil record of single-celled eukaryotes lacking skeletons is very poor, most of our ideas about the early evolution of this group are based on evidence derived from living organisms. The most remarkable of these ideas, which was not widely accepted until the 1970s, is that the eukaryotic cell arose from the union of two or more prokoryotic cells, at least one of which came to reside within another. According to this theory, the prokaryotic cell that first came to live inside another, was altered in minor ways to form a structure called a *mitochondrion*, one or more of which is present in nearly all eukaryotic cells. Mitochondria are, in fact, the structures that allow cells to derive energy from their food by means of respirations. In mitochondria, complex compounds are broken down by oxidation, which yields energy and, as a by-product, carbon dioxide. A mitochondrion is apparently the slightly modified evolutionary descendant of a small bacterium that became trapped within a larger one, as evidenced by the presence in mitochondria of DNA and RNA, both of which are essential to the independent existence of any cell. It is assumed that the smaller cell that became a mitochondrion was eaten by the larger one but proved resistant to the digestive processes of the predator cell.

There is one very primitive single-celled organism that seems to illustrate an early stage in the evolution of the eukaryotic condition. This is *Pelomyxa palustris*, a species that inhabits the bottom of freshwater ponds. Individuals of this species have nuclei, which constitute a eukaryotic trait, but the DNA within these nuclei is not arranged into chromosomes. Furthermore, mitochondria are lacking in *Pelomyxa* cells; in their place are bacteria that perform

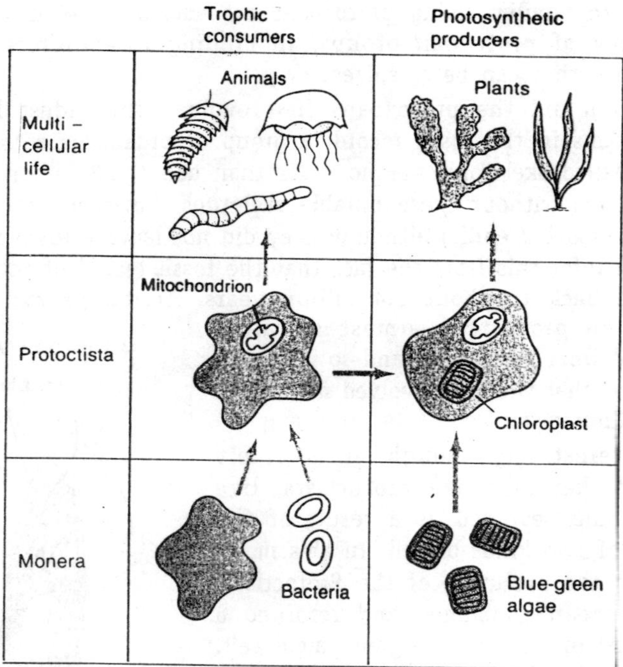

Fig. 3.5. The probable sequence of major events leading from Monera to multicellular animals and plants. The first protoctist apparently evolved when one moneran engulfed but failed to digest another, which then became a mitochondrion. A plantlike protoctist evolved when an animal-like protoctist engulfed but failed to ingest a blue-green algal cell, which then became a chloroplast.

essentially the same respiratory function. *Pelomyxa* dies if the bacteria living inside it are killed with an antibiotic. Although *Pelomyxa* itself may not have been ancestral to normal eukaryotes, it serves to illustrate the condition that may have preceded the transformation of intracellular bacteria into mitochondria. As the host, pelomyxa benefits from having its food broken down to supply its energy, while its intracellular bacteria benefit both from the abundance of food available and from the protection afforded by the host.

Single-celled eukaryotes and a few simple multicellular organisms constitute the kingdom Protoctista, and single-celled animal-like members of this kingdom (that is, consumers rather than producers) are called *protozoans*. It is believed that the earliest protoctists were protozoans, since their immediate ancestors were consumers that had eaten bacteria that became mitochondria. Among the more familiar living protozoans are amoebas and ciliates. A minority of protozoan groups, in cluding foraminifers, have skeletons that can be fossilized.

When did the protozoans first evolve? The oldest known protozoans in the fossil record, a group of organisms with rigid, vase-shaped skeletons, are no older than about 0.8 billion years. Protozoans without these durable skeletons, however, are known to have existed earlier although they did not leave a fossil record. We can infer this from the fact that the fossil record of acritarchs extend back to about 1.4 billion years. Acritarchs and other plantlike protoctists almost certainly evolved from protozoans and so it can be assumed that the later evolved more than 1.4 billion years ago.

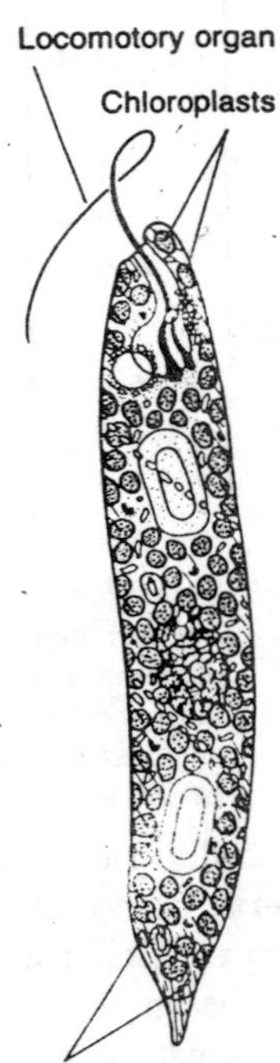

Fig. 3.6. Euglena, a plantlike (photosynthetic) protoctist, which, like many animal-like protoctists, is mobile. A Euglena cell contains large numbers of chloroplasts.

Interestingly enough, it is widely agreed that plantlike protoctists, like protozoans, evolved as a result of the union of two kinds of cell. In this major step in the evolution of the Protoctista, a protozoan consumed and retained a spherical or oblong blue-green algal cell. This cell then became an intracellular body known as a *chloroplast,* which served as the site of photosynthesis both in plantlike protoctists and in higher plants, which evolved from them. The similarity between blue-green algae and chloroplasts is striking. In both, for example, the pigment chlorophyll, which absorbs sunlight and permits photosynthesis, is located on layered membranes.

In many different kinds of protoctists, photosynthesis is conducted within chloroplasts, and it is generally believed that plantlike protoctists evolved many different times when protozoans retained within their cells the blue-green algae that they had eaten. Thus mobile protozoans may have evolved into mobile photosynthetic forms such as *Euglena* while immobile protozoans evolved into immobile photosynthetic forms.

Multicellular Algae

Algae (single, alga) is a term that encompasses a variety of forms. Photosynthetic protoctists, for example, are called algae, as are the prokaryotic blue-green algae and a large group of multicellular plants. The multicellular plants that are known as algae differ from more advanced land plants in that they lack multicellular reproductive structures to protect their eggs and embryos.

Today algae that are composed of many connected cells can be found in both freshwater and marine environments. Some groups form lettucelike carpets along rocky seashores, whereas others comprise the great kelps that rise up from some sea floors in twisted ribbons tens of meters long. Even these large forms, however, are fleshy structures that decay easily and are thus unlikely to be fossilized. Some ribbonlike fossils of Proterozoic age are probably segments of blue-green algal mats that only superficially resemble multicellular algae. Carbonaceous structures from 1.3 billion-years-old rocks of the Belt supergroup of Montana, on the other hand, may well be true multicelluler algae, since they are branched. Ribbons and oval remains from the 0.8- or 0.9-million-year-old Little Del Group of northwestern Canada almost certainly fall into this category.

Multicellular Animals

Multicellular animals evolved from animal-like protists rather than from multicellular plants. It is unlikely that any kind of stationary multicellular alga would evolve into an animal that must gather food. On the other hand, it is likely that certain mobile, predatory animal-like protists evolved into higher animals simply by developing multicellular body forms.

Before there were single-celled or multicellular eukaryotic consumers, the earth's ecosystem was relatively simple. Because the primary photosynthetic producers—blue-green algae and eukaryotic algae—did not suffer predation, they multiplied in aquatic settings to an extent that was limited only by the supply of nutrients essential to their growth. Seas and lakes were, in effect, saturated with algae—a situation that may have slowed evolution by leaving very little room for the origin of new species. Although the first organisms to feed on algae must have been animals like protists, today these forms are feeble in this role in comparison to multicellular animals, which because of their size can consume algae rapidly. Thus, the origin of eukaryotic consumers, especially multicellular animals, added a new trophic level to many aquatic ecosystems. Exactly when multicellular animals, evolved remains uncertain, but we now have evidence that it was not until late in Proterozoic time that these organisms first appeared. For many years, Paleontologists have searched well back into the Precambrian interval for skeletons of multicellular animals, but their search has been unsuccessful. The only skeletons that have been found have come from rocks situated close to or above the precambrian-Cambrian boundary—which, as we will see, cannot be recognized precisely in most areas.

Even during the last century, it was acknowledged that fossil skeletons appear in the stratigraphic record quite suddenly near the base of the Cambrian System. The complexity and variety of fossilized Cambrian life gave rise to speculation that multicellular animals had a long Precambrian history during which they lacked hard parts and therefore left no fossil record. One highly effective means of testing this hypothesis is based on the fact that soft-bodied multicellular animals—those that lack hard parts—have always crawled over the sea floor or burrowed into it and, in doing so, have often left trace fossils in the form of tracks, trails, and burrows in sedimentary rocks. It was therefore reasoned that if soft-bodied invertebrate animals had existed for a long Precambrian interval, some of them would have left such trace fossils. A search for trace fossils in Precambrian rocks has since turned up a striking pattern—such fossils have been found only in rocks less than about 1 billion years old. For example, 1.3 billion-year-old sedimentary rocks of the Belt Supergroup of Montana, which has yielded some of the multicellular algae of exhibit no tracks or trails. Belt mudstones are often strikingly well layered in comparison to younger deposits, in which laminae are often disrupted or destroyed by burrowing animals.

Although the oldest undisputed Precambrian trace fossils are thought to date back less than 1 billion years, the age of these fossils has never been precisely determined. Such fossils do, however, display a general evolutionary pattern. The oldest trace fossils are geometrically simple structures; often they are tubes made by wormlike animals that burrowed through the sediments.

In several region of the world, as stratigraphic section progress upward toward and into the Cambrian System, simple trace fossils are replaced by more and more complex types whose shapes also vary increasingly. This pattern of increase in both complexity and variety seems to represent the initial evolutionary diversification of multicellular animals in the world's oceans. In many regions the oldest trace fossils are found above the youngest Precambrian glacial sediments, which suggests that little mutlicellular animal life existed before the final glacial episode.

Additional evidence supporting the presence of multicellular life in the latest Precambrian interval has come to light since 1950. In many parts of the world, imprints of soft-bodied animals have been found in sandstone below the oldest rocks containing fossil skeltons. Many of these fossils have been thought to represent jellyfishes or sea pens, although this assignation has recently been questioned. Jellyfishes and sea pens are living groups that belong to the phylum *Coelenterata* (the same phylum as corals). Whereas jelly fishes float in the water, sea pens are stalked creatures that stand upright on the sea floor. The Ediacara fauna of Australia is the most famous of late Precambrian "soft-bodied" faunas and was also the first to be recognized. In addition to the possible coelenterates, the Ediacara fauna includes several additional kinds

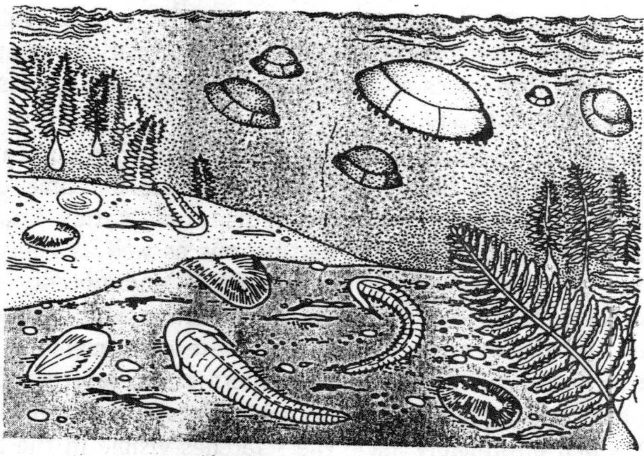

Fig. 3.7. A reconstruction of the Ediacara fauna based on the interpretation that some Ediacara fossils represent sea pens and some, jellyfishes. The stalked forms standing upright on the seafloor are sea pens, and the floating, belt-like animals are jellyfishes. Unlike the other forms represented here. these two groups of coelenterates have close living relatives.

of animals that cannot be related with certainty to any living group. Fossils of soft-bodied animals such as those of the Ediacara fauna are rare in rocks of Phanerozoic age. It seems likely that the late Precambrian faunas that were preserved escaped destruction only because few predators or scavengers at that early time were capable of devouring their carcasses quickly.

It is puzzling that multicellular animals, including those that form the Ediacara fauna, seem not to have undergone substantial evolutionary diversification until late in proterozoic time. In fact, the first multicellular animal may not have originated until considerably after 1 billion years ago. We have just seen that the presence of large cells in the fossil record suggests that the eukaryotic cell evolved more than 1.4 billion years ago. Thus, an interval of several hundred million years seems to have separated the origin of the eukaryotic cell from the diversification of multicellular animals. What could account for this evolutionary delay? It has been postulated that protozoans were too primitive to give rise to the multicellular animals before very late in Proterozoic time, perhaps because these organisms lacked sexual reproduction or other advanced eukaryotic features. This idea appears plausible but is unsupported by biological or geologic evidence. An alternative hypothesis is that atmospheric oxygen accumulated slowly, failing to reach the levels necessary to support higher life until late in Proterozoic time. This notion is also feasible, but to date it, too, lacks factual support. Certainly, however, there is geologic evidence—such as the initiation of widespread red-bed formation—supporting the hypothesis that levels of atmospheric oxygen were increasing by about 2 billion years ago.

What is clear is that animals more advanced than coelenterates were well established before the end of Proterozoic time. Among these advanced groups were the segmented worms known as *annelids*, which include modern earthworms as well as many kinds of marine and freshwater species. Annelids undoubtedly formed many of the tubelike fossil burrows of late Proterozoic age, perhaps including the simple ones. Also present were early members of the phylum *Arthropoda*, which includes modern crabs, lobsters, insects, and spiders. Arthropods have external skeletons and jointed legs, a set of which presumably made the scratches visible in the burrow. Coelenterates, annelids, and arthropods not only continued to flourish into Paleozoic time but have also remained important up to the present time.

Proterozoic Cratons

Although geologists have long attempted to determine how the continents of the modern world originated, the histories of these continents have been traced back only into Proterozoic time. Indeed, the configurations and relative positions of most older, Archean microcontinents will probably never be known. At the same time, uncertainties remain even about the histories of large Proterozoic cratons. The difficult is that depositional patterns and structural trends of rocks more than half a billion years old are often obscured by erosion, metamorphism, or burial; Paleomagnetic data are also sparse and difficult to interpret. Nonetheless, geologists have reconstructed partial histories for most large blocks of Proterozoic crust and have thus gained some knowledge of how they became part of the Phanerozoic world that forms the subject of the final seven chapters of this book.

In the present title, we will learn how North America grew episodically during Precambrian time and how this continent also appears to have lost terrane through continental rifting. We will then review what is known of the ancient landmass of Gondwana-land, which, during Proterozoic time, included the oldest block of continental crust that has been identified in the modern world. Finally, we will examine the Precambrian terranes that during Phanerozoic time became assembled into the great northern continent of Eurasia.

Before we discuss the histories of individual cratons, however, let us consider how most cratons undergo changes in size. As we have seen, size increase on a large scale results from the *suturing* of major cratons, which takes place along a subduction zone and is usually accompanied by mountain building in the vicinity of the suture.

Cratonic growth on a smaller scale, which is known as *continental accretion*, also entails mountain building, but this process occurs at the margin of a single large craton. As noted earlier, marginal accretion can result either from the attachment of a *microplate* to a large craton along a marginal subduction zone, or from the compression and metamorphism of sediments that have accumulated along a continental shelf. The latter is some times referred to as *stabilization*, since it thickens the crust and hardens both unconsolidated sediments and soft sedimentary rocks.

Stabilization is a cannibalistic process inasmuch as some of the sediment that is deposited and stabilized along a continental

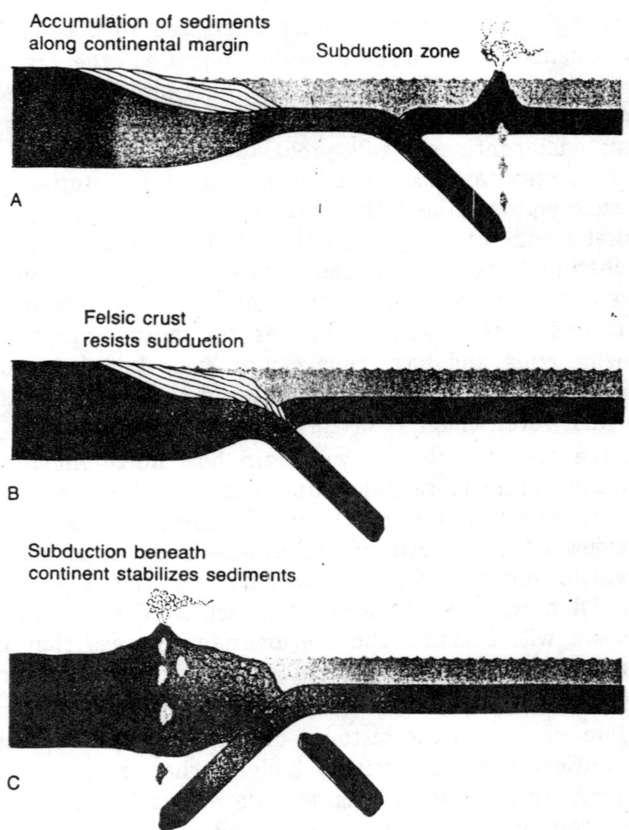

Fig. 3.8. Stabilization of sediments that have accumulated along the margin of a continent (A). The continent comes to rest along a subduction zone. Because the continent is of low density and resists subduction, the direction of subduction switches (B). Igneous activity then adds rock to the continental margin and metamorphoses the sediments that have collected there (C).

margin is derived from the interior of the continent by erosion. On the other hand, limestone that accumulates along continental margins is precipitated from sea water or secreted by organisms and thus represents an external contribution to the mass of the continent—as do the igneous rocks and oceanic crust that become welded to a continental margin resting along a subduction zone.

Orogenic processes do not simply add material to continents; they also alter preexisting crust. Regional metamorphism, for example—which sometimes operates in conjunction with structural deformation—*remobilizes* continental crust. As statied already, metamorphism often alters the character of preexisting rocks beyond recognition and resets rediometric clocks so that the age of the crust can no longer be determined. In reviewing the Precambrian history of individual Proterozoic cratons, we will encounter many examples of crustal remobilization.

How do continents decrease in size? They can shrink by erosion, but this process operates so slowly that it is of little overall significance. Far more important is the process of continental rifting, which operates on many scales. It can remove a small sliver of crust, or it can divide a large craton in half. Although continental rifting that took place more than half a billion ago is difficult to document, there is evidence suggesting that certain major rifting events occurred late in Proterozoic time.

North America, Greenland and Northern Britain as Laurentia

Although it lies close to North America, Greenland today is a continent in its own right. Until late in Phanerozoic time, however, Greenland was attached to North America, and these two continents were primary components of a large landmass known as Laurentia. As we will see, the crust that now forms northern Great Britain was also part of Laurentia during Proterozoic time, and it may be that Siberia and Scandinavia were also attached to this great landmass until mid-Proterozoic time. Forming the core of Laurentia was the crustal block that now represents most of the North American craton. This ancient block is well exposed today as the largest Precambrian shield in the world: the Canadian Shield.

The Canadian Shiel constitutes a large portion of the North American craton, including a small part of the northern united States. Precambrian rocks also underlie the interior of the continent to the south, where they are overlain by relatively thin veneer of Phanerozoic sedimentary rocks. Rocks obtained from wells that penetrate the Phanerozoic cover have provided a good picture of the general distribution of buried Precambrian rocks. These rocks, together with the exposed rocks of the Canadian Shield, reveal that in the course of Proterozoic time Laurentia gained territory by continental accretion, but also lost territory through continental rifting.

Precambrian Provinces and Continental Accretion

The Appalachian and Ouachita orogenic belts flank the North American craton on the east and south, while the Cordilleran belt

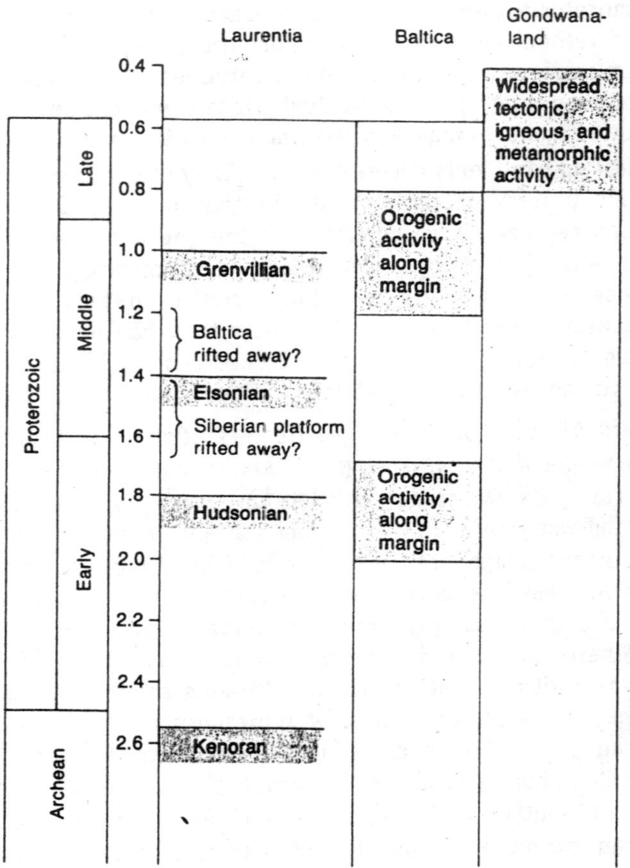

Fig. 3.9. Major geologic events in Laurentia, Gondwanaland, and Baltica after 2.7 billion years ago.

flanks it on the west. We have also seen that new crust was added to the margin of the continent during the development of the Appalachian mountain system in the Palezoic Era. As we will see in later chapter, the late Paleozoic events that produced the Ouachita mountain system also increased the size of North America, as did the Paleozoic, Mesozoic, and Cenozoic events that produced the Cordilleran system.

Evidence that North America was growing by accretion during Proterozoic time began to appear decades ago, when the recognition of structural trends and regional lithologic patterns permitted geologists to recognize natural geologic provinces within the Canadian Shield. More recently reliable radiometric dates for rocks of the Canadian Shield and for subsurface rocks bordering the shield have yielded a much more detailed picture. These dates are obtained from crystalline rocks, and thus they represent episodes of igneous and metamorphic activity. As it turns out, such activity occurred throughout Precambrian time but was concentrated primarily in four major intervals, along with comparable intervals on other cratons. The igneous and metamorphic rocks that were produced during each interval are not spread randomly throughout the North American craton, however; instead, they tend to be localize in descreted belts that are formally recognized as geologic provinces.

The arrangement of the Precambrian provinces tells the story of continental accretion. To understand this story, however, it is important to recognize that each date represents the youngest interval of widespread activity within a province. How might we expect provinces of various ages to be arranged within the North American craton? In general, we would expect older provinces to lie in the center of the craton and younger provinces to lie near the margin. In a general way, this is the pattern that we find within the North American craton. Note for example, that the ancient Superior Province which has remained stable since latest Archean time (about 2.6 billion years ago), is flanked by younger provinces. The Central province, which has remained largely stable for the last 1.4 billion years, borders it on the south. Both the Superior and Central provinces are bordered on the east by the Grenville Province, which represents a long marginal or ogen that was active about 1.0 billion years old. The Grenville Province, in turn is bounded by the younger Appalachian orogenic belt, which is of paleozoic origin.

The nature of the Grenville and Appalachian orogens illustrates that while an orogenic episode may add crust to the margin of a craton, it usually also deforms and metamorphoses older cratonic crust that lies alongside the orogenic belt. Thus while the *Grenville* orogenic episode probably accreted some new crust to the eastern margin of North America, it also metamorphosed a large segment of crust that had previously represented terrane of the Superior and Slave provinces. The Appalachian orogenic activity, in turn, overlapped the Grenville Province. In fact, the Blue Ridge and similar uplifts to the north and south consist of crystalline rocks of Grenville age that were caught up in the Appalachian deformation.

Stabilization of Archean Crust: The Kenoran Interval

Recall that podlike bodies of greenstone characterize many Archean terranes throughout the world. We observed that the largest group of Archean greenstones of the Canadian Shield occurs in the Superior Province. Within this province are numerous bodies of crystalline rock whose radiometrically determined ages cluster around 2.6 billion years. Widespread igneous activity produced these rocks during the *Kenoran* orogenic interval.

The Kenoran orogenic events at the close of Archean time appear to have united many small Archean crustal blocks to form a large, stable craton that now constitutes the Superior Province. The smaller Wyoming province is characterized by uplifts such as the Beartooth Mountains and the Black Hills, a domelike structure of Precambrian basement rocks that is elevated high above the Great Plains; here the giant presidential busts of Mount Rushmore are sculpted from Archean rocks.

In the Southern Province, which lies primarily Minnesota, are rocks that yield some of the oldest radiometric ages ever obtained on earth: nearly 3.7 billion years. This small province was continuous with the Superior Province until about 1.8 billion years ago, when the two provinces were separated by a narrow zone of metamorphic activity.

The Archean rocks that form the Nain Province of north-eastern Canada and Greenland are separated from the Superior terrane by a narrow band of remobilized rocks. Greenland, of course, is now a discrete continent, but its Precambrian terranes align with those of Canada on the opposite side of Baffin Bay. Greenland did not separate from North America until late in the Cretaceous Period, less than 100 million years ago. The southern tip of Greenland is considered to be part of the Nain Province, which means that it was continuous with the North American segment of this province at the end of the Kenoran orogenic interval. The Slave Province to the northwest would appear to be the remnant of another landmass that existed at the end of the Kenoran interval.

The Hudsonian Interval

The preceding section explains why geologists believe that sizable landmasses had formed from small Archean crustal blocks by the beginning of Proterozoic time, about 2.5 million years ago. What followed in early Proterozoic time was an interval of relative tectonic quiescence that lasted until slightly less than 2 billion years ago. The came another pulse of widespread orogenic activity in Canada that has been labeled the *Hudsonian* interval. Mountain building in the Wopmay orogen occurred at this time, accompanied by orogenic activity in a horseshoe-shaped belt that now encircles the peninsular region east of Hudson bay. In the area of this second orogen, which is now represented in the Labrador Trough, banded iron formations are among the most conspicuous sedimentary units. Also deformed and metamorphosed during the Hudsonian interval was the Animikie sequence near Lake superior. Mining of banded iron formations here has provided the United States with most of its iron ore.

Many crystalline rocks within the Churchill Province of western North America yield radiometric dates of about 1.8 billion years, which indicates that Hudsonian orogenic activity occurred here as well. Magnetic anomalies and deviations in the earth's gravitational field in this province reveal structural trends in Precambrian rocks lying beneath younger sediments. Distinctive north-south trends within the Wyoming and Superior provinces differ from those of the Churchill Province, suggesting that the latter provinces were united in the intervening zone (which now lies within the Churchill Province) when orogenesis occurred here about 1.8 billion years ago.

Largely spared from the deformation and metamorphism of the Hudsonian interval was the Huronian sequence of sediments east of Lake Superior. Here, Preserved with little alteration, are magnificent varved deposits that represent the Proterozoic glacial episode that took place more than 2 billion years ago.

The Elsonian Interval

Most of the present Canadian Shield had accreted by the end of the Hudsonian orogeny about 1.8 billion years ago. The stability that resulted is reflected in the fact that igneous and metamorphic rocks with dates appreciably younger than 1.5 billion years are all but unknown in the central provinces of the Canadian Shield (the Superior, Churchill, and Slave provinces). Rocks ranging in age from 1.4 to 1.5 billion years, however, are common in large areas along the eastern and southern flanks of the early Proterozoic craton. Drill cores show that much of the midcontinent of the

United States is also underlain by Elsonian-age rocks. Thus, although details are not yet cleat, it would appear that, soon after the close of the early Proterozoic interval, the craton grew toward the east and south.

Middle Proterozoic Rifting in Central and Eastern North America

The greatest disturbance of the central North American craton during the last 1.4 billion years was an episode of continental rifting that took place between about 1.3 and 1.0 billion years ago. This episode, which ultimately failed, was characterized by faulting and by mafic igneous activity in which lavas poured into downwarped basins along a belt that extended from the Great Lakes region to Kansas. Some of these lavas, which are known as Keweenawan basalts, are exposed near the southern border of the Canadian Shield. They contain ore deposits of native copper, a mineral consisting of the element copper uncombined with other elements. Similar basin basalts lie to the southwest beneath the sedimentary cover of the Midwest, as has been revealed both by the examination of rock cuttings taken from deep wells and by the detection from the earth's surface of strong magnetism. Because these basalts are rich in iron and magnesium and are therefore of high density, their presence is also associated with a feature known as the Midcontinental Gravity High, which is a local increase in the earth's gravitational field as measured from the surface. While the basalts were forming, numerous basic dikes were also emplaced across the Canadian Shield to the north, and a large plutonic igneous body, the Duluth Gabbro, was intruded within the area of the Keweenawan eruptions.

The configuration and composition of Keweenawan rocks and their subsurface counterparts are reminiscent of the Triassic Newark Series, whose configuration indicates the presence of a failed rift in the eastern United States. The Keweenawan volcanics, for example, are associated with red siliciclastic rocks and alluvial-fan conglomerates in what appear to be downfaulted troughs - configurations that tend to occur in newly forming continental rifts. Thus, it would appear that about 1.3 billion years ago, a spreading center formed beneath the late Precambrian craton and began to rift it apart. It is possible that this spreading center intercepted the eastern margin of the ancient craton, but this remains uncertain. In any event, it is obvious that the Keweenawan rifting was abortive. Rifting ceased before the continent was split but left its mark in the enormous volumes of mantle-derived lavas that were disgorged along a belt more than 1500 kilometers (~900 miles) long and 100 kilometers (~60 miles) wide!

We have seen that most of the present Canadian Shield had accreted by the end of the Hudsonian orogenic interval about 1.8 billion years ago. The emplacement of large dikes within the Churchill Province between 1.3 and 1.0 billion years ago, together with the rifting and the mafic extrusions of the Keweenawan Supergroup, indicates that the craton was rigid at this time. Paleomagnetic data provide further evidence that what we now Cell the Canadian Shield has behaved like a rigid block since the Hudsonian interval or even earlier: Rocks within the various provinces of the shield show similar paths of apparent polar wander for the last 2 billion years.

Let us now examine what happened along the eastern and western margins of the large North American craton which had formed by mid-Proterozoic time.

The Grenville Interval in Eastern North America

Recall that the Grenville orogenic belt formed along the east coast of North America. The igneous and metamorphic activity of the Grenville orogeny ended about 1 billion years ago. The resulting crystalline rocks and other are best exposed in the Canadian portion of the Grenville Province. To the south most crystalline rocks of Grenville age are buried, but some crop out here and there-especially in the Adirondack uplift of New york State and, as noted above, in the Blue Ridge Mountains and other uplifts lying between the valley and Ridge Province and the Piedmont. In the Gulf Coast region, most of the Grenville Belt lies beneath Phanerozoic cover but in central Texas crystalline rocks of Grenville age appear at the surface as a small, isolated prominence known as the Llano uplift.

Much evidence indicates that between about 1.4 and 1.0 billion years ago, a major episode of continental rifting took place in the eastern half of North America, creating a new continental margin in the area of the present northern Appalachian Mountains. The enormous failed rift of the midcontinent region, where the Keweenawan rocks accumulated, represents one piece of evidence, and another is a failed rift of similar age in the vicinity of Montreal. The rocks of the Montreal failed rift include volcanics situated in northeasterly trending dikes. These failed rifts must have projected

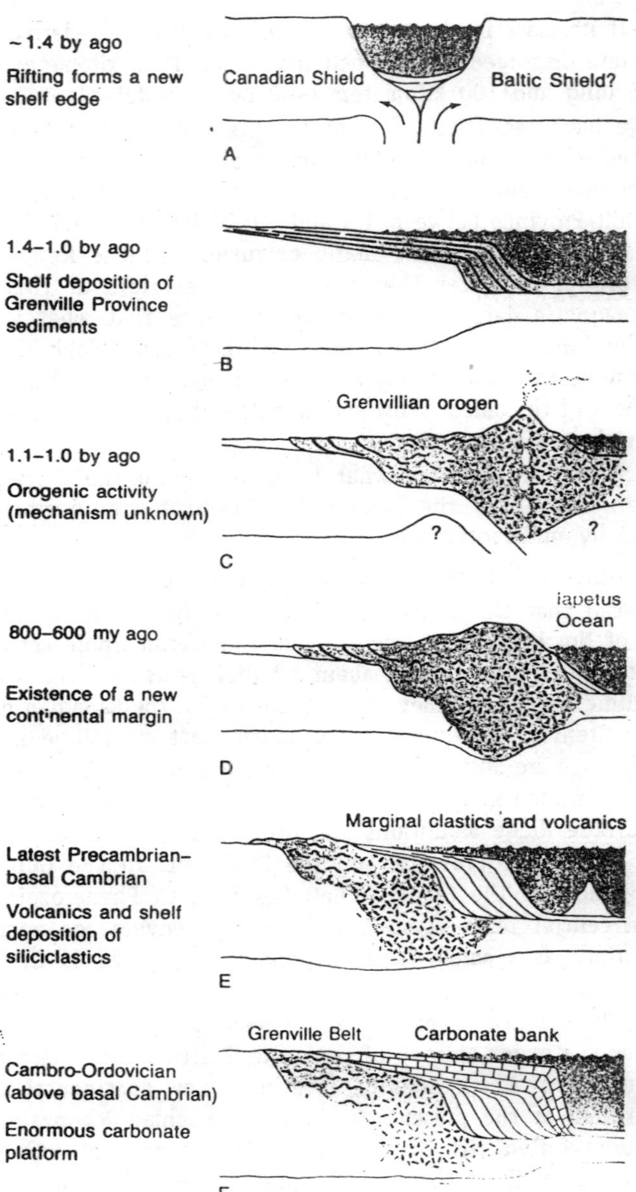

Fig. 3.10. Sequence of events along the eastern part of the North American craton from middle Proterozoic time until early Paleozoic time.

from a continental margin that probably came into existence about 1.4 or 1.3 billion years ago, precisely when the failed rifts began to form. The newly formed continental shelf must have lain to the southeast of the present Grenville Province, since sediments laid down along this shelf are now buried beneath the Piedmont Province and the Coastal Plain.

What landmass was rifted away from eastern North America to form the new continental margin? One possibility is that this landmass, or at least part of it, was the Baltic Shield. There is paleomagnetic evidence indicating that Scandinavia may have lain close to eastern North America about 1 billion years ago, and, as we will see in the next chapter, there is also a Precambrian province of Grenville age in eastern Scandinavia.

Figure 3.10 Summarizes the events that produced the rocks of the Grenville Province. As this figure illustrates, rifting was followed first by deposition along the newly formed shelf and then by the formation of the Grenville orogen. Deposition and stabilization of the sediments of the Grenville belt added a marginal increment of unknown proportion to eastern North America. It is not known, however, whether the Grenville orogeny between about 1.1 and 1.0 billion years ago, involved continental collision or simply the rafting of eastern North America against a subduction zone.

In any event, after the orogeny a new depositional margin formed along the eastern margin of North America. This margin seems to have developed earlier in the north, where marine deposition began about 800 million years ago, than in the southern United states, where it did not commence until about 600 million years ago. The body of water that bordered eastern North America during this interval of deposition was the lapetus Ocean. Recall that the latest Precambrian and early Paleozoic sediments of the first tectonic cycle of he Appalachian Mountain system accumulated along the margin of the lapetus Ocean and that in Cambro Ordovician time, these sediments were carbonates deposited along a vast carbonate platform.

The British Isles

Modern geology was born in Great Britain more than a century ago, and it was there that most of our formal goelogic systems were originally defined. Thus, it is lucky accident that this small group of islands happens to contain rocks representing every Phanerozoic period, together with a variety of rocks of Precambrian age. Precambrian rocks of the British Isles are best exposed in northern Scotland where James Hutton did his pioneering work in uniformitarian geology, as well as in the northern part of Ireland. In these northern areas, Precambrian rocks form part of an ancient shield that during late Precambrian and early Paleozoic intervals was connected to the Canadian and Greenland shields as part of Laurentia. At this time, southern Great Britain was a separate landmass. Northern and southern Great Britain became attached to each other during the middle Paleozoic Caledonian orogeny. As we will learn in later chapters, all of Britain (as well as the rest of Europe) remained attached to Greenland until early in the cenozoic Era, less than 60 million years ago.

Figure 3.11 illustrates the attachment of northern Britain to Greenland during late Proterozoic time and the general sequence of Precambrian rocks in northern Britain. The older rocks in this area, which are known collectively as the Lewisian sequence, range in age from late Archean to early Proterozoic (about 2.9 to 1.7 billion years), although metamor-phism may have disguised even older Archean rocks. Lewisian rocks are separated from younger Precambrian rocks by an unconformity. Metamorphic rocks (or gneisses), which make up the bulk of the Lewisian complex, crop out spectacularly throughout northern Scotland in what is known as the Northwest foreland.

Southeast of the main Lewisian outcrops in Scotland metamorphosed sediments known as Moinian are exposed over a large area. These rocks, which are about 1.2 billion years old, were originally poorly sorted siliciclastic deposits of the Grenville Belt, deposited when Scotland was attached to Greenland and North America. They were then metamorphosed and deformed during the Grenville orogeny about 1 billion years ago and were ultimately transported northwestward along what is known as the Moinian thrust fault during the Caledonian orogeny. In the Northwest foreland, another package of sediments, the Stoer Group, was deposited on top of the Lewisian slightly later than the Moine sequence was laid down as a consequence of which the Stoer Group is less severely deformed.

Between about 900 and 800 million years ago, an erosion surface developed on the deformed Moine and Stoer groups, exposing the older Lewisian rocks in some places. On top of this irregular surface were deposited thick sequences of sediment that

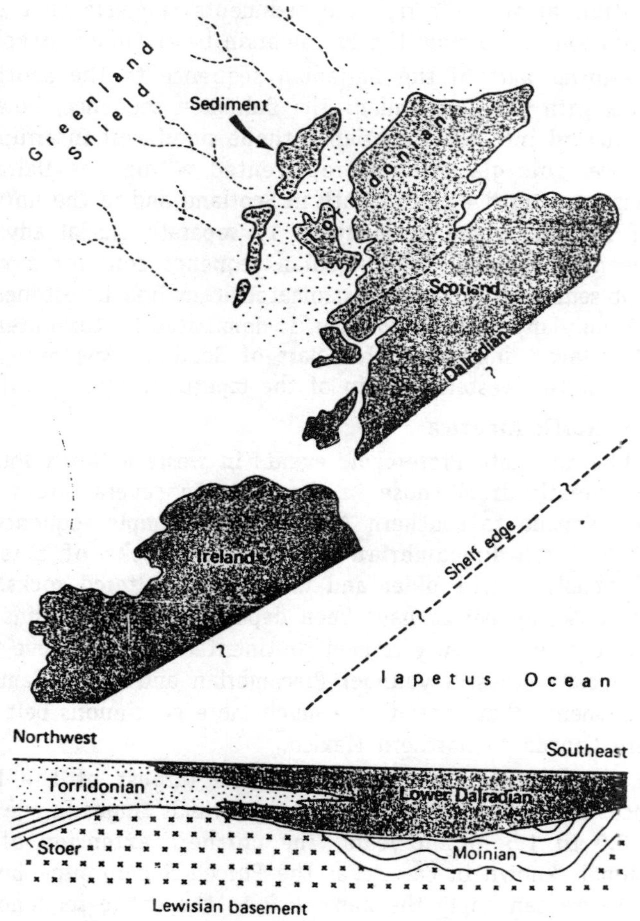

Fig. 3.11. Stratigraphic cross section of Precambrian rocks in northern Scotland and Ireland (above) and the generalized geography of northern Scotland and Ireland when Torridonian and Dalradian sediments were deposited near the end of the Precambrian (below); these regions were attached to Greenland along the margin of the Iapetus Ocean.

have traditionally been labeled Torridonian in the north-west and Dalradian in the southeast. The Torridonian sequence, which accumulated between 800 and 650 million years ago, is often separated from underlying rocks by a striking angular unconformity and includes several distinct packages of sedimentary rocks, the most conspicuous of which consist of red arkoses. The mineral composition of some Torridonian sediments suggests that these sediments were shed from the Precambrian basement of Greenland.

The lower part of the Dalradian sequence to the southeast correlates with the Torridonian. The Dalradian sequence, however, ranges upward into the Cambrian without significant interruption. Late Proterozoic glaciation is represented within the Dalradian sequence by extensive tillites both in Scotland and in the northern part of Ireland, where as many as 17 separate glacial advances have been recognized. The Dalradian sequence contains a varied group of sediments, including stromatolitic marine limestones. Its upper (Cambrian) portion, however, is dominated by turbidites and by pillow lavas, indicating that part of Scotland was once deep sea floor at the western margin of the lapetus Ocean.

Western North America

Middle and Late Proterozoic events in western North America differed greatly from those in the east. In several areas from southern Canada to southern Arizona, for example sequences of relatively fresh Precambrian sedimentary rocks of this age unconformably overlie older and more highly altered rocks. The younger rocks appear to have been deposited in local basins that probably lay near a newly formed continental margin. Above these basin deposits are still younger Precambrian and Lower Cambrian shelf sediments that extend in a much more continuous belt from northern Canada to northern Mexico.

In the northern United States, the largest of the basin sequences is the famous Belt Supergroup, which ranges in age from about 0.9 to 1.5 billion years. The northern extension of this formation is known in Canada as the Purcell Supergroup—but for simplicity we can apply the name *Belt* to the entire sequence. In general, the Belt thickens toward the west, sometimes reaching a thickness of 16000 meters.

The Belt formed in a northwesterly trending basin resembling a failed rift. Sandstones increase in abundance toward the western part of the sequence, while limestones increase toward the east, where deposition occurred in shallower water. In general, however, mudstones predominate. The Belt apparently formed as a result of sediment accumulation in very shallow water during rapid subsidence. Salt crystals and mud cracks, both of which indicate a drying up of shallow bodies of water, are present in Belt sediments, and shallow-water stromatolites are also abundant in the limestones.

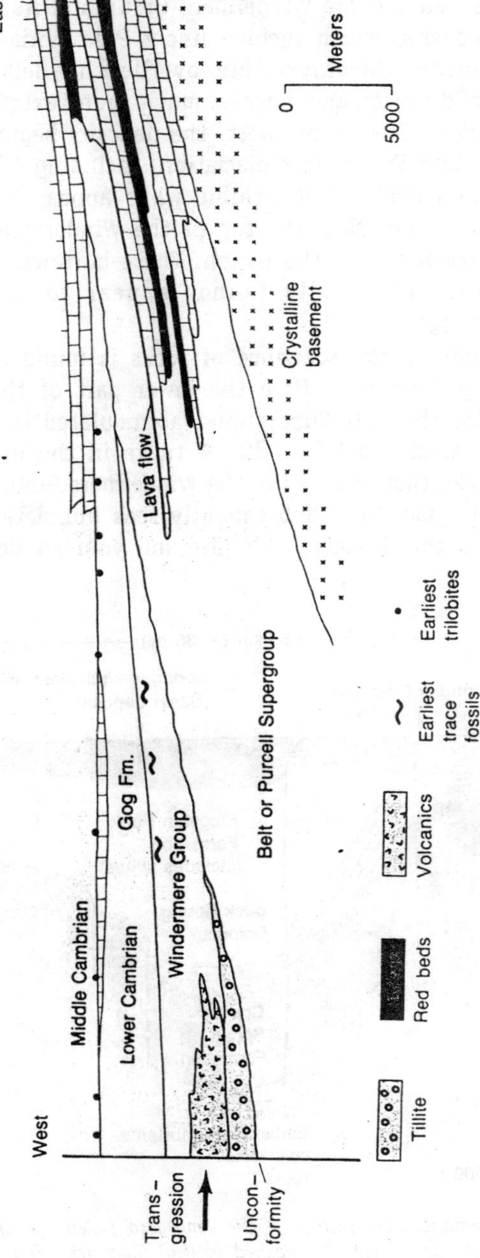

Fig. 3.12. Stratigraphic sequence in Canada just north of Montana, ranging from the middle Proterozoic through the Middle Cambrian. The Belt Supergroup accumulated in a large basin.

Above the Belt are the Windermere Group and its equivalents. These shelf deposits, which include upper Precambrian and lower Cambrian deposits, unconformably overlie the Belt and other localized basin deposits and have a much more extensive north-south distribution. Toward the west, the deposits begin with rocks that represent late Proterozoic glaciation, including tillite as well as pebbles and cobbles that exhibit what appear to be glacial striations or scratches. Near the top of the Windermere are found the earliest trace fossils of the region. These burrows, which occur well below the first shelly faunas appear to be very late Precambrian in age.

A remarkably similar sequence of rocks is found as far south as Death Valley, California. Here the lower part of the sequence, which resembles the Belt Supergroup, accumulated in a structure known as the Amargosa failed rift. A tillite in the upper part of the sequence, like that recorded in the Windermere Group, represents late Proterozoic glaciation. Subsequently seas inundated a broader area over which the Noonday Dolomite and younger deposits were laid down.

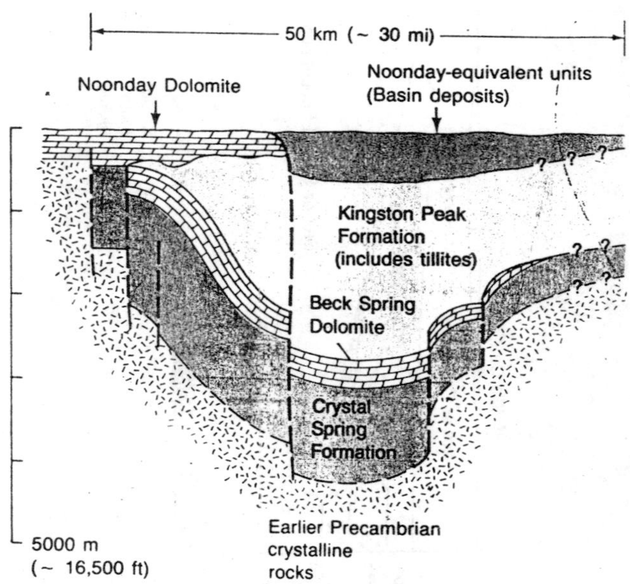

Fig. 3.13. Diagrammatic cross section of the Amargosa failed rift after deposition within it had ceased. The dashed vertical lines are normal faults.

Additional localized sedimentary units were deposited in late Proterozoic time to the south and east. Thus, the interval represented by the Belt Supergroup (approximately 1.5 to 0.9 billion years ago) was a time of localized basin deposition in western North America. In many areas, however—including the Belt basin—sediments deposited during this interval were soon weakly deformed. The more widespread latest Precambrian and Early Cambrian sequences—the windermers and its equivalents—were then deposited during a marine transgression along the full length of the continental margin.

It seems evident that a rifting episode preceded the interval of Belt deposition, since a new continental margin was produced, and another failed rift cut into the continent in the Death Valley region. The other new marginal basins may have represented failed rifts as well. In this regard two major questions remain to be resolved. First, what produced the modest deformation of these units in many areas before the younger Windermere and its equivalents were spread over them? And second, what happened to the landmass that was rifted away about 1.5 billion years ago, leaving a new continental margin along which the Belt and other marginal basin sequences formed.

The first question remains unanswered; we simply do not know what tectonic mechanism disturbed the Belt and its equivalents. For the second question, however, geologists have a likely answer. The Siberian platform, which now lies far across the Pacific, exhibits a number of features indicating that it may have been the landmass that was rifted away. Rotation of the Siberian platform through less than 90° orients it so that it fits rather neatly against the margin of the north American continent that existed when the Belt and its equivalents were deposited. In this arrangement, the structural trends of the two cratons also align. The Aldan and Anabar massifs, which are Siberian blocks of Archean age, then become partners of the Slave and Superior of provinces North America and of the Wyoming Province. (*Massif* is a name commonly used for durable, crystalline crustal blocks that are smaller than Precambrian areas recognized as shields.) The region between the Anabar and Aldan massifs, which aligns with the Churchill Province, experienced orogenic activity between 1.7 and 2.1 billion years ago—at about the time when Hudsonian orogenesis affected the Churchill Province. The crystalline rocks of the Siberian platform are also overlain by middle and late Proterozoic sedimentary rocks that resemble those of the North American craton. In the interior of the Siberian craton basal conglomerates and sandstones are followed by stromatolitic carbonates. Thicker deposits of the shelf margin begin with basal sandstones and tillites that pass upward into alternating siliciclastics and stromatolitic carbonates. In places, marginal failed rifts have been recongnized. In a general way, this configuration is a mirror image of what we find in western North America and it thus tends to support the hypothesis that the Siberian craton is the missing continent. Still, the so-called Siberian connection remains unproven.

Early History of Gondwanaland

Godwanaland was the great southern continent that fragmented during the Mesozoic Era to form South America, Africa, peninsular India, Australia, and Antarctica. Similarities among the life, strata and geologic structures of these five landmasses formed the strongest early arguments in favor of continental drift.

We have seen that Laurentia experienced considerable marginal accretion during the Elasonian, Grenville, and Appalachian orogenic episodes. Gondwanaland, too, experienced marginal accretion, which was largely confined to belts along what now constitute the margimns of South America, Antarctica, and Australia. Gondwanaland remained intact throughout the Paleozoic era, and Laurentia and Gondwanaland became united as the dominant elements of the supercontinent Pangaea near the beginning of the Mesozoic Era more than 200 millioin years ago. Subsequently, the fragmentation of Pangaea produced the continents of the modern world, including North America and Greenland, which broke off from the part of pangaea that had once been Laurentia.

When Pangaea broke apart, the large portion that had once been Gondwanaland suffered much more severe fragmentation, fracturing into five fragments—four of which represent large continents today. The reassembly of these five fragments into Gondwanaland reveals interesting patterns of orogenesis and metamorphism. When continents are assembled into the hypothetical landmass of Gondwanaland, those mountain belts in South America, Antarctica, and Australia that formed after early Paleozoic and early Paleozoic time line up neatly to form what was known as the Samfrau chain.

Inland from the Samfrau mountain chain in Gondwanaland was an older zone of Proterozoic and early paleozoic age that represented a vast network of igneous, metamorphic, and tectonic activity in Gondwanaland between about 800 and 400 million years ago. What is striking about this network of activity is that much of it seems to have taken place as regional metamorphism within rather than along the margins of Gondwanaland; thus far, no evidence of continental suturing has been found along the interior metamorphic zones. If suturing did not occur however, the metamorphic zones cannot be explained by conventional plate-tectonic processes, and their origin remains a mystery. In Figure, the zones representing igneous, metamorphic, or tectonic activity between 800 and 400 million years ago have been collectively labeled mobile belts rather than orogenic belts because those belts that display only metamorphism may not have formed parts of conventional mountain systems.

It is interesting to note that some 400- to 800-million-year-old mobile belts outline continental blocks that coincide with the present-day continents derived from Gondwanaland. In other words, when Gondwanaland finally broke apart during the Mesozoic Era, many of the rift zones tended to follow the mobile belts of the earlier interval of instability. It would appear that zones of crustal weakness from this earlier interval persisted for hundreds of millions of years.

The earlier history of Gondwanaland is difficult to decipher indeed, it is not even certain that Gondwanaland existed much earlier than 800 million years ago. Thus, for the purposes of our discussion, we will consider the Precambrian rocks of Gondwanaland in terms of the five modern continents where they are found today. The precambrian record of Africa will be our first subject.

Africa

As described in the previous chapter, the southern part of modern Africa includes the oldest cratonic block of large proportions now recognized in the world. This block, which is more than 3 billion years old, may have been the crustal core around which Gondwanaland grew. Radiometric dating reveals that several regions in modern Africa consist of crustal elements older than 3 billion years; by early Proterozoic time (about 2 billion years ago), nearly all of the modern craton was in existence.

During mid-Proterozoic times, deformation continued on a smaller scale. Then, from the end of the Precambrian interval to Paleozoic time, a series of orogenic events rejuvenated not only a large percentage of Africa but much of Gondwanaland as well.

Approximately 400 million years ago, Africa became a remarkably stable landmass, and it remained so until very recently. This stability can be attributed to the fact that ,much of Africa lay in the interior of Gondwanaland and was thus protected by South Africa on one side and by India and Antarctica on the other. Africa's northwestern margin and southern tip, which were expased along the perimeter of Gondwana land, were deformed during the Paleozoic Era: and, as is often the case, these continental margins were rafted against subduction zones. Today the surface of Africa stands high above sea level, the continent is fractured by rift systems that developed relatively recently. It is possible that a few million years hençe, Africa will be divide into two or more smaller continents. On the other hand, the rifting may fail, as did the rifting that began to divide North America about 1.3 billion years ago.

South America

South America, Africa and Antarctica, which were attached to each other for hundreds of millions of years as part of Gondwanaland, have histories that are interrelated in many ways. There are five general areas of Precambrian terrane in South America; the two in the north are the large Guyana and Brazilian shields. Because Archean rocks are uncommon throughout South America, the relative ages of these shields are difficult to assess. The most widespread Archean units are found in the northern part of the Guyana Shield, and a small Archean region lies near the Atlantic margin of the Brazilian Shield. One of the reasons Archean dates are relatively scarce here is that both shields were extensively metamorphosed about 2 billion years ago. Along the coast of the Brazilian Shield is a younger orogenic zone, the Brazilian orogen. This area has been radiometrically dated at 650 to 450 million years and thus represents one of the many orogenic episodes that affected Gondwanaland between 800 and 400 million years ago.

There is a striking difference between the Phanerozoic behaviour of Precambrian terranes in North and South America. Even at the beginning of Phanerozoic time, it would appear that the Precambrian rocks of North America had been eroded to low relief, and certainly the Canadian Shield is a low-lying area today. In contrast, the South American shields and massifs have remained positive topographic features throughout Phanerozoic time. To the east of the tall Andes Mountains, which have been uplifted intermittently throughout most of Phanerozoic time, depositional basins lie between the precambrian massifs. Although some marine sediments in this area accumulated during Paleozoic time, most of the sediments here are much younger Cenozoic calstics that have been shed from the Andes. The three southern Precambrian massifs lie within this sedimentary belt, but these durable massifs have tended to stand above the other terrane east of the Andes and, as a result, have received relatively little sedimentary cover.

These high-standing massifs provide only one example of the strong influence that the Precambrian shields and massifs have exerted over the Phanerozoic topography of South America. Another example concerns the Amazon Basin, which is one of the most interesting regions of the continent. The course of the Amazon River did not develop by accident. The elongate basin in which the great river now flows has a history tracing back at least to late Precambrian time, when it was the depositional site of extensive sediments derived from the stable Guyana and Brazilian cratons. The basin persisted through much of the Paleozoic interval, although it was periodically invaded by marine waters that separated the Guyana and Brazilian shields. Thus, the passage of the modern Amazon through the less resistant sedimentary terrane between the shields is only the most recent evidence of weakness in this region. To the east and south, three other shallow basins—the Parnaiba, the Sao Francisco, and the Parana-received sediment during part of the Phanerozoic despite the fact that they formed within the Brazilian Shield. Here, however, sedimentary accumulation id relatively thin, and after Triassic time only nonmarine deposits accumulated.

Looking farther back into the history of the Andean orogen, we find patterns indicative of continental accretion; for example the thickest Paleozoic sequences of sediment lie seaward of the thickest late early Paleozoic sequences, and the thickest Mesozoic sequences lie even closer to the Pacific margin. More generally, it is important to note that when Gondwanaland is reconstructed, the Andes form a segment of the Samfrau mountain belt.

Australia

The continent of Australia reveals a striking pattern of accretion along the margin of what was once Gondwanaland. The only large area of exposed Archean rocks in Australia is the Yilgarn Shield which lies near the western margin of the continent. Here and in the neighbouring Pilbara Shield are found the familiar greenstones and granites that characterize Archean terranes throughout the world. Metamorphic Precambrian belts to the east—the Arunta, Musgrave, and Gawler belts—probably include altered Archean deposits as well, indicating that the western two-thirds of the present continent existed at the end of Archean time or soon thereafter. Before we consider the pattern of eastward accretion that followed in Proterozoic and Paleozoic time, however, let us examine the vast accumulations of Proterozoic sedimentary deposits that have been found in Australia.

Sedimentary successions of Proterozoic age are more extensively exposed in Australia than anywhere else in the world, and this is why Australia has been and will continue to be an especially rich hunting ground for fossils of Proterozoic age. The widespread distribution of sedimentary deposits for three intervals of Proterozoic time has been shown. Early in the Proterozoic interval, approximately 2.3 to 1.8 million years ago, extensive marine sedimentary deposits accumulated in northern Australia. In the west there developed banded iron formations similar to those that formed in many other parts of the world about 2 billion years ago. Then, about 1.8 billion years ago, many areas were disturbed by orogenic activity. Sedimentation during the remainder of Proterozoic time was most extensive in what is now the central part of Australia. It is interesting to note, however that Proterozoic deposits younger than about 1.8 billion years include no turbidites or other deposits indicative of deep water. For more than a billion years, deposition of shallow-water marine sediments, including stromatolites, prevailed over large areas of the Australian craton. In addition, tillites, varved deposits, and scoured pavements offer evidence of late Proterozoic glacial activity in many areas.

The stabilization of late Proterozoic sediments in the Adelaide Trough, which lay near the eastern margin of the Proterozoic craton, led to considerable eastward accretion of the craton. Among the sedimentary units found here is the Pound Quartzite, which has yielded the Ediacara fauna of soft-bodied marine organisms.

The orogenic stabilization of the Adelaide sediments occurred in Cambrian time, representing yet another segment in the network of orogenic activity that took place in Gondwanaland between 800 and 400 million years ago.

Eastward continental accretion continued with the late Paleozoic uplift of the Tasman orogenic belt adjacent to the Adelaide Belt. Like the Andes, this zone represents part of the Samfrau mountain system of Gondwanaland. Volcanism here has continued into the Neogene Period.

Antarctica

Because about 99 percent of its surface is covered by ice, Antarctica is less well understood from a geologic stand-point than any other continent. Small outcrops have nonetheless revealed the general geologic history of this continent, which entails asymmetrical marginal accretion resembling that of Antarctica's Gondwanaland neighbour, Australia.

East Antarctica consists of large Precambrian shield in which several Archean regions have been indentified. The Ross orogen to the west of the shield contains some Precambrian rocks of uncertain age together with Cambro-Ordovician sedimentary units. The latter were deformed, intruded, and metamorphosed during Ordovician time, shortly after the deformation of the Adelide belt of Australia. Like the Adelaide episode, the ordovician Ross episode was one of the many events that affected Gondwanaland between 800 and 400 million years ago.

Later in Phanerozoic time, Antarctica underwent further accretion through stabilization of crust in what is known as the Western orogen. Like the Andes of South America and the Tasman Belt of Australia, this orogen represents part of the Samfrau system that was fragmented by separation of the three continents during the Mesozoic Era.

India

The peninsula of India, though now joined to Asia, was once wedged between Antarctica and Africa as part of Gondwanaland. When Gondwanaland broke apart during the Cretaceous Period, the Indian peninsula began a long northward journey. As we have seen, this journey ended in the Cenozoic Era with collision against the Asian continent and uplift of the Himalayan Mountains.

It is possible that almost all of the present craton of peninsular India existed in Archean times. We cannot prove that this is the case, however, because extensive Proterozoic metamorphism has left only a few areas of recognizable Archean terrane. The largest of these is the Dharwar Shield, which forms much of the southern portion of the peninsula; smaller areas of Archean crust lie to the north. The Archean terrane of India, like that elsewhere, is typified by areas of greenstones and granites as well as by areas of high-grade metamorphic rocks.

Most of the Proterozoic rocks of Indian are crystalline. The unaltered sedimentary rocks of India are primarily nonmarine and are therefore difficult to date by means of fossils, but they are known to span an interval of time beginning about 1.4 to 1.2 billion years ago and extending into the Cambrian. These rocks were deposited near the margin of Gondwanaland during a period in which the tip of what is now the modern Indian Peninsula experienced tectonism and plutouism as part of the widespread mobilization to which we have repeatedly referred.

Eurasia: A Composite Landmass

The vast modern continent of Eurasia is a patchwork quit of ancient continental blocks that have been joined together during the Phanerozoic Eon. What is now Eurasia was an array of small continents at the end of Precambrian time. Thus, while Gondwanaland broke apart during Phanerozoic time, Eurasia was assembled into a large continent.

We have already seen how two peninsulas became attached to Eurasia late in the assembly of this great landmass. These are the Arabian and Indian peninsula, which had been part of Gondwanaland. Like the Indian peninsula, the Arabian peninsulas was attached late in the Cenozoic Era. This western suturing event took place when Africa collided with Eurasia and mountain system, including the Alps, were uplifted in the Mediterranean region.

The greatest suturing event in the assembly of Eurasia however, occurred much earlier, at the end of the Paleozoic Era, when the Siberian and Russian platforms, which had been separate cratons, were united along the Ural Mountains. The assembly of eastern Asia was more complex and is poorly understood. It would appear however, that this region formed from the union of several microplates late in the Paleozoic Era.

Today, the total area of Precambrian rocks exposed in Eurasia is relatively small since a considerable amount of Precambrian crust has been blanketed by Phanerozoic sedimentary rocks. The exposed Archean terranes of Eurasia are especially meager, and they do not exhibit greenstone belts as well developed as those found in other parts of the world. Nonetheless, there remain four principal shields that include Archean rocks. Two of these the Anabar and Aldan massifs, lie Within the Siberian platform. Although much of the platfrom is how covered with younger sediments. The Anabar and Aldan massifs belong to a single large block of Precambrian crust. As we have seen, this block may have been attached to western North America during the Proterozoic Eon.

The part of Eurasia that lies west of the Ural suture also displays two sizable shield areas that contain Archean rocks. One of these, the Ukrainian massif, lies in the southern part of the Russian platform. More than half of the surface of this massif is Archean. Furthermore, it is continuous with the Baltic Shield beneath Phanerozoic sediments. As the largest Precambrian shield area in all of Eurasia, the Baltic deserves a closer examination.

The Baltic Shield, like the Canadian Shield, has been scoured by Pleistocene glaciers. Its eastern protion constitutes a sizable province of Archean rocks. The shield and, Presumably, the entire Archean crustal block underlying the surface deposits of the Russian platform grew by continental accretion on what is now a westerly direction. The eastern (Archean) portion of the shield is flanked on the west by a zone of rocks that was stabilized 1.7 to 2.0 billion years ago; to the west of this lies a zone that was stabilized 0.8 to 1.2 billion years ago.

As we noted earlier, the Baltic Shield may represent at least part of a crustal block that was rifted away from eastern North America between about 1.3 and 1.0 billion years ago. The youngest Precambrian province of the Baltic Shield is about the same age as the Grenville Province of North America and like the Grenville it may have formed through an episode of continental accretion shortly after Baltica and Laurentia split apart.

4

Early Paleozoic Era

Prelude

Anyone who has experienced a course in history is likely to appreciate the way historians group seemingly endless facts into tidy packages bearing such labels as "The Renaissance" or "The Hellenistic Period." With the help of such groupings, one is better able to retain an overall impression of the period. This is also a reason for dividing the history of the Paleozoic Era into "late" rather than providing a period by period account. The scheme is also convenient in that each of the two increments of time is of approximately equivalent duration. The Cambrian, Ordovician, and Silurian Periods constitute the Early Paleozoic, which lasted about 190 million years. In the subsequent 180 million years, the Devonian, Mississippian, Pennsylvanian, and Permian Periods ran their course. Each of the two parts of the Paleozoic was characterized by very general similarities in the historical sequence of events. For example, on most continents the Early Paleozoic began with gradual marine invasions of low-lying regions of the interior. These wide expanses of shallow epicontinental seas moderated climate and provided habitats for a multitude of marine organisms.

Subsequently, the continents began to stir. Either by uplift of the lands or by decline in sea level, the inland seas began an oscillatory regression back toward the major ocean basins. Increasing thicknesses of terrestrial sediments and volcanic rocks appeared in the stratigraphic sequence, and contorted strata attest to the growth of mountain ranges. It was a time of land, just as the preceding episode was a time of seas. However, gradually the lands were reduced and the seas returned as we pass from the Early into the Late Paleozoic.

After the total physical history of the Paleozoic Era is examined, it becomes evident that long periods of quite sedimentation were punctuated at intervals by severe change involving earth movements and mountain building. European geologists have recognized three such mountainbuilding, or orogenic, events. They refer to them as the *Caledonian*, *Hercynian*, and *Alpine* Orogenies. You will recall our discussion of how Pangaea 1 disaggregated. The return of the errant lands to form Pangaea II may very well be associated with the Caledonian and Hercynian Orogenies. The Alpine Orogeny records a similar but much later event.

Lands, Seas, And Mobile Belts

Clues to Paleogeography

Because of the many processes that have ceaselessly worn away or altered the older parts of the geologic record, it is difficult to reconstruct details of Early Paleozoic geography. From paleomagnetic studies we have learned that most of North America during the Cambrian lay along the paleoequator, which was aligned nearly perpendicular to present parallels of latitude from north-central Mexico to Ellesmere Island in the Arctic. The Ordovician, equator extended slightly more northeastward, from what is now Baja California to Greenland and thence across the British Isles and Central Europe. These paleoequatorial positions are substantiated by paleoclimatic indicators. Salt and gypsum accumulations, for example, are found in Ordovician rocks of Arctic Canada at a paleolatitude of about 10° from the paleoequator. Great thicknesses of carbonate rocks bearing corals and other relatively warm water marine invertebrates are located within 30° of the paleomagnetically determined equator.

Both paleomagnetic evidence and paleontologic evidence suggest either the existence of a supercontinent at the very beginning of the Paleozoic or that the continents were gathered into a relatively compact assemblage. The land area that in a previous chapter we termed Pangaea I was situated primarily in the southern hemisphere. The Gondwana region was turned abut 180° from its present orientation, whereas what is now North America and Europe lay somewhat east of South America. Very early in the Cambrian, or possibly late in the Precambrian, Period ended, the lines of separation had widened to admit the seas. These elongate tracts

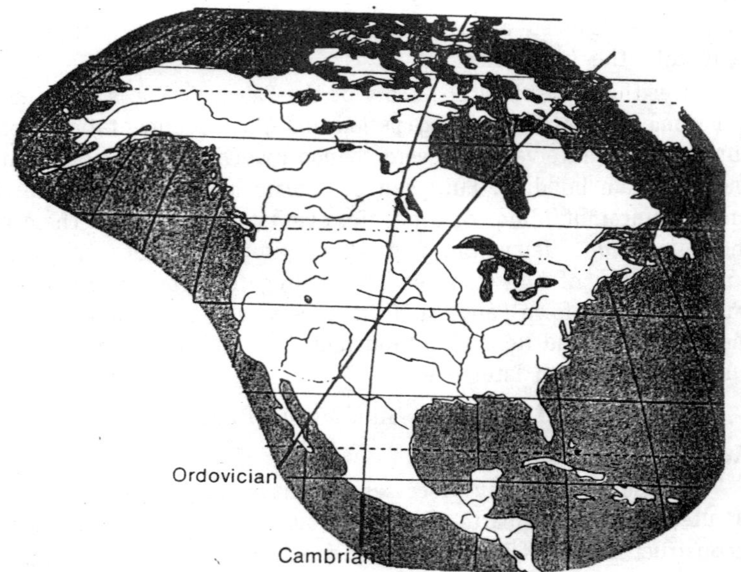

Fig. 4.1. Inferred position of Cambrian and Ordovician paleoequator relative to North America.

became great collecting troughs for the sediments that are now metamorphosed and folded in the highlands of northwest Europe, in the Urals, and in the Appalachians. By Ordovician time large continental fragments were drifting away from the parent mass.

The earth scientists James Valentine and Eldridge Moores which are further have described the sequence of events. These configurations show Pangaea I divided into four continents by Late Cambrian and Early Ordovician. Then, during Late Ordovician and Silurian time, Europe west of the Uralian trough pushed into North America, forming a chain of ancient mountains called the *Caledonides* and welding the two continental segments into a "Euramerica." The next major collision occurred about 300 million years ago when Euramerica joined Gondwana land, forming the ancient Hercynian Mountains, which have since been largely eroded away and covered with younger strata over much of Europe. Finally, late in the Permian, Asia was joined to Euramerica. The sediments caught and crushed in the closing ocean formed the Urals. Pangaea II was born.

It must be remembered that these theories of continental splitting and drift, and our methods of locating the positions of continents long ago, are fraught with possibilities for error or misinterpretation. We are dealing with speculations based on limited data, and we should not be surprised if in a decade hence our views of Pangaea I and II are altered.

The Continental Framework

The Cambrian Period of the Paleozoic Era began approximately 570 million years ago. By that time most of the continental masses, even though clustered together, could be considered to have two parts that differed in their history and structure. These two major geologic divisions have been termed *cratons* and *mobile belts.*

Cratons

Cratons form the ancient central nuclei of continental masses. Cratons may include either or both of two kinds of surficial geology. Cratonic regions consisting of surficial Precambrian igneous and metamorphic rocks are termed *shields*. Most shield areas were exceptionally stable during the Paleozoic, in contrast to their Precambrian history, which included intense orogenic activity. The shield portion of the craton may be partially clothed in a veneer of relatively flat post-Precambrian sedimentary strata. This portion is called the *platform*. Like shields, platforms were generally stable throughout most of Phanerozoic geologic history. Platform strata consist of wave-washed sandstones and carbonates deposited in shallow seas that periodically flooded continental regions of low relief. The strata platforms are not, however, entirely horizontal. Here and there they are gently warped into broad synclines, basins, domes, and arches. The resulting tilt to the strata is so slight that it is usually expressed in feet per mile rather than degrees. In the course of geologic history, the arches and domes stood as low islands or barely awash submarine banks.

Domes and arches seem to have developed in response to vertically directed forces quite unlike those that formed the compressional folds of mountain belts. Whether domes or basins, structures of the platform can be recognized by their pattern of outcropping rocks. Erosional truncation of domes exposes older rocks near the center and younger rocks around the periphery. Sequences of strata over arches and domes tend to be thinner. Also, because these structures were periodically above sea level, they characteristically exhibit a greater frequency of erosional unconformities. Basins, on the other hand, were more persistently covered by inland seas, have fewer unconformities, and develop greater thickness of sedimentary rocks. In erosionally truncated basins, younger rocks are centrally located, and older strata occur successively farther toward the periphery.

Mobile Belts

The North American craton is bounded on four sides by great elongate tracts that have been the site of intense deformation, igneous activity, and earthquakes. Such tracts are called *mobile belts*, and one or more are present on all continents. Most, like the North American Cordilleran Belt, are located along the margins of continents. It is also possible to have an intraoceanic mobile belt, such as occurs along the Urals. Systems of deep sea trenches with associated volcanic island arcs are intraoceanic mobile belts. A *geosyncline* is a mobile belt that, in its early history, received a great accumulation of sediment. As described already, mountain ranges are related to geosynclines in that they are thought to have evolved through several stages of development, beginning with a period of sedimentation and ending with an episode of severe crustal deformation.

Early Paleozoic Events

The beginning of the Paleozoic Era was a relatively quiet time for North America, as if the continent rested from the rigors of orogeny and severe climate that had occurred near the end of the preceding era. We know little of the Paleozoic ocean basins, for those old sea floors have long since been consumed at subduction zones. However, oceans did exist and gradually spilled out of their basins onto low regions of the continents. Although a large part of the sedimentary record has been removed by erosion, enough remains to indicate that practically all of the Canadian Shield was inundated at times. Initially, the dominant deposits were sands and clays derived from the weathering and erosion of igneous and metamorphic rocks of the shield. Later, as these quartz- and clay-producing terrains were reduced, limestones and dolostones became increasingly more prevalent. These rocks contain abundant remains of lime-secreting marine organisms and indicate that shallow seas were common throughout much of the equatorial region of the earth during the Early Paleozoic. Indeed, the advances and retreats of these shallow seas were the most apparent events in the Early Paleozoic history of the continental interiors.

The North American Craton

It is convenient to discuss the geologic history of the continental mass that now constitutes North America in terms of its cratonic and geosynclinal regions. It is also of advantage to subdivide the Early Paleozoic of the craton on the basis of advances (transgressions) and retreats (regressions) of the inland or epeiric seas. Regressions, of course, exposed the old sea floors to erosion and produced extensive unconformities that are used as stratigraphic markers.

Although the cratonic rocks of North American can be correlated with those of Europe on the basis of fossils, most of the stratigraphic breaks in deposition do not correspond to those that define the European period boundaries. Indeed, if the relative time scale had been worked out in North America in stead of Europe, there would probably have been only two periods in the "Early Paleozoic" rather than three. The American stratigrapher Laurence Sloss has suggested a workable remedy for this disparity. He suggested that the Paleozoic rocks of the North American craton be divided into "sequences" of deposition. For the Paleozoic, he named these the *Sauk, Tippecanoe, Kaskaskia,* and *Absaroka* sequences. The boundaries of these sequences are marked by widespread unconformities. The Sauk and Tippecanoe are the Early Paleozoic sequences, whereas the Kaskaskia and Absaroka are Late Paleozoic.

During the earliest years of the Sauk Sequence, the seas were largely confined to geosynclinal troughs, so that most of the craton was exposed and undergoing erosion. No doubt it was a bleak and barren scene, for vascular land plants had not yet evolved. Uninhibited by protective vegetative growth, erosional forces gullied and dissected the surface of the land. Over a span of at least 50 million years, the Precambrian crystalline rocks were deeply weathered and must have formed a thick, sandy "soil. " Eventually, marine waters spilled out of the geosynclines and began a slow encroachment over the eroded and weathered surface of the central craton.

By Late Cambrian, seas extended across the southern half of the craton from Montana to New York. A vast apron of clean sand was spread across the sea floor for many miles behind the advancing shoreline. This sandy aspect, or facies, of Cambrian deposition was replaced toward the south by carbonates. Here the waters were

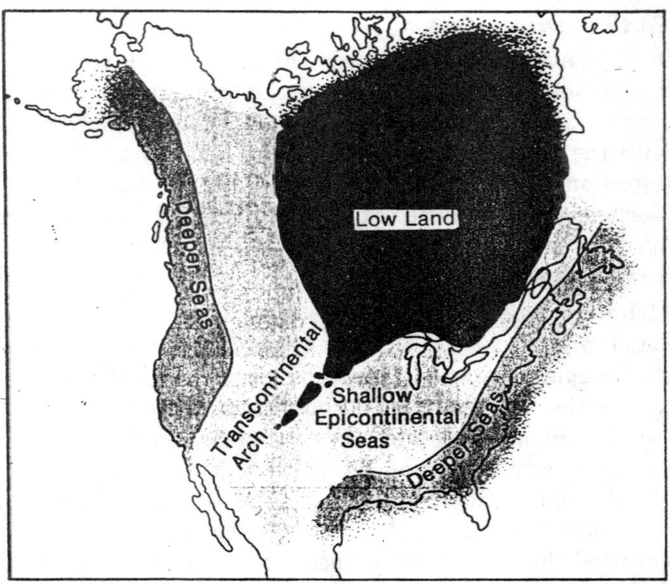

Fig. 4.2. An interpretation of Late Cambrian paleogeography. During the Cambrian, the seas transgressed the craton until by late in the period marine waters covered most of the United States. The Canadian Shield remained as a lowland area, and large islands existed along the trend of the Transcontinental Arch.

warm, clear, and largely uncontaminated by clays and silts from the distant shield. Marine algae flourished and contributed to the precipitation of calcium carbonate. Invertebrates, although present, did not contribute to the volume of sediment to the degree they did in later periods.

Along the Cordilleran Geosyncline, the earliest deposits are sands, which graded westward into finer clastics and carbonates. An excellent place to study this Sauk transgression is along the walls of the Grand Canyon of the Colorado River. In this region one finds the Lower Cambrian Tapeats Sandstone as an initial strandline deposit above the old Precambrian surface. The Tapeats can be traced both laterally and upward into the Bright Angel Shale, which was deposited in a more offshore environment. Next is the Muav Limestone, which originated in a still more seaward environment. As the sea continued its eastward transgression, the early deposits of the Tapeats were covered by clays of the Bright Angel Formation, and the Bright Angel was in turn covered by clays of the Bright Angel Formation, and the Bright Angel was in turn covered by limy deposits of the Muav Formation. Together these formations form a typical transgressive sequence, recognized by coarse deposits near the base of the section and increasingly finer (and more offshore) sediments near the top.

Cambrian rocks of the Grand Canyon region not only provide a glimpse of the areal variation in depositional environments as deduced from changing lithologic patterns but also illustrate that particular formations are not usually the same age everywhere they occur. Detailed mapping of the Tapeats, Bright Angel, and Muav units combined with careful tracing and correlation of trilobite occurrences indicates that deposition of the highest facies (Muav) had already begun in the west before deposition of the lowest facies (Tapeats) had stopped in the east. Correlations based on the guide fossils provided reliable evidence that the Bright Angel sediments were largely Lower Cambrian in age in California and mostly Middle Cambrian in age in the Grand Canyon National Park area. This example illustrates the *principle of temporal transgression*, which stipulates that sediments deposited by advancing or retreating seas are not necessarily of correlative geologic age throughout their areal extent.

The episode of carbonate deposition that was so characteristic of the southern craton continued into the Early Ordovician almost without interruption. Then, near the end of the Early Ordovician, the seas regressed, leaving behind a landscape underlain by limestones that was subjected to deep subaerial erosion. That erosion produced a widespread unconformity which geologists use as the boundary between the Sauk and the Tippecanoe sequences.

The second major transgression to affect the platform occurred when the Tippecanoe Sea flooded the region vacated as the Sauk Sea regressed. Again the initial deposits were great blankets of clean quartz sands. Perhaps the most famous of these Tippecanoe sandstones is the St. Peter Sandstone, which is nearly pure quartz and is thus prized for use in glass manufacturing. Such exceptionally pure sandstones cannot be developed in a single cycle of erosion, transportation, and deposition. They are the products of chemical and mechanical processes acting on still older sandstones. In the St. Peter Formation, waves and currents of the transgressing Tippecanoe Sea reworked Late Cambrian and Early Ordovician sandstones and spread the resulting blanket of clean sand over an area of nearly 7500 sq km. As described, sandstones can be described by their texture and composition as either mature or immature. These textural and compositional traits are, of course, related to the processes that produced the sandstone. Sandstones with a high proportion of chemically unstable minerals and angular, poorly sorted grains are considered immature, whereas aggregates of well-rounded, well-sorted, highly stable minerals (like quartz) are considered mature. The St. Peter Sandstone is so pure as to be geologically unusual and perhaps should be termed an "ultramature sandstone."

Another noteworthy basal sandstone of the Tippecanoe Sequence is the Harding Sandstone of the Rocky Mountain Front ranges. Within this formation are found worn fragments of bone and bony scales that are the earliest fossil record of fishes in North America.

The sandstone depositional phase of the Tippecanoe was followed by the development of extensive limestones, often containing calcareous remains of brachiopods, bryozoans, mollusks, corals, and algae. Some of these carbonates were chemical precipitates, many were fossil fragment limestones (bioclastic limestones), and some were great organic reefs. Frequently, the deposited lime underwent

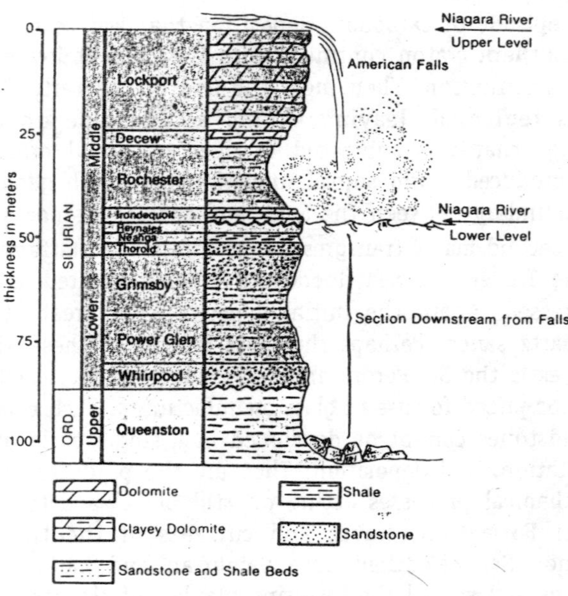

Fig. 4.3. The stratigraphic section at Niagara Falls.

chemical substitution of some of the calcium by magnesium and in the process was converted to the carbonate rock called dolostone.

In the region east of the Mississippi River, the dolostones and limestones were gradually supplanted by shales. As we shall see, these, clays are the peripheral sediments of the Queenston Delta, which lay far to the east. The geologic section exposed at Niagara Falls is a classic locality to examine these rocks.

Near the close of the Tippecanoe Sequence, landlocked, reef-fringed basins developed in the region now occupied by the Great Lakes. The evaporation of these basins caused precipitation of salt and gypsum on a vast scale. Salt from the Salina Group of the Michigan Basin is about 500 meters thick and is extensively mined. The Tippecanoe Seas were well into their regressive phase when these evaporites were being precipitated.

The Western Geosyncline

The great Cordilleran Geosyncline, which covered a wide belt from Alaska down through western Canada and the United States, was the dominant feature of western North America during the Early Paleozoic. During much of this time, the Pacific Ocean extended inland all the way to the *transcontinental arch*: an elongate, ridgelike structure extending from Southern California to the Lake Superior region of Canada.

The Cordilleran Geosyncline appears to have originated during the Proterozoic, when North America began to separate from a segment of the continent that extended westward beyond the present continental margin. The geosyncline formed along the passive trailing edge of the continent. One would expect to find evidence of rifting if such a continental separation occurred, and indeed such evidence does exist. The Belt Supergroup of Montana, Idaho, and British Columbia, the Uinta Series of Utah, and the Pahrump Series of California were all deposited in deep fault-controlled basins formed during an episode of rifting about 0.85 to 1.4 billion years ago.

One of the grandest sections of Cambrian rocks in the world outcrops in the Canadian Rockies of Alberta. The formations have been erosionally sculptured into magnificent mountain scenery. Lower Cambrian rocks include ripplemarked quartz sandstones, probably derived from the Canadian Shield. By Middle Cambrian time the seas had transgressed farther eastward, and shales and carbonates became more prevalent. An interesting section of these

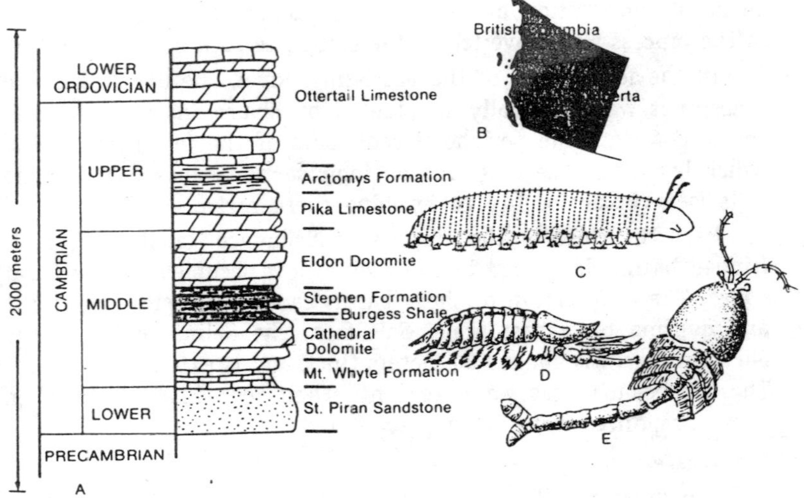

Fig. 4.4. Cambrian stratigraphic section (A) at Kicking Horse Pass, British Columbia (B). According to a now famous story, this is where a slab of fossiliferous shale was kicked over by a pack horse so as to catch the eye of paleontologist Charles Walcott, who followed the trail of rock debris up the side of the mountain to discover the Burgess Shale Beds with their rich fauna of Middle Cambrian fossils. Among these were Aysheaia (C), an "onycophoran" believed to be intermediate in evolutionary position between segmented worms and arthropods. Among the thousands of specimens were trilobite-like arthropods such as Leanchoilia superlata (D) and Waptia fieldensis (E).

Middle Cambrian rocks is exposed along the walls of Kicking Horse Pass in Alberta. One of the units in this section, the Burgess Shale, has excited the keen interest of paleontologists around the world because it contains abundant remains of Middle Cambrian soft-bodied animals, most of which have never been discovered elsewhere.

By viewing the patterns of sedimentation in the Cordilleran region, it is apparent that the geosyncline can be naturally divided into an eastern miogeosynclinal belt, in which continental shelf sediments predominate, and an adjacent eugeosyncline, which received great thicknesses of continental rise siliceous shales, cherts, graywackes, and volcanics. The stratigraphic sequence of the miogeosyncline resembles carbonate and quartz sandstone associations of the platform. The thickness of sediment, however, was much greater. Because of the large numbers of calcareous invertebrates fossilized in the miogeosynclinal limestones, they are often referred to as the "shelly facies." Eugeosynclinal deposits, on the other hand, often include dark shales containing carbonized remains of colonial organisms called *graptolites*. For this reason the eugeosynclinal sediments have been dubbed the "graptolite facies."

The Ordovician of the Cordilleran Geosyncline is represented by several thousand meters of fossiliferous carbonates that were deposited along the miogeosyncline. These strata contrast markedly with the dark graptolitic siltstones and graywackes of the eugeosyncline to the west. The eugeosynclinal facies are best exposed in great thrust sheets in central Nevada, where the section measures over 5000 meters in thickness. Similar eugeosynclinal rocks outcrop in central Idaho, in eastern Washington, and in the central Sierra Nevada Mountains of California.

Silurian deposits of the Cordilleran are generally similar to those of the Ordovician. Eugeosynclinal facies can be recognized in southeastern Alaska, California, and Nevada. Nevada also contains exposures of Silurian miogeosynclinal carbonates.

The Northern Geosyncline

The geosyncline at the top of North America has been named the Franklinian Geosyncline. It extended over a distance of 2250 km from northernmost Canada to northeastern Greenland. Cambrian rocks are not exposed in this frigid region, but Ordovician formations of the shelly facies are widespread. Ordovician limestone reefs and evaporites are also prevalent. On the Ellesmere and Axel

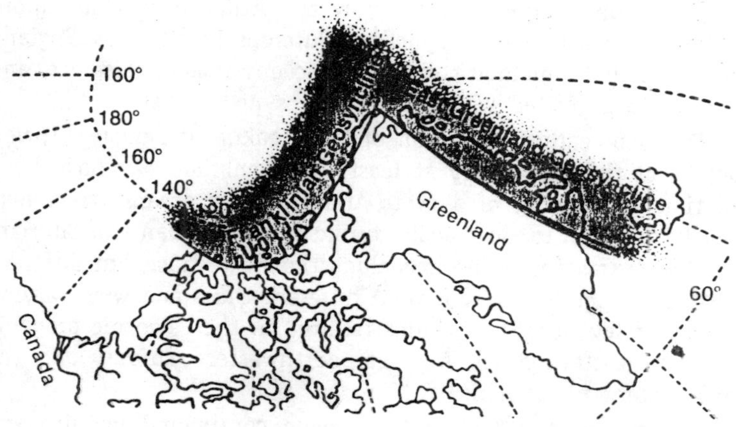

Fig. 4.5. Location of the Franklinian Geosyncline.

Heiberg Islands, thick sequences of graywackes, black shales, conglomerates, and volcanics indicate the existence of a well-developed eugeosyncline.

The Silurian section in this region is one of the thickest in the world. Here, too, it contains extensive reef deposits and thick accumulations of evaporites. Such sedimentary rocks imply former tropical conditions, an interpretation strengthened by paleomagnetic data indicating that the Franklinian Geosyncline was located about 15° from the equator during the Early Paleozoic.

The Eastern Geosyncline

Bordering the interior platform of North America on the east from Newfoundland to Georgia was the great Appalachian Geosyncline. Its southwestward extension is called the Ouachita Geosyncline. Unlike the Cordilleran Geosyncline, which was a relatively passive plate margin during most of the Early Paleozoic, the Appalachian belt was subjected to frequent episodes of intense orogenic activity resulting from plate collisions. Sediment was poured into this great arcuate trough from both the craton on the one side and lands and volcanic islands that existed just beyond the present continental boundaries on the other. Once again, Cambrian sands derived from shield source areas are the initial deposits of the miogeosyncline. These sandy deposits are followed by relatively shallow water Middle Cambrian and Ordovician carbonates that have great thicknesses (in excess of 3000 meters in the Ouachita part of the geosyncline). Eastward of the miogeosyncline, along what is now the Atlantic coastal region, lay the more active eugeosyncline. Outcrops in the New England states and Newfoundland consist of siliceous shales, volcanics, and graywackes deposited within that eugeosynclinal bel.

From our earlier discussion of the breakup of Pangaea, I, you may recall that, in theory at least, the Cambrian may have been the time of opening of a proto-Atlantic. That oceanic tract may then have begun to close again during the Ordovician and Silurian. The closure was not continuous along the entire tract, but different parts came together at somewhat different times and with varying degrees of severity. The closures resulted in the Paleozoic orogenic events of eastern North America, northwestern Europe, and the northwestern tip of Africa.

The coming together of the broken continental margins was not without a depositional record. With increasing frequency one

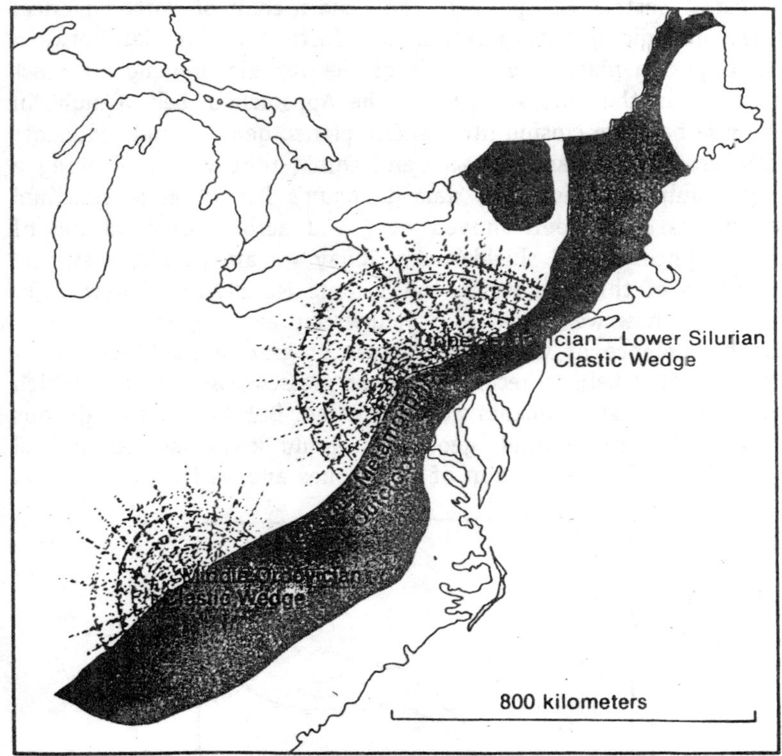

Fig. 4.6. Great wedges of clastic sediments spread westward as a result of erosion of mountain belts developed during the Early Paleozoic.

finds ever thicker and coarser clastics that signify orogenic activity in the eugeosyncline during the Early Paleozoic. The clastics covered most of the miogeo-syncline; they are coarsest on the eastern margin, grade westward to finer and more clayey materials, and ultimately merge into carbonate strata along the edge of the platform. Long ago, geologists viewing this great wedge of sediment that became thicker, coarser, and more volcanic toward the Atlantic speculated that some now vanished highland area must have supplied the detritus.

Pulses of orogenic activity became relatively frequent during the Early Ordovician. This preliminary unrest was followed by several more intense deformational events in Middle and Late Ordovician that comprise the *Taconic Orogeny*. The Taconic Orogeny was caused by the collision of another tectonic plate, probably part of ancestral western Europe, with the eastern coast of North America. Later orogenic episodes resulted from further compression between the opposing plates. The effects of the Taconic Orogeny are most apparent in the northern part of the Appalachian belt. Caught in the vise between closing lithospheric plates, geosynclinal sediments were crushed, metamorphosed, and thrust northwestward along a great fault. In this fault, named *Logan's Thrust*, eugeosynclinal sediments have been shoved over and across some 48 km of miogeosynclinal and shield rocks. Today we are able to view the remnants of this activity in the Taconic Mountains of New York. Ash beds, now weathered to a yellow clay called *bentonite*, attest to the violence of volcanism. Masses of granite now exposed in the Piedmont help to record the great pressures and heat to which the geosynclinal sediments were subjected. But even if the igneous rocks were never found, geologists would know mountains had formed, for the great apron of sandstones and shales that outcrop

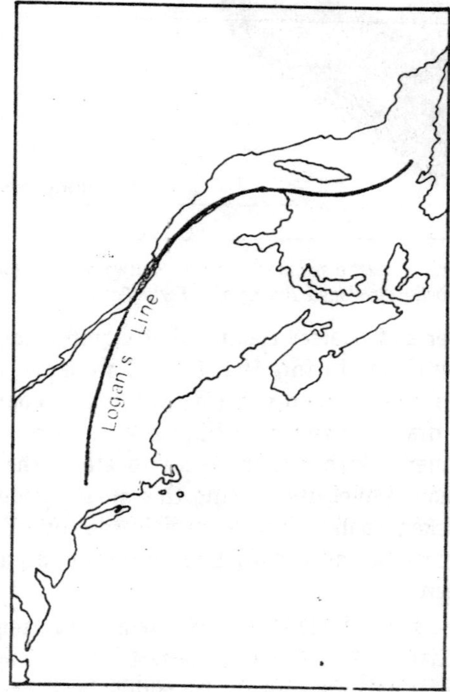

Fig. 4.7. "Logan's line." As a result of folding and thrust-faulting, the New England region of North America was compressed by several hundred kilometers.

across Pennsylvania, Ohio, New York, and West Virginia must have had their source in the rising Taconic ranges.

From a feather edge near Cincinnati, this baren wedge of rust-red terrestrial clastics called the *Queenston Delta* becomes increasingly thicker and coarser toward the ancient source area to the east. It has been estimated that over 600,000 cu km of rock were eroded to produce the enormous volume of sediment in the Queenston Delta. It is not unlikely that the Taconic ranges exceeded elevations of 4000 meters.

During the Silurian, the locus of most intense orogenic activity seems to have shifted northeastward into the Caledonian belt. Meanwhile, erosion of the Taconic Highlands continued. Early Silurian beds are often coarsely clastic, as evidenced by the Shawangunk Conglomerate of New York. These pebbly strata give way upward and laterally to sandstones such as those found in the Cataract Group of Niagara Falls. Silurian iron-bearing sedimentary, deposits accumulated in the southern part of the geosyncline. The greatest development of this sedimentary iron ore is in central Alabama, where there are also coal deposits of Pennsylvanian age. The coal is used to manufacture coke, which is needed in the process of

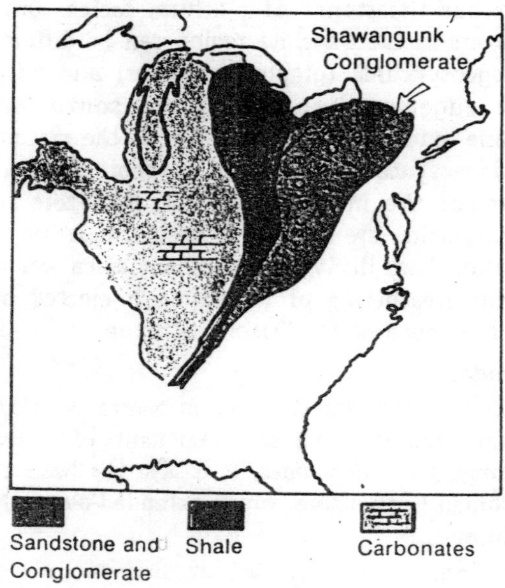

Fig. 4.8. Lower Silurian lithofacies.

smelting. Limestone, used as blast furnace flux, is also available nearby. The fortunate occurrence of iron ore, coal, and limestone in the same area accounts for the important steel industry of Birmingham, Albama.

Extending across the southern margin of the North American craton, and continuous with the Appalachian Geosyncline, is the Ouachita Geosyncline. Although this segment of the mobile belt is over 1500 km long, only about 300 km of its folded strata are exposed. Additional information about the distribution of Paleozoic rocks in this region is derived mostly from oil well drilling operations.

Altogether, nearly 10,000 meters of Paleozoic sediment fill the Ouachita Geosyncline. However, of this thickness, only about 1600 meters were deposited in the Early Silurian. This indicates more rapid subsidence and deposition in the trough during the Late Paleozoic. Unfossiliferous metamorphosed graywackes, quartzites, and cherty beds of probably Early Cambrian age are the oldest Paleozoic sediments known from the Ouachita Geosyncline and indicate eugeosynclinal deposition. The next oldest rocks are Middle Cambrian igneous bodies that yield dates of 525 to 535 million years. These are overlain unconformably by basal sandstones and fossiliferous limestones of platform facies. Ordovician and Silurian deposits in the Ouachita region can be differentiated into distinctly eugeosynclinal (graptolitic shales) and miogeosynclinal facies. The eugeosyncline lay to the south, whereas the miogeosyncline extended along the edge of the craton. Northward from the miogeosyncline, the Early Paleozoic section thins as it becomes part of the interior platform. Paleozoic rocks of the Ouachita Geosyncline are noted for their unusually siliceous and cherty character. Very likely, the abundant silica was derived from the submarine weathering of volcanic ash ejected by volcanoes that lay to the south of the Ouachita trough.

The Caledonides

Scotland's ancient name, still used in poetry, is *Caledonia.* Eduard Suess used the name for the Scottish remnants of an Early Paleozoic mountain range, the *Caledonides*. It is also the basis for the name of the Caledonian Geosyncline, which extended along the northwest border of Europe.

The Caledonian and Appalachian Geosynclines have had a generally similar history. This is not surprising, for both are really part of one greater Appalachian-Caledonian system. The geosyncline

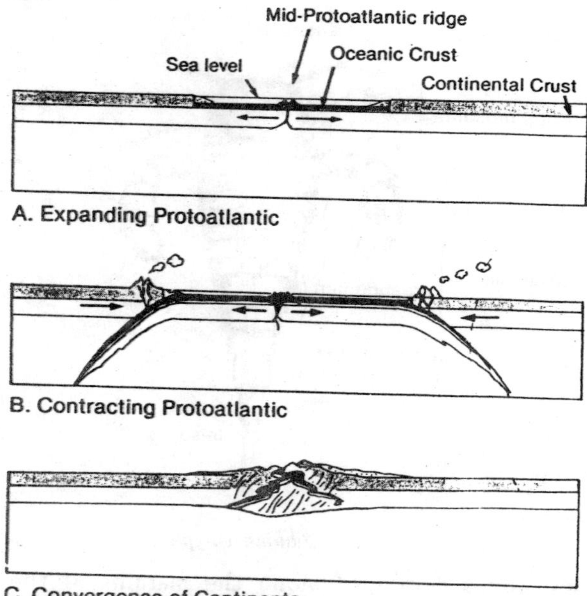

Fig. 4.9. The cycle of oceanic expansion and contraction that controlled the depositional and deformational history of the Appalachian and Caledonian Geosynclines.

evolved as a result of a cycle of ocean expansion and contraction such as is schematically illustrated in Figure 4.9. The cycle began with a Late Precambrian to Middle Ordovician episode of spreading, as the Proto-Atlantic widened to admit new oceanic crust along a spreading center. On the continental shelves and rises of the separating blocks, thick lenses of sediment accumulated. This depositional phase is marked by the development of two distinct facies. The graptolite facies consists of more than 6000 meters of volcanics, graywackes, and shales. Graptolites are prevalent in this facies and are used for subdividing the Ordovician and Silurian into biostratigraphic zones. A shelly facies with clean sandstones and fossiliferous limestones help us to identify the miogeosynclinal or continental shelf facies. Here and there on the margins of the Caledonian Miogeosyncline are found freshwater deposits of Late Silurian age that are noteworthy for fossil remains of early fishes and strange arthropods called *eurypterids*.

The closure of the Proto-Atlantic and crumpling of the Caledonian Geosyncline began in Middle Ordovician time, when

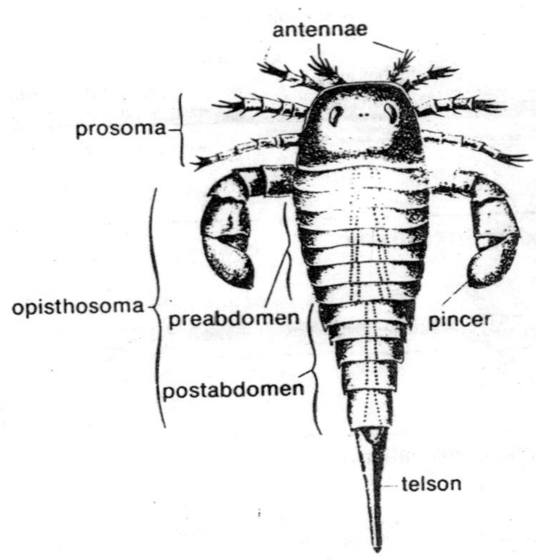

Fig. 4.10. A Silurian eurypterid.

subduction zones developed along the margins of the formerly separating continents. This event is recognized from the volcanic rocks that occur in the Canadian Maritime Provinces, northwestern England, northeastern Greenland, and Norway. Little by little, the Proto-Atlantic closed until the opposing continental margins converged in a culminating mountain-building event termed the *Caledonian Orogeny*. The orogeny reached its climax in Late Silurian to earliest Devonian time. It was most intense in Norway, where Precambrian and Lower Paleozoic sedimentary rocks are drastically metamorphosed and thrust-faulted. Southwestward, in the British Isles, the effects were not as severe, although mountainous terrains also developed there.

South of the unruly Caledonian belt there is evidence of a land mass that during Late Silurian and Devonian received sandy detritus from the growing Caledonides. This region, on which was deposited clays and sands of the Old Red Sandstone Formation, is often referred to as the "Old Red Continent." It was a thickened wedge of nonmarine clastics not unlike the Queenston Delta.

Rocks of the Caledonian Geosyncline are well exposed not only in the British Isles but also across northeastern Greenland and Spitzbergen. The geologic succession in the Caledonides of East Green land begins with a thick (100,00 meter) sequence of Precambrian rocks. The uppermost unit in the Precambrian sequence contains tillites, indicating an episode of glaciation just prior to the beginning of the Paleozoic. In eastern Greenland, Cambrian deposition began with sandstones followed by shales, dolostones, and limestones. Carbonates continued to be prevalent in the Early Ordovician, but thereafter sedimentational patterns began to change in response to the closing of the Proto-Atlantic as it was manifested in the Caledonian Orogeny. Erosion of the metamorphosed and folded geosynclinal rocks led to the accumulation of the continental Old Red Sandstone facies in basins adjacent to the growing mountains.

The Uralian Geosyncline

The Uralian Geosyncline extended southward from the Russian Arctic along the course of the present Ural Mountains and eastward to central Mongolia. A eugeosynclinal belt developed here also and was bordered on either side by miogeosynclines. However, the eastern miogeosyncline is poorly known because of a thick cover of Cenozoic strata. This is not so for the western miogeosyncline, where carbonates and sandstones grade eastward into thick sequences of shale, coarse arkosic sandstones, and volcanics.

Although the Uralian orogenic climax was not to come until the Late Paleozoic, folded strata, unconformities, and mafic intrusions attest to plate convergence and subduction in the region.

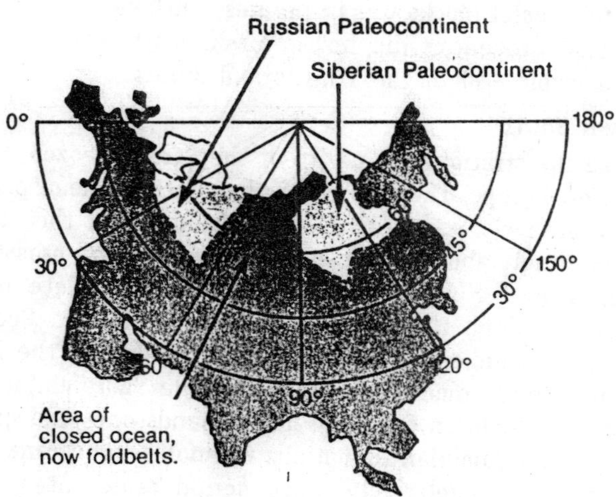

Fig. 4.11. Present positions of the Russian and Siberian continental blocks relative to the Uralian fold belts.

The history of the geosyncline seems to have followed an ocean expansion and contraction cycle in some ways similar to that of the Caledonian Geosyncline. Paleomagnetic studies indicate that the Russian and Siberian platforms were separated by wide oceanic tract during the Early Paleozoic. During the Silurian and Devonian, subduction zones developed between the two opposing continental masses. As the oceanic crust was being subducted, the intervening ocean narrowed, and the continents began to converge upon one another. By the end of the Paleozoic, and possibly continuing into the Triassic, the geosyncline experienced its orogenic climax as the two continental blocks collided.

Asian Begins to Take Form

Asia east of the Urals is a very ancient and complex terrain formed from the convergence and accretion of several crustal blocks. The sutures along which the continents collided are marked by ophiolite belts, narrow zones of high-pressure metamorphism, and volcanic rocks that characterize crustal tracts above subduction zones that descend at continental margins. Some geologists believe as many as nine separate crustal blocks or microcontinents may have combined to form Asia. In the Early Paleozoic, faunal and paleomagnetic evidence suggests that most of these blocks were separated by considerable widths of oceanic crust. By Middle Ordovician, narrowing of these oceans and convergence of the continental crustal blocks was in progress, although major episodes of collisional orogenesis did not occur in most regions until late in the Paleozoic and in the Triassic and Jurassic.

Way Down South

Unlike the stratigraphic record of the Early Paleozoic for North America and Europe, the record for the southern hemisphere land masses is relatively poor. This may be the result of the continents being prevailingly above sea level, which would favor erosion rather than deposition. However, there is a fairly complete record of sedimentation along the north-south-trending Tasman Geosyncline of Australia. Sedimentation there began just before the Paleozoic in a time period sometimes designated the "Eocambrian." These uppermost Precambrian rocks are mostly sandstones and grade into overlying Lower Cambrian formations without a significant erosional unconformity. The relatively quite period represented by these sediments was followed by a full-scale orogeny in the Late Cambrian. Thick graywackes and volcanics record the orgenic disturbances.

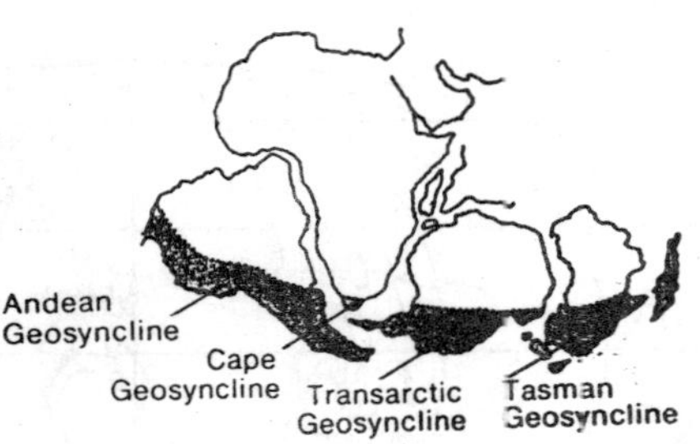

Fig. 4.12. Gondwanaland geosynclines.

During the Ordovician, a broad shelf spread across central Australia and received deposits of well-sorted sandstones and limestones.

The African record of the Early Paleozoic is far more obscure than that of Australia. The most complete sequences of strata are located north of the equator, where one finds Cambrian through Silurian rocks that were laid down in shallow embayments of a seaway that was the precursor of the Tethys Geosyncline. Africa south of the equator was persistently above sea level and was being eroded during most of the Early Paleozoic. Strata of this age are indeed uncommon.

One feature in the geologic record of the northern part of Africa is rather astonishing: There is remarkable evidence of an Ordovician episode of glaciation in the Sahara. One can only surmise that this part of Africa was close to the Ordovician South Pole or was at sufficient altitude to sustain glaciation.

Lower Paleozoic rocks in South America are also poorly known and difficult to interpret. Those strata deposited along the Andean Geosyncline have been greatly deformed, and frequently obliterated, by Mesozoic orogenic activity. Elsewhere, the Paleozoic sequence lies buried beneath the great blanket of sediment resulting from the erosion of the Andes. We do, however, have scattered bits of the record. We know that Cambrian and Ordovician seas did occupy the western margin of the continent and that sediments of both graptolitic and shelly facies were deposited there. Platform deposits were laid down in a shallow basin in what is now the Amazon region. In Silurian time, and elongate seaway covered Ecuador and Peru and then swung eastward across Argentina in apparent response to uplift in the southern Andean mobile belt. Finally, we note a short period of glaciation that left its traces in Ordovician rocks of western Argentina.

Early Paleozoic Global Geography

We are so accustomed to the shapes and locations of today's continents that earth's Early Paleozoic geography seems alien to us. Geologists have attempted to reconstruct the positions of

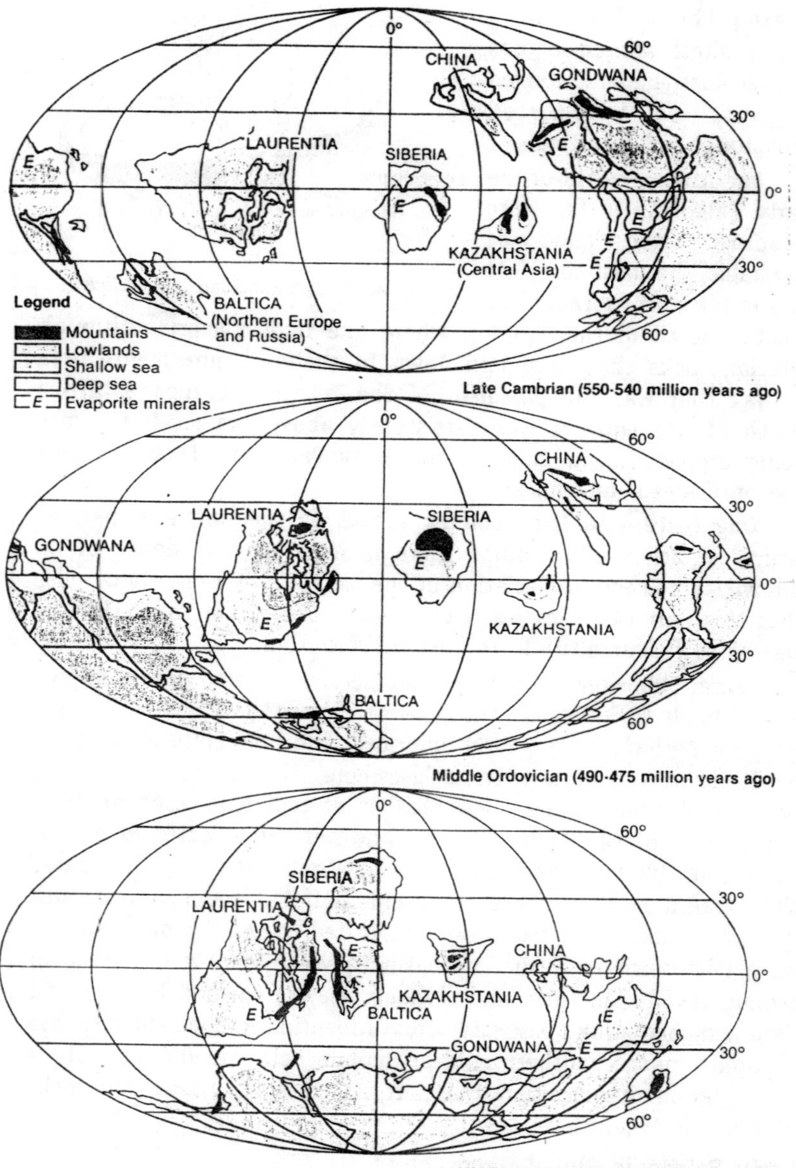

Fig. 4.13. Global paleogeography of the Early Paleozoic.

continents during this ancient time by combining all relevant paleomagnetic, paleoclimatologic, biogeographic, and tectonic information. This synthesis of data has resulted in global paleogeographic maps such as those provided in Figure. These maps indicate that by Late Cambrian time the major land areas were pasitioned at now latitudes as rather isolated continents in a well-interconnected world ocean. The main continental blocks at this time were Kakzakhastania (a piece of crust now located in Central Asia north of Iran), Siberia, Laurentia, Baltica, and Gondwana. There was no land north or south of 60° latitude, and the poles were therefore in the middle of oceanic areas. During the Ordovician and Silurian, the most dramatic change in global geography was the movement of Gondwana from its equatorial position southward to a polar location. Baltica (northern Europe and Russia), meanwhile, gradually circled and set a collision course toward eastern Laurentia. The actual collision took place in the Devonian as part of the Acadian Orogeny and the assembly of Pangaea II.

Aspects of Early Paleozoic Climates

During the Early Paleozoic, the earth had a north and south pole, was encircled by an equator, and must have been characterized by latitudinal and topographic variations in climate just as such variations exist today. As we have seen, during episodes of great marine transgressions and low-lying terrains, climates were probably somewhat milder; and during those periods when continents stood high and great ranges diverted atmospheric circulations patterns, climates were more diverse and extreme, much as is the case today. However, there were also factors that would have made those climates of long ago rather unique. During the Early Paleozoic, the earth turned faster and the days were shorter. Tidal effects were stronger, and there was no green covering of vegetation to better absorb the sun's radiation. There may even have been changes in solar radiation or atmospheric composition that left no unambiguous clues. The total climatic effects of these possibilities can only be surmised.

At the very beginning of the Cambrian, the climate was probably somewhat cooler than the average for the entire Early Paleozoic. The earth was still recovering from a great ice age that had culminated near the close of the Precambrian. Soon, however, climates became warmer. Epeiric seas slowly began to encroach upon the continental interiors. Paleogeographic reconstructions suggest that North America, Europe, and even Antarctica may have lain astride or near the equator during the Early Paleozoic. These geophysical data are compatible with the depositional record of thick limestones and extensive reefs, which today form only in the warmer regions of the oceans. However, some evaporites and fossiliferous limestones have been found in rather high paleolatitudes, indicating that worldwide climates may have been somewhat warmer than average for a time. In the Ordovician stratigraphic sequences of Northern Canada, for example, thick deposits of gypsum and salt are found at paleolatitudes (latitudes determined by paleomagnetic studies) of about 10° S. The North American reefs of Ordovician age generally fall within 30° of the paleoequator, whereas the richly fossiliferous carbonate rocks of the craton would have been within about 40° of the paleoequator. The study of cross-bedding in sandstones suggests that eastern North America lay in a zone of trade winds that blew from northeast to southwest with respect to the present directional grid. These winds thus moved in a direction opposite to those prevalent in eastern North America today.

That there were severe as well as equable climates in the Early Paleozoic is indicated by extensive Late Ordovician glacial deposits in the region of the present Sahara Desert. These deposits are sufficiently widespread that they must have been emplaced by continental glaciers and certainly must have been accompanied by a lowering of mean annual temperatures at middle and high latitudes. Study of the distribution of the glacial deposits, directions of glacial striations, and paleomagnetic data indicate that in Ordovician time the north African part of Gondwana was positioned astride the North Pole.

During the Silurian, there were latitudinal variations in climate somewhat similar to those of today. Glacial deposits of this age are found only in higher locations, usually in excess of 65° from the paleoequator. Coral reefs, evaporites, and desert sandstones are found within 40° of the Silurian equator. Certainly there were regions of marked aridity during the Silurian. Epeiric seas became relatively less extensive as the Silurian drew to a close. We can speculate that the "Proto-Atiantic," formed by the division of that hypothetic continent Pangaea I, had already begun closing and that the Caledonian crunch was about to take place. The curtain was ready to rise on the world of the Late Paleozoic.

Life of the Early Paleozoic

In the previous chapter we noted that fossil indications of life during most of the time that preceded the Paleozoic consisted largely of algae, fungi, and bacteria. Unquestioned animal remains appear only in uppermost Precambrian strata—"Eocambrian" strata—like those of Ediacara Hills, Australia. The Ediacara fauna includes some rather complex animals. They may have been the products of a previous lengthy period of evolution for which we have scant fossil evidence, or they may possibly have evolved rapidly just before the Cambrian. The Eocambrian, as mentioned, was a time of generally emergent lands, worldwide glaciation, and limited epeiric seas. It was a period that favoured neither fossilization nor development of diversity among marine invertebrates. It is likely that the ability of invertebrates to secrete calcium carbonate shells was limited.

In passing from the Late Precambrian to the Early Cambrian, the record gradually improves. Here and there around the shallow margins of the continents, isolated groups of invertebrates had begun to establish themselves. Soon the great inland seas began to creep across the cratons, providing as they did so a multitude of opportunities for the diversification and expansion of the phyla.

Certainly the improvement in the fossil record following the Eocambrian can be partly attributed to the spread of shell-building abilities. During the Cambrian, shell-building brachiopods and trilobites were abundant, but the most rapid expansion and diversification of shell-bearing groups began later, in Early Ordovician time. Many theories have been proposed to account for this expansion of shell-builders, including some that suggest pre-ordovician seas were chemically unsuitable for the secretion of calcium carbonate by organisms. However, Cambrian trilobites and brachiopods apparently had no difficulty in manufacturing exoskeletons of calcium carbonate. In modern shell bearing invertebrates, the shell serves as protection and provides support for soft tissues. It seems reasonable that these were also the functions served by the shells of Early Paleozoic invertebrates. Because there is little fossil evidence of numerous predatory animals until about Late Cambrian, support may have initially been the more important function. However, with the advent of abundant and vigorous carnivores (such as the predatory cephalopods) during the Late Cambrian, active predation may have provided the selective pressures that would have favored rapid evolutionary expansion of protective exoskeletons. The result was the multitude of shellfish that have populated the oceans down to the present day.

Plants

The history of plants has an obscure beginning among the bacteria and algae of the Precambrian. Indeed, the Precambrian sedimentary iron ores are presumed to have been produced by bacterial activity. However, as has been noted, fossils of such tiny, soft-bodied creatures are very rare.

An exception to the poor fossil record of earliest plants is provided by the stromatolites which are found frequently in Late Precambrian rocks and continue to be found in limestones of all younger ages. Laminations are the most apparent feature of stromatolites. They may vary in form from massive and incrusting to branching or fingerlike. Present-day colonies of blue-green algae cause the construction of similar masses. As a result of the photosynthetic activity of that algal mat, calcium carbonate is precipitated in the surrounding sea water, settles onto the algal colony, and adheres to the gelatinous film. Successive additional layers result in the laminations. Stromatolitic reefs were widespread during the Cambrian but were more restricted during the Ordovician, perhaps because of the rise of grazing marine invertebrates.

The green algae also produced some interesting fossils during the Early Paleozoic. A limesecreting family, the Dasycladaceae, includes fossils known as *receptaculids*. When first encountering one of these fossils, one is reminded of the appearance of the seed-bearing central area of a large sunflower; this has caused fossil collectors to designate them "sunflower corals." They are, however, neither sunflowers nor corals. Because the fossils contained spicule-like rods and circulatory passages, receptaculids were for many years classified as ancient varieties of sponges. Although most frequently found in Ordovician rocks, the receptaculids also occur sparsely in Silurian and Devonian strata.

The Late Precambrian and Early Paleozoic seas harbored a variety of unicellular, solitary, or colonial phytoplankton. Life in the oceans then, as today, would have been dependent upon those plants, which are known primarily from the chemically resistant coverings of their spores.

But what of life on land? The appearance of our landscapes today owes much to a verdent covering of forests and grasslands. At the beginning of the Paleozoic, however, there were no forests or grassy meadows. Most living organisms were confined to the seas. It is very likely that some species of algae invaded freshwater streams and lakes, and some may even have evolved means of moisture control independent of their environment, moved onto damp areas of the land, and begun the evolutionary progression toward more complex land plants.

Most land plants are *vascular*. Because they are not continuously immersed in water, they require systems of tiny tubes to transport fluids and nutrients. The tissues of land plants are tougher and more resistant to decay than those of the simple, nonvascular plants, and thus preservation is enhanced. The most reliable clues to the early presence of vascular plants are microscopic fossil spores discovered in mid-Ordovician rocks from Libya. The basic configuration of these microfossils corresponds to that in simpler types of spores known from Lower Devonian land plants of the

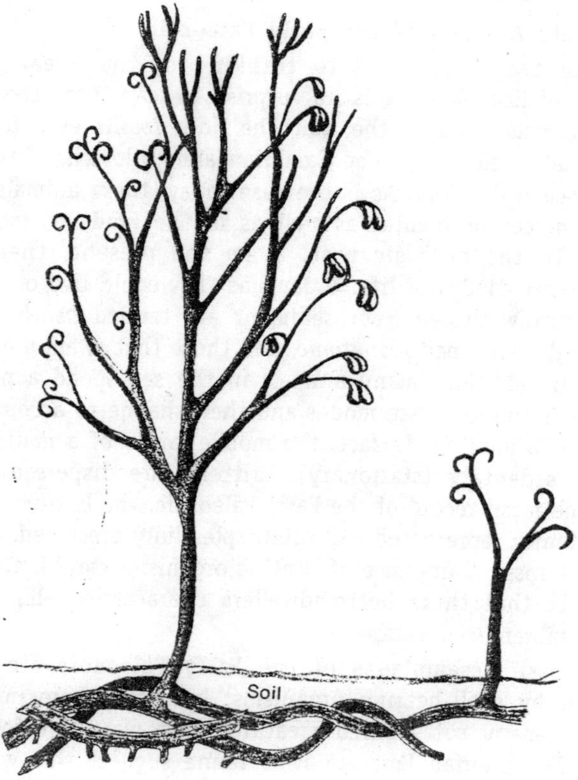

Fig. 4.14. A psilophyte from the Lower Devonian of Gaspe.

Family Psilophytaceae. Actual remains of spores from these so-called *psilophytes* have been found in Middle Ordovician Strata.

The psilophytes were small plants characterized by a horizontal stalk that grew just beneath the ground in moist soil and from which grew short, vertical "stems" bearing smaller branches and spore sacs. Although they had neither true leaves nor roots, psilophytes were more advanced than the algae and mosses in that they did have simple vascular tissue and even delicate strands of wood cells. The vascular system permitted part of the plant to exist underground, where there was water but no light, and another part to grow where the water supply was uncertain but sunlight for photosynthesis was insured. The psilophytes paved the way for the evolution of large, upright trees that covered large regions of the continents during the Late Paleozoic.

Invertebrate Animals of the Early Paleozoic

People are accustomed to thinking of the ocean as the birthplace of life. Thus, it is no surprise to learn that the earth's earliest animals lived in the sea. The only fossils ever found of animals that lived during the Cambrian and Ordovician have been those of ocean dwellers. As is the case today, these animals varied in their choices of habitat as well as in the kinds of food they required. In the geologic past, as in the present, there were basically three modes of life. Organisms that could live on the sea floor or burrow in sea floor sediment are termed *benthic*, those that floated are termed *planktonic*, and those that swam are termed *nektonic*. In addition, many animals in the sea spend a phase of their lives in one of these modes and then change to accommodate themselves to another. In fact, the mobile larvae of a multitude of otherwise sedentary (stationary) "critters" are dispersed in this way. In the fossil record of the Early Paleozoic, the bottom-dwellers are abundantly represented and often splendidly preserved. At least part of the fossil abundance of benthic organisms can be attributed to the fact that these bottomdwellers characteristically develop shells or mineralized carapaces.

The fossil assemblages of the Early Paleozoic are clearly dominated by shell-bearing organisms; however, there must also have lived many soft bodied creatures that were not fossilized. The Late Precambrian Ediacara Hills fauna suggest this was true, but more convincing proof derives from the discovery of the Middle Cambrian Burgess Shale fauna. This un-usual find was made by

Charles D. Walcott in 1910 while he was on an expedition in the Canadian Rockies. Near Field, British Columbia, while ascending the southwestern slope of Mt. Wapta, Walcott happened to notice some slabs of black shale that contained shiny, jetlike impressions of a variety of arthropods, worms, jellyfish, sea cucumbers, and sponges. The organisms are preserved as flattened, carbonaceous films with such perfection that fine, hairlike appendages, intestinal organs, and delicate dermal patterns are revealed. Walcott described 130 species from this locality. The Burgess Shale fauna is an excellent example of Early Paleozoic marine biota. It also reminds us of the bias of the fossil record, which only occasionally provides fossils of soft-bodied animals.

The diversification of marine invertebrates during the Early Paleozoic appears to have proceeded in two stages. An initial "experimental" evolutionary radiation during the Cambrian established most of the invertebrate phyla with preservable hard parts (the bryozoa do not appear until Ordovician). Many taxonomic groups within this original radiation did not expand widely, and many became extinct. Those that were best adapted survived and formed the ancestral stock for a grander radiation that took place during the Ordovician. Indeed, most of the classes of marine invertebrates with representatives in modern seas did not appear until Late Ordovician. Nearly 170 million years of evolution was necessary for this diversification.

The buildup to a complete marine fauna involved the expansion of life into a multitude of habitats and lifestyles. The many factors involved in evolutionary adaptation produced *epifaunal* animals that lived on the surface of the sea floor, as well as *infaunal* creatures that burrowed beneath the surface. There were borers as well as burrowers, attached forms and mobile crawlers, swimmers, and floaters. Some were *filter-feeders*, which strained tiny bits of organic matter or microorganisms from the water; others were *sedimentfeeders*, which passed the mud of the sea floor through their digestive tracts in order to extract the nutrients within. There were animals that grazed upon algae that covered parts of the ocean bottom, and carnivores that consumed these grazers. A host of scavengers processed organic debris and aided in keeping the seas suitable for life. Some classes of invertebrates maintained a single mode of life, whereas groups within other classes adapted to several different life styles. Snails, for example, included scavengers, herbivores, and carnivores. With the passage of geologic time, many evolutionary changes and extinctions were to occur among the invertebrates, but these changes tended to occur *within* the taxonomic classes that had been established during the Early Paleozoic.

The Unicellular Animals

The postulated biochemical steps leading to the first unicellular animals were discussed in the previous chapter. The first unicellular organisms were, of course, plants, but among modern unicells are creatures that defy classification as either animal or plant. Fortunately, there are also some single-celled forms that can with confidence be identified as animals. Two groups of creatures in the last category that are capable of building shells or supportive structures are the *foraminifers* and *radiolarians.*

Foraminifers build their tiny shells by adding chambers single, in row, in coils, or in spirals. Some species construct the shell, or test, of tiny particles of silt; others secrete tests composed of calcium carbonate. The test is characteristically provided with holes through which extend "tentacles" of protoplasm for feeding. It is

Fig. 4.15. Forminifers that are found in Early Paleozoic strata. A—Ammodiscus. B—Turritellella. C—Lituotuba. D—Bathysiphon. E—Lagenammina.

from these holes, or foramina, in the test that foraminifers take their name. Foraminifers range from the Cambrian to the present, but they are rare and poorly preserved in Lower Paleozoic strata. During these early stages of their history they were mostly simple, saclike, tubular, or loosely coiled benthic creatures with tests composed of sedimentary particles. In modern oceans the tests of calcareous foraminifers rain down continuously on the sea floor, much like an unending limy snow. The accumulated debris forms, in part, a deep sea sediment called globigerina ooze because of the prevalence of the remains of species of *Globigerina*.

Radiolarians are also single-celled planktonic organisms that have been present on earth at least since the Paleozoic began. Like the foraminifers, they have thread like pseudopoelia that project from an ornate lattice-like skeleton of opaline silica or a proteinaceous substance. In some regions of the oceans today, radiolarian skeletons accumulate to form deposits of radiolarian or siliceous ooze. Although radiolarians do occur in the Early Paleozoic, they are rare and not yet useful for stratigraphic correlation. They are more abundant during the Mesozoic and

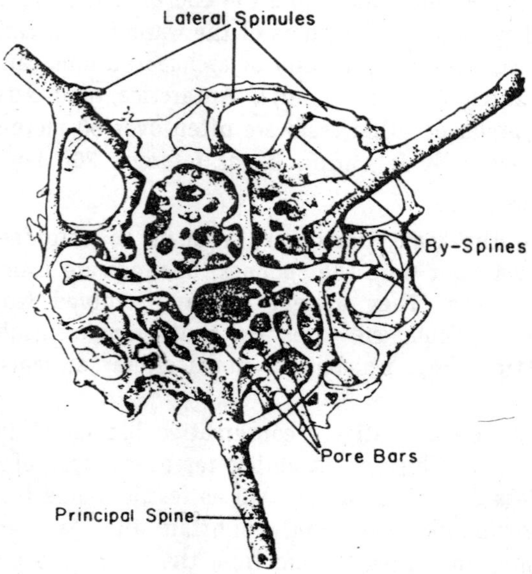

Fig. 4.16. A Late Ordovician radiolarian from the Hanson Creek formation of Nevada. It is unusual to find well-preserved radiolarians in rocks of the Early Paleozoic.

Cenozoic and have been shown to be particularly useful in correlating Pleistocene deep sea deposits. Of special interest is the observation that Pleistocene radiolarian extinctions appear to be associated with dates of reversal in the earth's magnetic field.

The Cup Animals

Archaeocyatha means "ancient cups." It is an appropriate name for a somewhat enigmatic group of Cambrian organisms that constructed conical or vase-shaped skeletons out of calcium carbonate. Archaeocyathids hold two paleontologic records. They are the earliest abundant reef-building animals on earth, and they are members of a major phylum that suffered extinction. Although quite abundant during the Early Cambrian, they had entirely died out before the end of that period. Gregarious in habit, they carpeted the floors of the warm inland Early Paleozoic seas. Low reefs formed primarily of archaeocyathids can be studied today in North America, Siberia, Antarctica, and Australia. The Australian archaeocyathid reefs are often over 60 meters thick and extend horizontally in narrow bands for over 200 km.

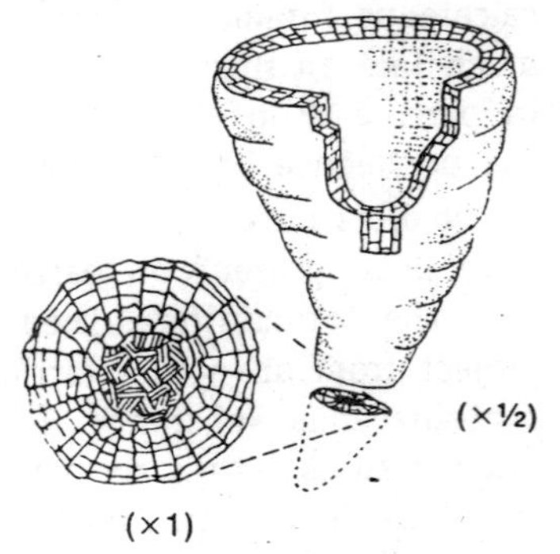

Fig. 4.17. An archaeocyathid.

The Pore-Bearers

Among the many stationary animals to colonize the Early Paleozoic sea floor were the sponges. Sponges are members of the Phylum *Porifera*. They appear to have evolved from colonial flagellated unicellular creatures and thus provide insight into how the transition from unicells to multicellular animals may have occurred.

Sponges are a relatively conservative branch of invertebrates that have a long history. Cambrian representatives of all but one modern class of Porifera are known as fossils. Some fossil sponges, like *Protospongia* from the Cambrian and *Astraeospongium*, *Microspongia*, and *Astylospongia* from the Silurian, are well known to geologists as index fossils.

Sponges have always been predominantly marine creatures, although a few modern species live in fresh water. Spicules formed

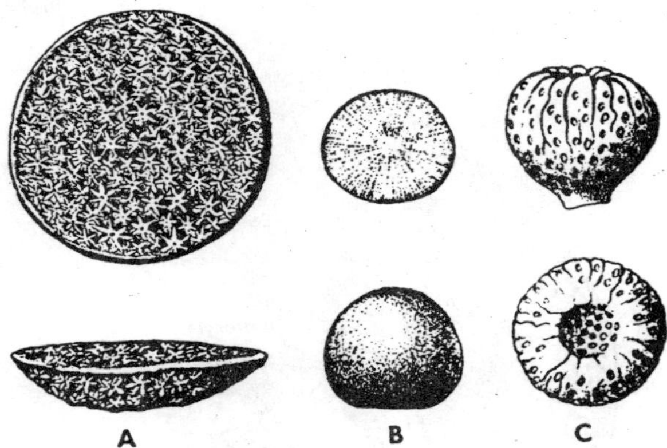

Fig. 4.18. Early Paleozoic (Silurian) sponges. A—Astraeospongea. B—Microspongea. C—Astylospongea.

in the walls and cell layers of sponges are distinguishing characteristics of the phylum and provide both protection and support. The spicules may be composed of silica, calcium carbonate, or a proteinaceous material termed *spongin*. Naturally, the mineralized spicules are more commonly preserved and are frequently found by geologists examining rocks that have been disaggregated for study. Spicules are also important in the classification of sponges. For example, the *Desmospongea* consist entirely or in part of spongin, which may be reinforced with siliceous spicules. *Hyalospongea* develop siliceous spicules of distinctive shape, and *Calcispongea* are characterized by spicules made of calcium carbonate.

Although sponges vary greatly in size and shape, their basic structure is that of a highly perforated vase modified by folds and canals. The body is attached to the sea floor at the base, and there is an excurrent opening, or osculum, at the top. The wall consists of two layers of cells. Facing the internal space is a layer of collar cells (choanocytes), and on the outside is a protective wall of flat cells that somewhat resemble the bricks of a worn masonry pavement. Between these two layers one finds a gelatinous substance called *mesenchyme*. Here amoeboid cells go about the work of secreting the spicules. Sponges lack true organs. Water currents moving through the sponge are created by the beat of *flagellae*. These currents bring in suspended food particles, which are ingested by the collar cells. In a simple sponge, water enters

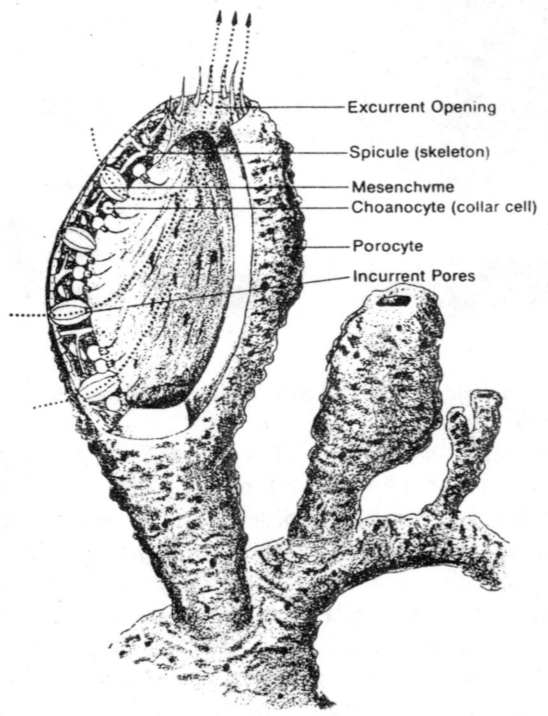

Fig. 4.19. Schematic diagram of a sponge having the simplest type of canal system. The path of water currents is indicated by arrows.

through the pores, flows across sheets of choanocytes in the central cavity, and passes out through the osculum.

One group of Early Paleozoic sponges that are particularly interesting because of their reef-building capabilities is the *stromatoporoids*. These organisms constructed fibrous, calcareous skeletons of pillars and thin laminae that can nearly always be found in reef-associated carbonate rocks of the Silurian and Devonian. Apparently, stromatoporoids grew profusely and in close association with corals, brachiopods, and other invertebrate reef-dwellers. Silurian coral-stromatoporoid reefs are well known to geologists primarily because of extensive study of reef exposures on Gotland Island off the Baltic coast of Sweden and in the region around and southwest of the Michigan Basin. Reef development during the Devonian was similar to that in the Silurian, with blanket-like, massive, and cylindric stromatoporoids forming a considerable part of the total mass of the reef structures.

Corals and Other Coelenterates

Sea anemones, sea fans, jellyfish, the tiny *Hydra*, and the reef-forming corals are all representatives of the Phylum *Coelenterata*, which is known for the great diversity and beauty of its members. The coelenterate body wall is composed of an outer layer of cells, the *ectoderm*, an inner layer, the *endoderm*, and a thin, noncellular intermediate layer, the mesoglea. In the endoderm are found primitive sensory cells, gland cells that secrete digestive enzymes, flagellated cells, and nutritive cells to absorb nutrients. A distinctive feature of many coelenterates is the presence of stinging cells, which, when activated, can inject a paralyzing poison.

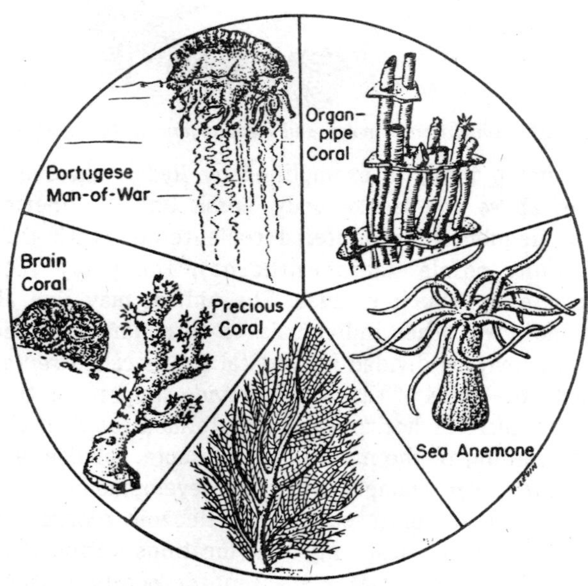

Fig. 4.20. Diversity among the coelenterates.

Body form in coelenterates may be either *polyp* or *medusoid*. The medusoid form is seen in the jellyfish, which somewhat resembles an umbrella in shape. Jellyfish have a concave undersurface that contains a centrally located mouth; for this reason, it is designated the oral surface. In jellyfish such as the living *Aurelia*, which is common along the eastern shore of the United States, the mouth is surrounded by four oral arms. All jellyfish have tentacles, and in *Aurelia* these are located around the margin of the umbrella. Most jellyfish swim by rhythmic contractions of the umbrella.

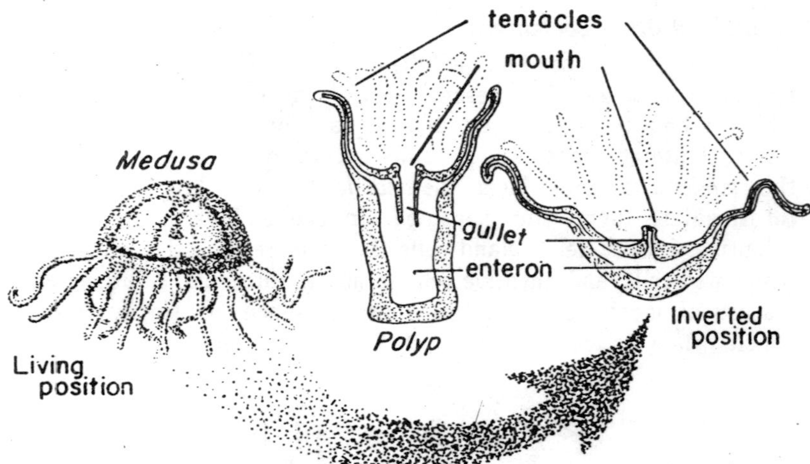

Fig. 4.21. Comparison of polyp and medusa form in coelenterates.

In the polyp form, as exemplified by *Hydra*, there is a circle of tentacles above the saclike body. Corals and sea anemones are among the frequently encountered coelenterates with this polyp form. In stony corals (Class Anthozoa), the polyp secretes a calcareous cup, in which it lives. The animal may live alone or may combine with other individuals to form large colonies. The cup, or *theca*, may be divided by vertical plates called *septa*, which serve to separate layers of tissue and provide support. As the animal grows, it also secretes horizontal plates termed *tabulae*. Corals are identified according to the nature of their septa, tabulae, and other skeletal features. For example, after the development on an initial embryonic set of six protosepta, the Paleozoic *rugose corals*, or *Rugosa*, insert new septa at only four locations during growth. In other Paleozoic corals, septa are absent or poorly developed, so that tabulae are the most important features in classification. These are the *Tabulata*. Tabulates include many interesting colonial forms, such as the honeycomb and chain corals. The rugose corals and tabulate corals became extinct at the end of the Paleozoic and were followed in the Mesozoic by anthozoans of the Order *Scleractinia*. In all scleractinian corals, septa are inserted between the mesenteries in multiples of six.

The fossil record for coelenterates begins with the discovery of fossil jellyfish impressions in the Late Precambrian. The phylum is still poorly represented in Cambrian rocks. In the succeeding

Ordovician the record improves dramatically, for the lime-secreting anthozoans begin to expand and diversify. The first stony corals were the tabulates, recognized by their simple, often clustered or aligned tubes divided horizontally by transverse tabula. During the Silurian, *Halysites*, the chain coral, and *Favosites*, the honeycomb coral, contributed their skeletons to extensive reef developments. The reefs in turn provided a variety of habitats for use by other marine invertebrates. Simple representatives of the horn-shaped rugose corals appeared in the Ordovician only slightly later than the tabulates. The tabulates, however, were the dominant corals of Silurian reefs. By Devonian time, tabulates began to be overshadowed by numerous and diverse members of the expanding rugose corals. These rugose corals joined with the declining tabulates and the ubiquitous stromatoporoids to form the reef that were common in the epicontinental seas that covered the craton. The reefs varied in size. Some were only a few meters thick, whereas other attained thicknesses of nearly 300 meters. Many of the reefs in Illinois and Ohio are covered by younger strata and were discovered during oil explorations. Because of their porous nature, buried reefs provide excellent structures for the accumulation of petroleum. Corals have also been remark-ably useful in local and regional stratigraphic correlations. Forms such as *Tetradium* were common Ordovician tabulates, whereas *Favosites* and *Heliolites* populated Silurian seas. Among Early Paleozoic rugosa *Favistella* and *Streptelasma* are used widely in correlation.

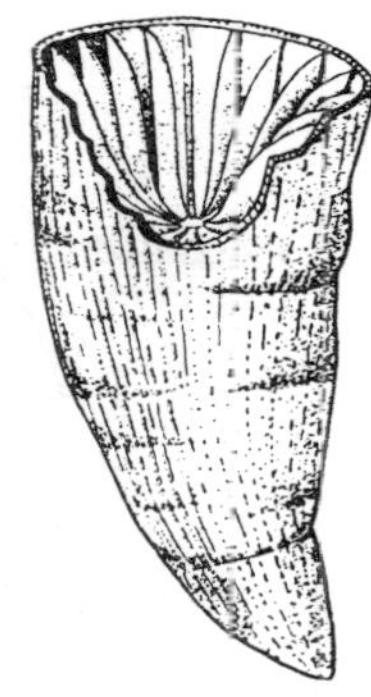

Fig. 4.22. An Ordovician rugose or "horn coral."

Moss Animals

Bryozoans are minute, bilaterally symmetric animals that grow in colonies that frequently appear twiglike when viewed without the aid of a magnifier. The individuals, called *zooids*, are housed in a capsule, or *zooecium*, that is often preserved as a result of its being calcified. The zooecia appear as pinpoint depressions on the outside of the colony, or *zoarium*. The zooid has a complete, U-shaped digestive tract with mouth surrounded by a tentacled feeding organ called the *lophophore*.

Today, there are more than 4000 living species, and nearly four times that number are known as fossils. Their earliest reliably

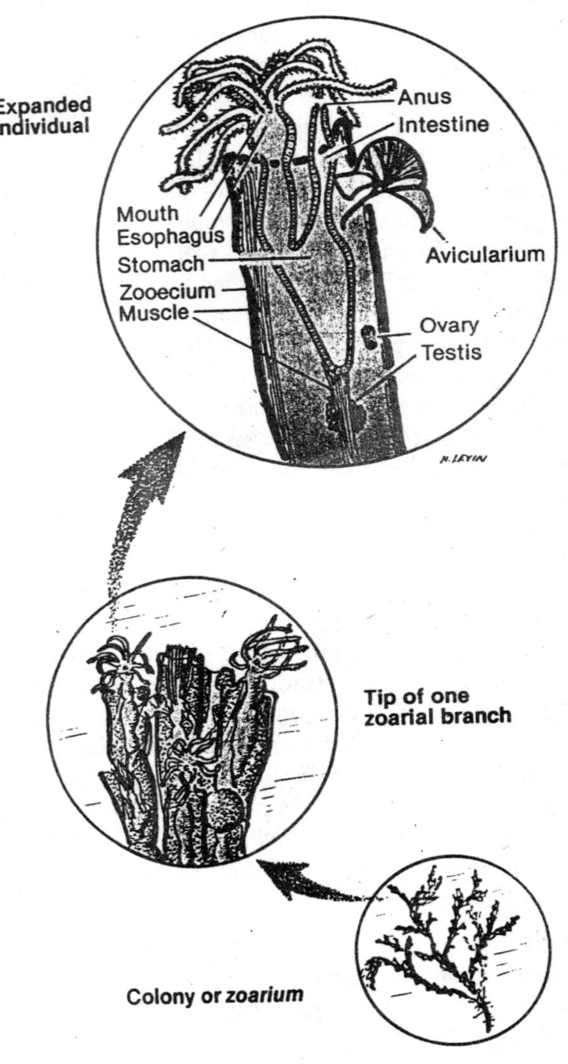

Fig. 4.23. Relation of bryozoan individuals to the zoarium.

reported occurrence is from Lower Ordovician strata of the Baltic, but they did not become abundant until Middle Ordovician and Silurian. In rocks of these ages, their remains sometimes make up much of the bulk of entire formations. Like the corals, bryozoa contributed to the framework of reefs. One of the commonest of all the Early Paleozoic bryozoans was *Fistulipora*, different species of which develop massive, incrusting, or arborescent colonies. The genus *Hallopora* is characterized by abundant nodes called monticules. Star-shaped patterns on the surface of *Constellaria* provided the inspiration for its generic name.

Like corals, bryozoa are useful in stratigraphic correlation. Biostratigraphic zones based on overlapping ranges of fossil bryozoans have made possible correlation of Ordovician strata from Oklahoma across the central United States to Ontario and from North American to the Baltic area of Europe.

Brachiopods

Second only to the trilobites, brachiopods are the most abundant, diverse, and useful fossils in Paleozoic rocks. They are characterized by a pair of enclosing valves, which together constitute the shell of the animal. They resemble clams in this regard, but, in symmetry of the valves and soft part anatomy, brachiopods are quite different from pelecypods. Brachiopod valves are almost always symmetric on either side of the mid line, and the two valves differ from each other in size and shape. The valves of a clam are right and left, whereas those of a brachiopod are dorsal and ventral. The valves may be variously ornamented with radial ridges or grooves, spines, nodes, and growth lines. Calcium carbonate is the usual hard tissue of brachiopods, although the valves of some families are composed of mixtures of chitin and calcium phosphate. This is particularly true of the group called *inarticulates*.

In the *articulate* brachiopods, the valves are hinged along the posterior margin and are prevented from slipping sideways by teeth and sockets. The less common inarticulates lack this definite hinge, are held together by muscles, and characteristically have simple, spoon-shaped or circular valves. Although brachiopod larvae swim about freely, the adults are frequently anchored or cemented to objects on the sea floor by a fleshy stalk (*pedicle*) or by spines. Some simply rest on the sea floor. One of the more conspicuous soft organs of brachiopods is the *lophophore*, a structure consisting of two ciliated coiled tentacles whose function is to circulate the water between the valves, distribute oxygen, and remove carbon dioxide. Water currents generated by *cilia* on the lophophore move food particles toward the mouth and short digestive tract.

Brachiopods still live in seas today, although in far fewer numbers than they did during the Paleozoic. Cambrian brachiopods were frequently the chitinous inarticulate varieties. The inarticulates

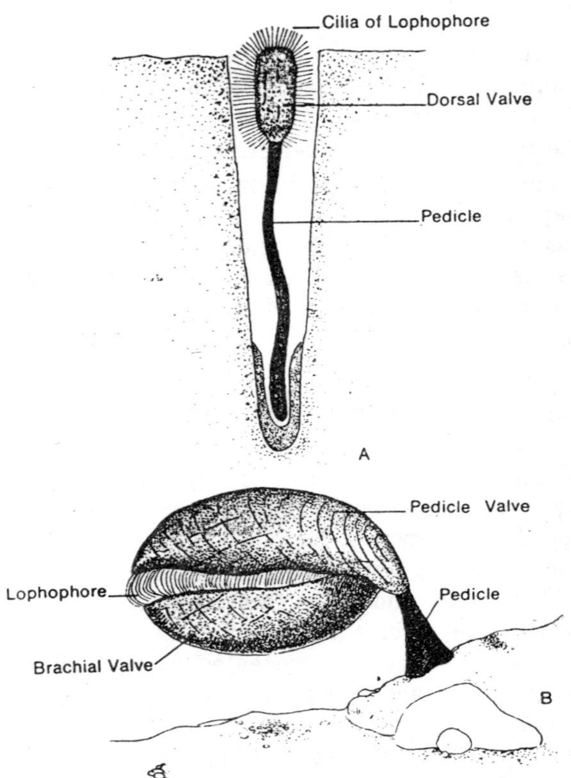

Fig. 4.24. Living positions of two modern brachiopods. The inarticulate brachiopod Lingula excavates a tube in bottom sediment and lives within it. The pedicle secretes a muscus at the end that glues the animal to the tube. The life position of a living articulate is shown in B, with the lophophore barely visible through the gape in the valves.

became somewhat more diverse during the Ordovician and then declined. A few species of only three families remain today.

The articulates also first appeared in the Cambrian but became truly abundant during the succeeding Ordovician Period, when there was a great expansion of all sorts of shelled invertebrates. Across the floors of Early Paleozoic epeiric seas, large and small aggregates of these filter-feeders could be found. Today, their skeletons compose much of the volume of thick formations and provide the stratigrapher with essential markers for correlation. Among the articulate brachiopods that were particularly abundant during the first three periods of the Paleozoic were the orthids, strophomenids, pentamerids, rhynchonellids and spiriferids.

The Mollusks

A stroll along almost any seashore will provide evidence that mollusks are today's most familiars marine invertebrates. The shells of most members of this phylum have been readily preserved and provide enjoyment to collectors and conchologists today. Such well-known animals as snails, clams, chitons, tooth shells, and squid are included within the *Phylum Mollusca*. Although most mollusks possess shells, some, like the slugs and octopods, do not. The various member classes of the Mollusca may differ considerably in external appearance, yet they have fundamental similarities in their internal structure. There is a muscular portion of the body called the *foot* that functions primarily in locomotion. In cephalopod forms, the foot is modified into tentacles. A fleshy fold called the mantle has the function of secreting the shell. In aquatic mollusks, respiration is accomplished by means of gills. Well-developed organs for digestion, sensation, and circulation attest to the advanced stage of evolution that mollusks have attained.

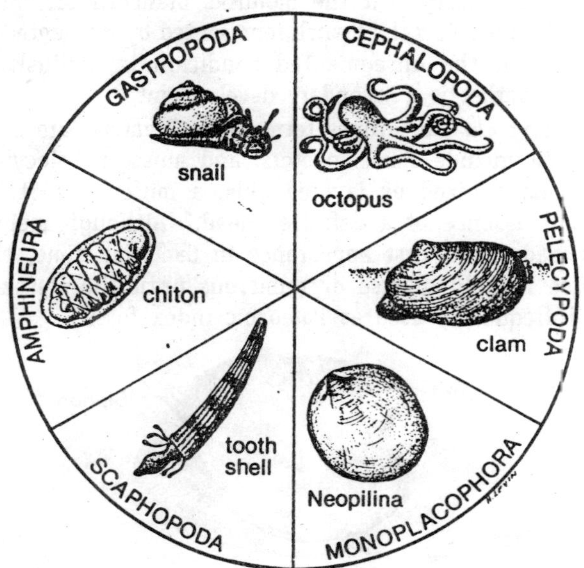

Fig. 4.25. Diversity among the members of the Phylum Mollusca.

As presently classified, the Phylum Mollusca consists of eight classes. Three of these can be collectively termed *placophorans*. They are relatively primitive mollusks with multiple paired gills.

The placophorans are divided into the *Monoplacophora*, which have simple cap-shaped shells; the *Polyplacophora*, represented by chitons, which possess eight overlapping shell plates; and the *Aplacophora*, which lack mineralized plates. The remaining classes of Mollusca include *Scaphopoda*, *Hyolitha* (small, conical mollusks with lids, or opercula), and the more familiar *Gastropoda*, *Pelecypoda*, and *Cephalopoda*.

Fossil monoplacophorans are known to occur in rocks as old as Lower Cambrian. This group persists today in the form of small, limpet-like mollusks named *Neopilina*.

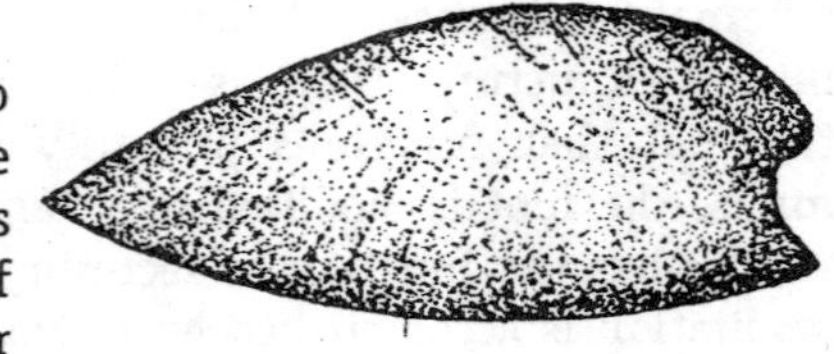

Fig. 4.26. The "living fossil" *Neopilina ewingi*.

Of great interest to biologists is the fact that the "living fossil" *Neopilina* displays a segmental arrangement of gills, shell muscles, and other organs. This evidence of segmentation provides some support for the theory that the mollusks branched off the family tree at some point below the branch represented by segmented annelid worms and that the unsegmented condition in mollusks is not primitive but rather is a secondary development.

The *Pelecypoda* (sometimes termed the Bivalvia) are a class of mollusks that include clams, oysters, and mussels. Pelecypods are generally characterized by layered gills, a muscular foot, bilobed mantle, and absence of a definite "head." Although members of the class made their first appearance in Cambrian time, they did not become notably diverse or numerous until the Mesozoic and thus are infrequently used as Paleozoic index fossils.

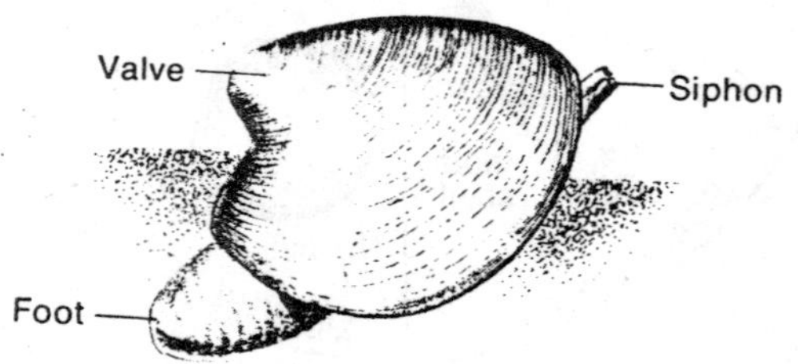

Fig. 4.27. Modern clam showing muscular foot, siphon (for water circulation) and one of two valves.

Gastropods first appear in Lower Cambrian strata. Earliest forms constructed small, depressed, conical shells. During later Cambrian

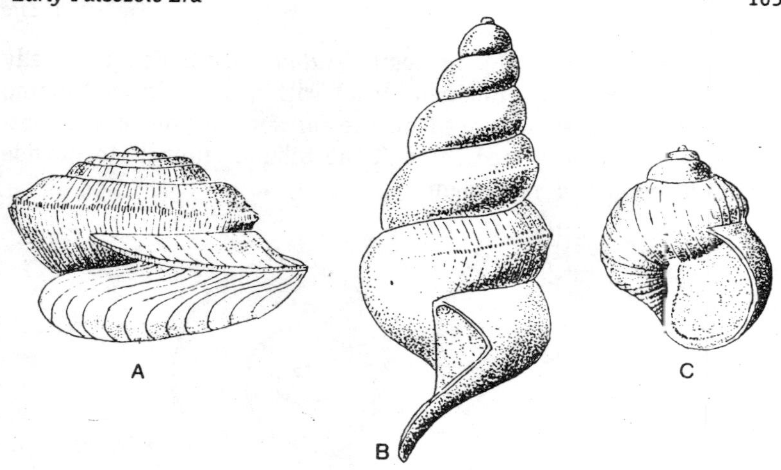

Fig. 4.28. Three wide-spread Ordovician marine gastropods. A—Eotomaria supracingulata. B—Hormotoma trentonensis. C—Cyclonema limatum.

and Ordovician time, gastropods with the more familiar coiled conchs became commonplace. It appears that coiling in a plane was not unusual initially but was supplanted by spiral coiling. Gastropods of the Ordovician and Silurian had shapes similar to those of living species. One presumes that their soft parts were also similar, with distinct head, mouth, eyes, tentacles, and a vertically flattened foot to provide for gliding movement. Certain species of gastropods had relatively short geologic ranges and are therefore useful as guide fossils.

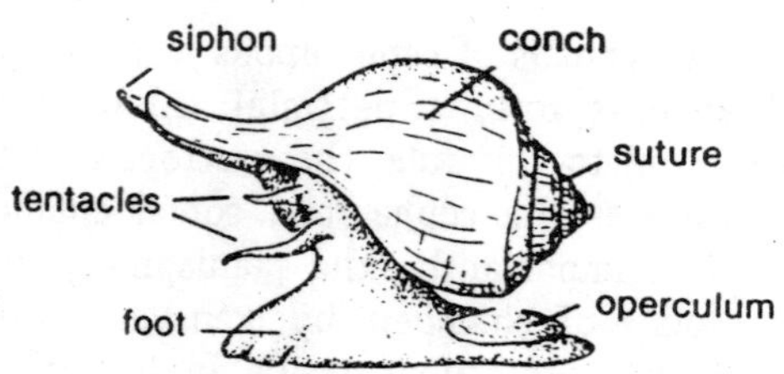

Fig. 4.29. Externally visible features of a marine gastropod.

The *cephalopods* may very well be the most complex of all the mollusks. Today, this marine group is represented by the squid, cuttlefish, octopods, and the lovely chambered nautilus. The nautilus in particular provides us with important information about the soft anatomy and habits of a vast array of fossils known only by their preserved conchs. In the genus *Nautilus*, one finds a bilaterally symmetrical body; a prominent head with paired, image-forming eyes; and tentacles developed on the forward portion of the foot. Water is forcefully ejected through the tubular "funnel" to provide swift, jet-propelled movement.

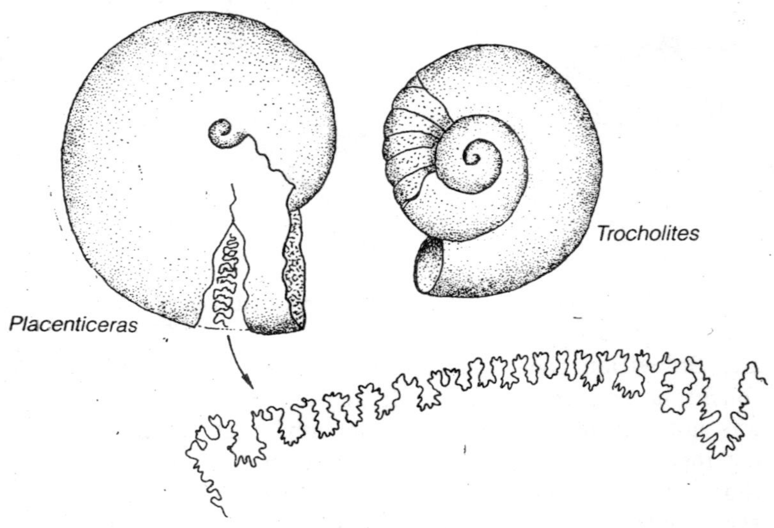

Fig. 4.30. Nautiloid cephalopods are characterized by relatively straight sutures, as shown in Trocholites from the Ordovician of New York. Ammonoids have complexly wrinkled sutures, as exemplified in Placenticeras from the Cretaceous.

At first glance the conchs of cephalopods resemble snail shells, but the resemblance is only superficial. Although we have mentioned an exception to this rule, the gastropod shell generally coils in a spiral, whereas the cephalopod conch characteristically coils in a plane. More importantly, the planispirally coiled conch is divided into a series of chambers by transverse partitions, or septa. The bulk of the soft organs reside in the final chamber. Where the septa join the inner wall of the conch, suture lines are formed. These lines are enormously useful in the identification and classification of cephalopods. For example, cephalopods placed in the *Nautiloidea* have straight or gently undulating sutures, whereas the *Ammonoidea* have complicated wriggly sutures. Ammonoids became widespread during the Mesozoic; for this reason well will delay our discussion of them until a later, more appropriate, chapter.

***Fig. 4.31.** Two specimens of straight-conch cephalopods from the Saluda Formation of Indiana. The specimens are Upper Ordovician in age.*

The oldest fossils to be classified as cephalopods were small, conical conchs discovered in Early and Middle Cambrian rocks of Europe. The class gradually increased in number and diversity and became ubiquitous inhabitants of Ordovician and Silurian seas. Indeed, by Ordovician time, a great variety of conch forms—from straight to tightly coiled had developed. Some of the elongate forms exceeded 4 meters in length. The first signs of a decline in nautiloid populations can be detected in Silurian strata. After Silurian time the group continued to dwindle until today only a single genus, *Nautilus*, survives.

Arthropods

The arthropod phylum is enormous. It includes such living animals as lobsters, spiders, insects, and a host of other "critters" that possess chitinous exterior skeletons, segmented bodies, paired and jointed appendages, and highly developed nervous system and sensory organs. Members of the Arthropoda that have left a particularly significant fossil record are the *trilobites, ostracodes*, and *eurypteriods*.

Trilobites were swimming or crawling arthropods that take their name from division of the dorsal surface into three longitudinal segment, or *lobes*. There are, for example, a central axial lobe and two lateral (Pleural) lobes. There was also a transverse differentiation of the shield into an anterior *cephalon*, a segmented *thorax*, and a posterior *pygidium*. The skeleton was composed of chitin strengthened by calcium carbonate in parts not requiring flexibility. As in many other arthropods, growth was accomplished by molting. Although many trilobites were sightless, the majority had either single-lens eyes or compound eyes composed of a large number of discrete visual bodies. The earliest trilobites lacked a pygidium and had a large number of thoracic segments. These characteristics have led paleontologists to speculate that trilobites may have evolved from annelid worms sometime during the Precambrian.

If one considers the entire fossil record for only the Cambrian, the trilobites are clearly the most abundant and diverse. More than 600 genera are known from the Cambrian. Their first appearance, marked by remains of the genus *Olenellus*, is used to identify the initial strata of the Paleozoic.

The earliest trilobites were apparently bottom-dwelling, crawling scavengers and mud processers. Some preferred limy bottoms of

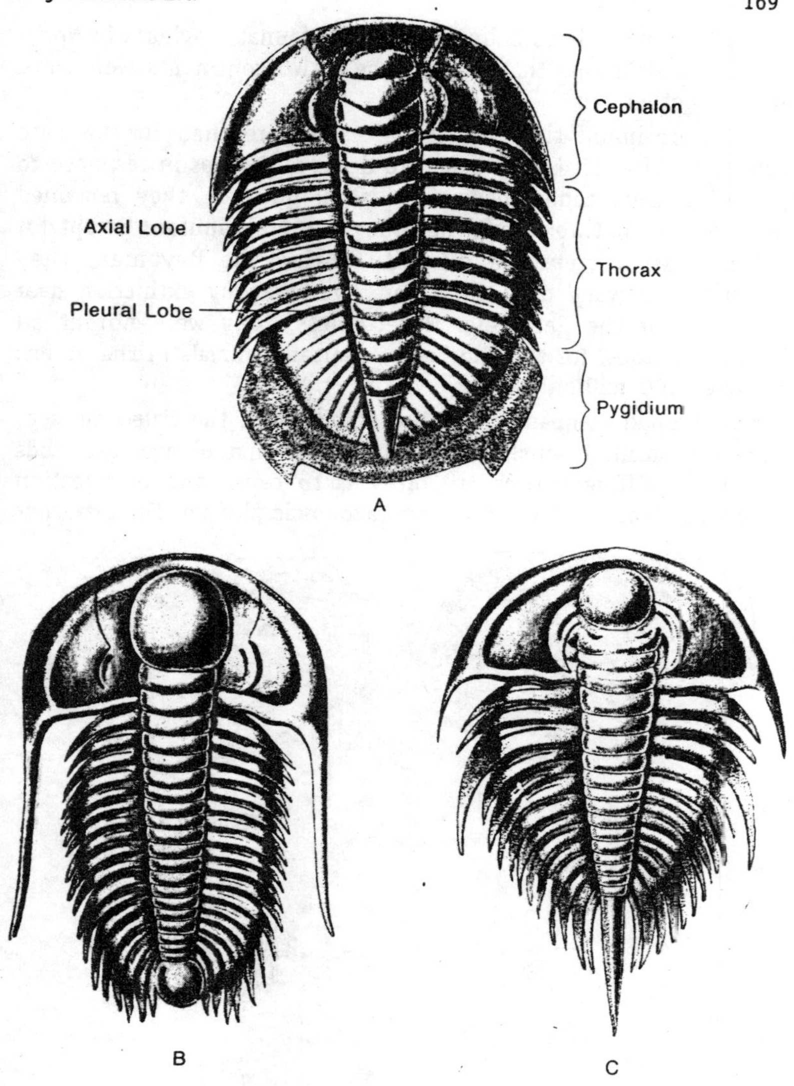

Fig. 4.32. Three well-known Cambrian trilobites. A—Dikelocephalus minnesotensis (Upper Cambrian), B—Paradoxides harlani (Middle Cambrian), C—Olenellus thompsoni (Lower Cambrian).

cratonic and miogeosynclinal areas, and others, like *Paradoxides* of the Middle Cambrian, inhabited the muddy or silty floors of eugeosynclinal tracts. As a result of such environmental preferences, it has been possible to delineate trilobite faunal provinces in North America and Europe that suggest these two continents were once close together.

The optimum time for trilobites was reached in the Late Cambrian. After that they began to decline, perhaps in response to predation from cephalopods and fishes; however, they remained fairly abundant throughout the Ordovician and Silurian. Except for a temporary increase in diversity during the Devonian, they continued to wane until they were overtaken by extinction near the close of the Paleozoic. Nevertheless, they were not at all biologic failures, for the had been important animals in the oceans for over 300 million years.

Arthropod companions to the trilobites in the Paleozoic seas were the small, bean-shaped ostracods. At first glance, ostracods appear so different from trilobites as to cause one to question their classification within the same taxonomic phylum. The ostracods

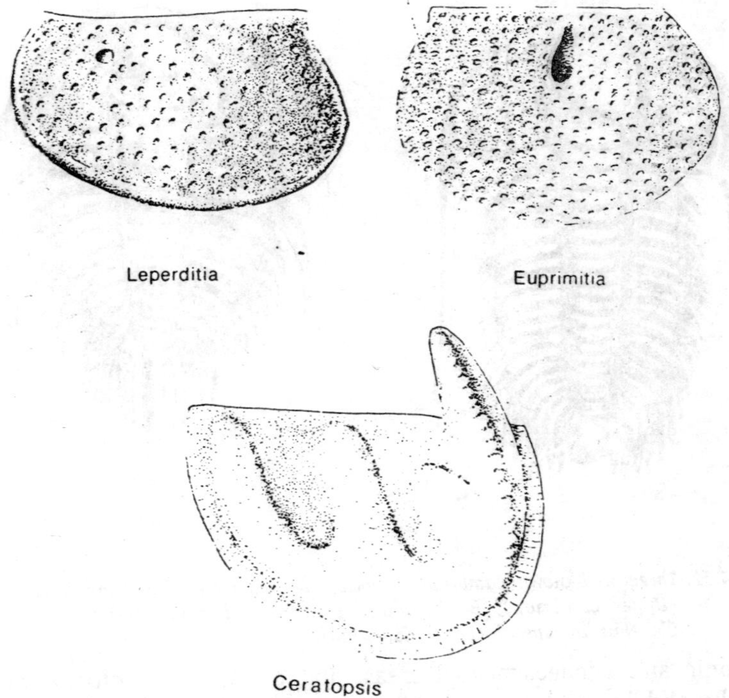

Fig. 4.33. Three common Ordovician ostracods.

have a bivalved shell vaguely suggestive of some sort of tiny clam. However, this bivalved carapace encloses a segmented body from which extend seven pair of jointed appendages. Adult animals are about ½ to 4 mm in length. The valves are composed of both chitin and calcium carbonate and are hinged along the dorsal margin.

Ostracods first appeared early in the Ordovician, and some limestones of this age are almost completely composed of their discarded carapaces. *Laperditia, Euprimitia,* and *Ceratopsis* are common Ordovician genera. Ostracods continue in relative abundance to the present day. They occur in both marine and fresh water sediments. Because of their small size, they are brought to the surface in wells drilled for oil and are used by exploration geologists in correlating the strata of oil fields.

Eurypterids are a group of arthropods that, because of their rarity, are less useful in stratigraphic studies than are either trilobites or ostracods. Nevertheless, they are among the most impressive of Early Paleozoic marine invertebrates. Although many were of modest size, some were nearly 3 meters long and, had they survived, would be suitable subjects for a Hollywood monster film. On their scorpion-like bodies were five pairs of appendages and a fearful-looking pair of pincers. They were also equipped with a venomous stinger. Eurypterids ranged across portions of the sea floor and brackish estuaries from Ordovician until Permian time but were especially abundant during the Silurian and Devonian.

Spiny-Skinned Animals

Just as modern seas abound with starfish, sea urchins, and seal lilies, so also were the oceans of the Early Paleozoic populated with members of the Phylum Echinodermata. Echinoderms are animals with pentamerous symmetry that masks an underlying primitive bilateral symmetry. They are well named "the spiny skinned," for spines are indeed present in many species. A unique characteristic of the phylum is the presence of a system of vessels—the water vascular system—which functions in respiration and locomotion. Members of the phylum are exclusively marine, typically bottom-dwelling, and either attached to the sea floor or able to more about slowly.

Of considerable interest is the evolutionary relationship between echinoderms and vertebrates. There are so many allied embryologic parallelisms among primitive chordates and echinoderms that some zoologists speculate that both may have arisen from similar ancestral

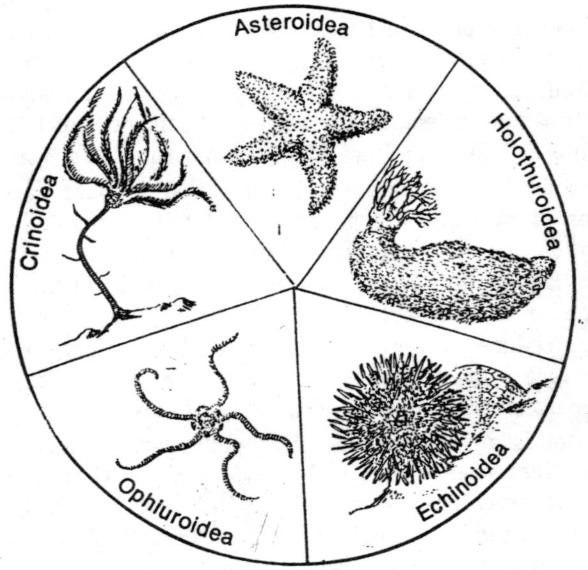

Fig. 4.34. Living echinoderms.

forms. For example the larvae of echinoderms closely resemble those of the living protochordates called acorn worms. Also, in embryologic development, the mesodermal layer of cells, as well as certain other elements of the body, arises in the same way. Biochemistry has provided additional evidence for a relationship between echinoderms and chordates by revealing chemical similarities associated with muscle activity and oxygen-carrying pigments in the blood.

Among the many classes of the Phylum Echinodermata, the Asteroidea (starfish), Ophiuroidea (brittle stars), Edrioasteroidea, Crinoidea (sea lilies), Blastoidea, and Cystoidea are the most abundant and useful in geologic studies. The last four of these classes constitute echinoderms that were sedentary and attached. They were particularly characteristic of the Paleozoic Era, whereas the other unattached and more mobile forms were more frequent in later eras. For this reason, we will discuss the stemless echinoderms in a later chapter probably with Mesozoic life.

The oldest probably echinoderm thus far discovered may be *Tribrachidium* of the Late Precambrian Ediacara Hills fauna. This peculiar echinoderm appears to be related to a group called *edrioasteroids*, which are among the oldest members of the phylum

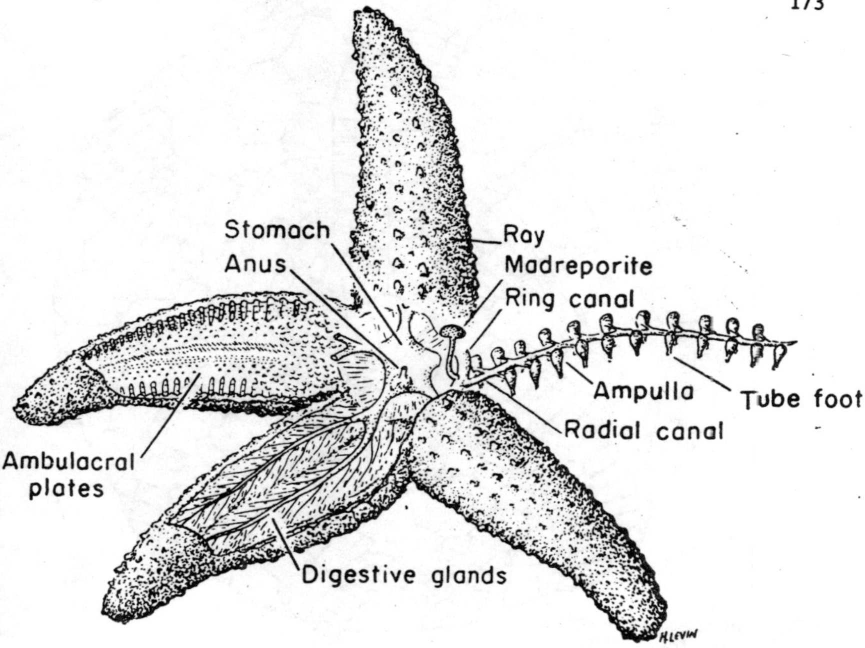

Fig. 4.35. Partially dissected starfish showing elements of the water vascular system and other organs.

and considered by many paleontologists to be ancestral to starfish and sea urchins. Edrioasters like *Edrioaster bigsbyi* had developed several general echinoderm characteristics, such as calcareous plates, a central mouth, radiating food grooves (ambulacra), and a water vascular system.

The stemmed or stalked echinoderms first occur in Middle Cambrian strata but do not become abundant until Ordovician and Silurian time. Stalked forms called *cystoids* are the most primitive among this group. The striking pentamerous symmetry that is evident in most echinoderms is often less well developed in cystoids. Beginning students of paleontology often recognize cystoids by the characteristic pores that occur on the plates of the calyx and that compose part of the water vascular system. Although cystoids range from Cambrian into Late Devonian, they are chiefly found in Ordovician and Silurian rocks.

Unlike some of the cystoids, stalked echinoderms known as *blastoids* have a beautifully symmetric arrangement of plates. The

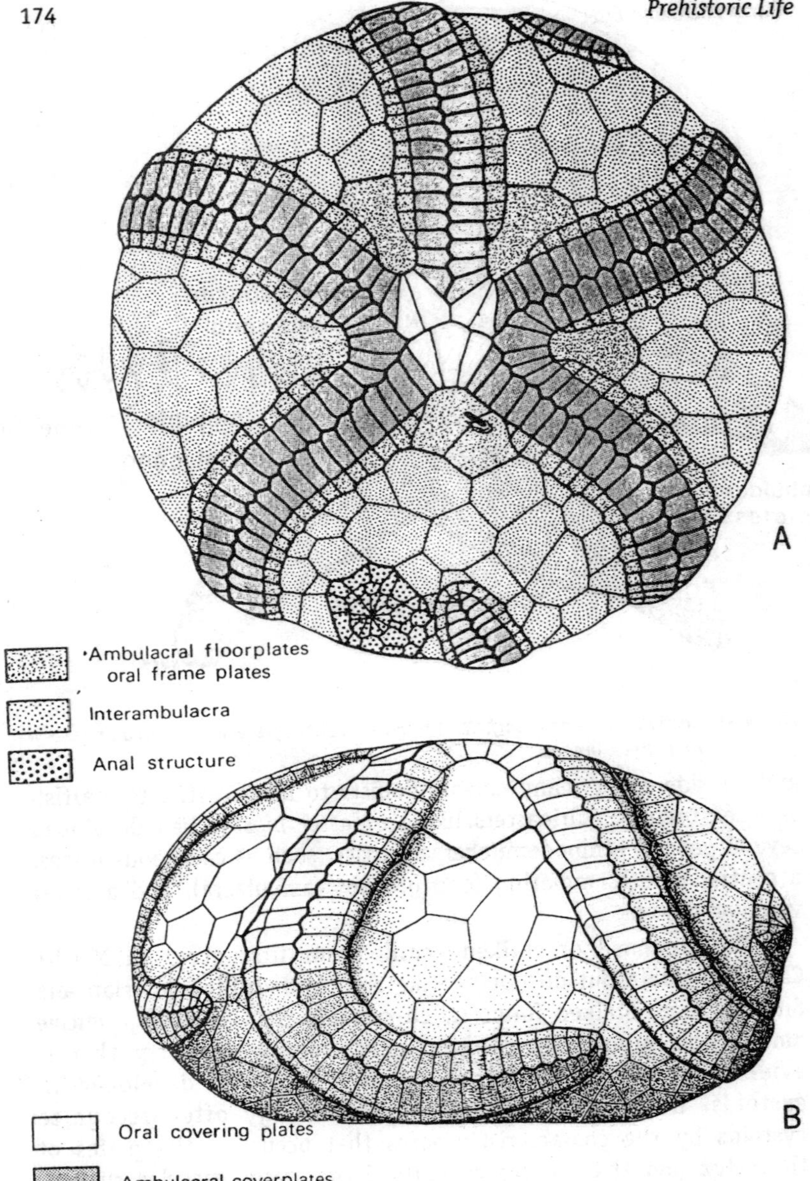

Fig. 4.36. Edrioaster bigsbyi, a Middle Ordovician edrioasteroid. Specimen is 45 mm in diameter. A—Oral surface. B—Lateral view of the globoid fossil.

ambulacra are prominent, bear slender branches or brachioles along their margins, and have a well-developed and unique water transport system. The blastoids first appear in Silurian time, expand in the Mississippian and decline to extinction in the Permian.

Crinoids, like most cystoids and blastoids, are composed of three main parts: the *calyx*, which contains the vital *organs*; the *arms*; and the *stem* with its rootlike *holdfast*. The arms bear ciliated

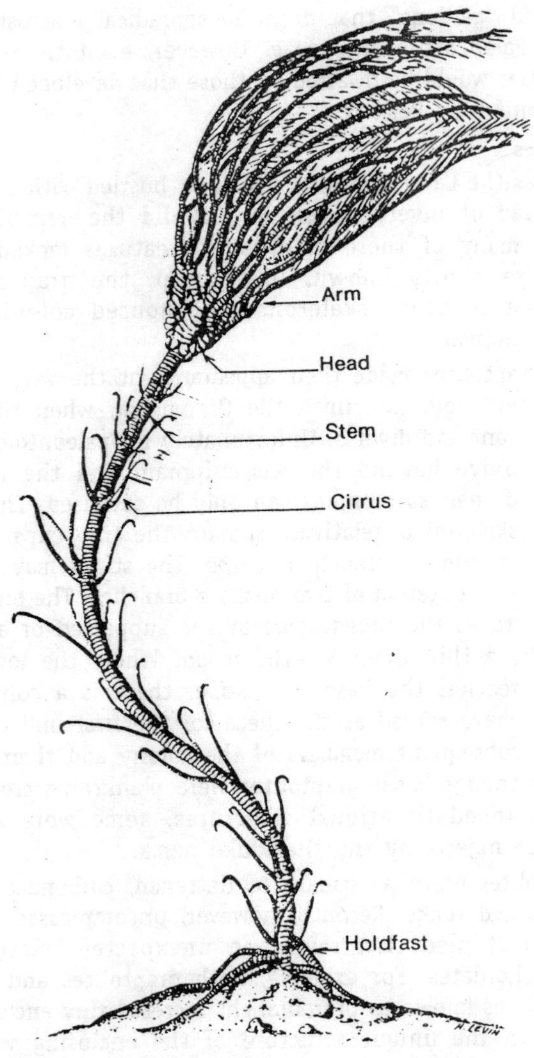

Fig. 4.37. Crinoid in living position on sea floor.

food grooves and, like the brachioles of blastoids, serve to move food particles toward the mouth. Crinoids are found in rocks that range in age from Ordovician to Recent. In some areas, Ordovician and Silurian rocks (and later Missisippian and Pennsylvanian) have such great quantities of the disaggregated plates of crinoids that they are formally named crinoidal limestones. One can visualize the crinoid "gardens" that occurred sporadically across the floors of Early Paleozoic epeiric seas. However, even these impressive communities would be dwarfed by those that developed later during the Carboniferous Periods.

Graptolites

Just as the Early Paleozoic sea floors bustled with the activities of a myriad of invertebrates, so also did the overlying waters. However, many of these planktonic creatures lacked skeletons and so are poorly known. One group, the graptolites, had preservable corneous skeletons that housed colonies of tiny individual animals.

The graptolites made their appearance at the very end of the Cambrian but were rare until the Ordovician, when they became quite abundant and diverse. Unfortunately for paleontologists, they did not survive beyond the Mississippian, and the nature and functions of their soft organs can only be surmised. The structure of the exoskeleton is relatively simple. The tiny cups, or thecae, are arranged along a branch, or *stipe.* The stipes may be solitary or formed into a system of two or more branches. The entire colony is referred to as the *rhabdosome* and is supported or attached at one end by a thin filament—the *nema.* Where the lower end of the nema reaches the base of a *stipe,* there is a conical *sicula,* which may have served as the theca for the first individual. From the sicula subsequent members of the colony add their thecae by budding Although most graptolites were planktonic creatures and often developed flotational structures, some were apparently attached to objects by the threadlike nema.

Graptolites occur as streaks of flattened, carbonaceous matter in fine-grained rocks. Recently however, uncompressed graptolites have been studied that reveal an unexpected relationship to primitive chordates. For example, both graptolites and the living protochordates known as pterobranchs secreted tiny enclosed tubes, and in both the unique structure of the enclosing wall of the sheath was so similar that a relationship is virtually certain.

During the Early Paleozoic, graptolites were at times so abundant that they have been visualized as forming sargasso sea-like floating masses. They were apparently carried about by ocean currents and thus achieved a worldwide distribution, which has enhanced their use as index fossils. There is, however, a limitation, in that they are seldom found preserved except in fine-grained, usually carbonaceous, shaly sediments. Perhaps these occurrences came about as masses of graptolite colonies floated into areas of toxic conditions, died, and settled to the sea floor. It is also possible that such accumulations resulted from fortuitously favourable preservation.

Conodonts

One group of well-known but enigmatic fossils may have some affinity to a prechordate animal. However, their true biologic relationships are still being hotly debated, and many paleontologists are convinced that they are really food-filtering parts of some ancient invertebrate animal. These problematic fossils are called conodonts. Some are plate like or cone-shaped, and all are composed of calcium phosphate.

At the base of conodont fossils, one finds cavities or attachment scars where they were presumably attached to a supportive part of the parent organism. Conodonts occur mostly as single disarticulated fossils, but occasionally, paired groupings are discovered that are suggestive of an animal's feeding mechanism. Conodonts are excellent guide fossils. They occur in a variety of marine sedimentary rocks and range from latest Precambrian into the Triassic.

Vertebrate Animals of the Early Paleozoic

A momentous biologic occurrence took place in the Early Paleozoic. It came gradually and in a manner that left no fossilized traces. The event, of course, was the birth of the chordate line. A chordate is an animal that has, at least at some stages in its life history, some kind of stiff, elongate supporting structure, a dorsal hollow central nervous system, gill slits, and blood that circulates forward in a main ventral vessel and backward in the dorsal. The supportive structure in primitive chordates is called a notochord and has been studied by generations of biology students in such animals as the lancelet *Branchiostoma*. In the taxonomic hierarchy, vertebrates are simply those chordates in which the notochord is supplemented or replaced by a series of cartilaginous or bony vertebrae.

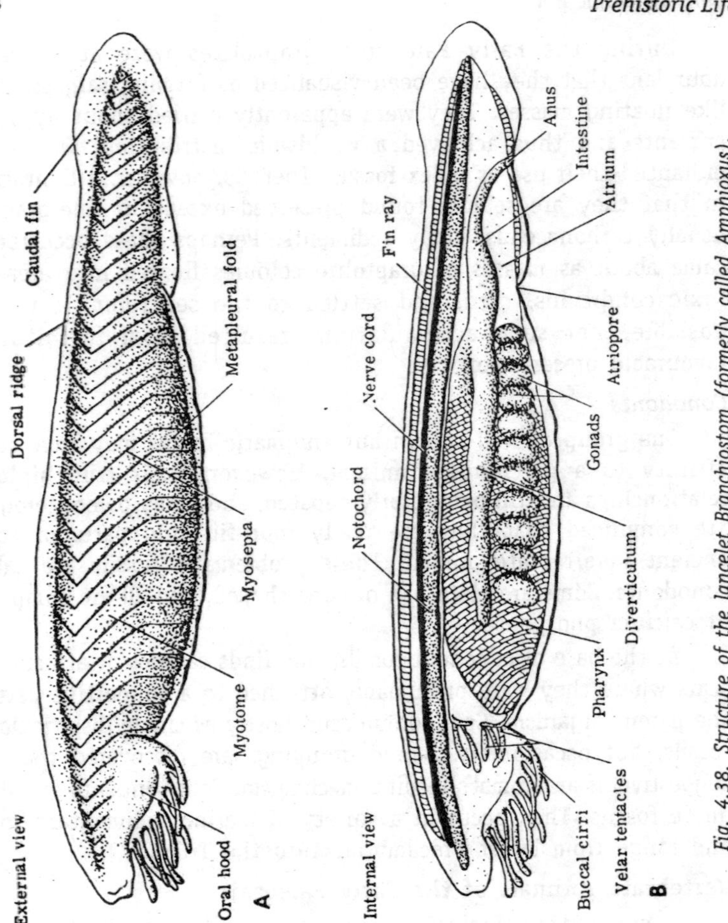

Fig. 4.38. Structure of the lancelet Branchiostom (formerly called Amphioxus).

As we have mentioned previously, the ancestors of the vertebrate lineage may lie somewhere among the echinoderms. The theoretic evolutionary progression that was to lead to vertebrates may have begun with sedentary, filter-feeding animals that had exposed cilia located along their arms, somewhat in the fashion of crinoids. From such a beginning there may have evolved filter-feeders with cilia brought inside the body in the form of gills. In a subsequent stage, the organisms may have become free-swimming, gilled animals, in superficial appearance not unlike small, simple fishes or frog tadpoles. With the advent of vertebrae, the evolutionary progression toward true vertebrates would have been achieved. Presumably, we can call those earliest vertebrates fishes.

The oldest fossils that are considered to be the possible remains of fishes are found in strata of Ordovician age. They consist of tiny mouth parts bearing still smaller, pointed teeth. Scales and scraps of bony armor are found in Ordovician rocks in both North America and Russia.

Recently, small fish scales and plates were recovered from Middle Ordovician Viola Limestone exposures in the Arbuckle Mountains of Oklahoma. The Viola also contains a rich assemblage of marine invertebrates, thus providing evidence that the earliest vertebrates now known lived in the sea rather than in freshwater bodies.

The vertebrates that we loosely call fishes are actually divided into at least five distinct taxonomic classes. There are the jawless fishes, or *Agnatha*; two groups of archaic jawed fishes, the *Acanthodii* and *Placodermi*; and the familiar *Osteichthyes* with their highly developed bony skeletons the cartilaginous fishes or chondrichthyes. It is the first three of these categories—namely, the Agnatha, Acanthodii and Placodermi—that frequented freshwater and salt water bodies of the Early Paleozoic.

The agnaths still live today in the guise of the inelegant hagfish and lamprey. However, these specialized survivors are quite unlike

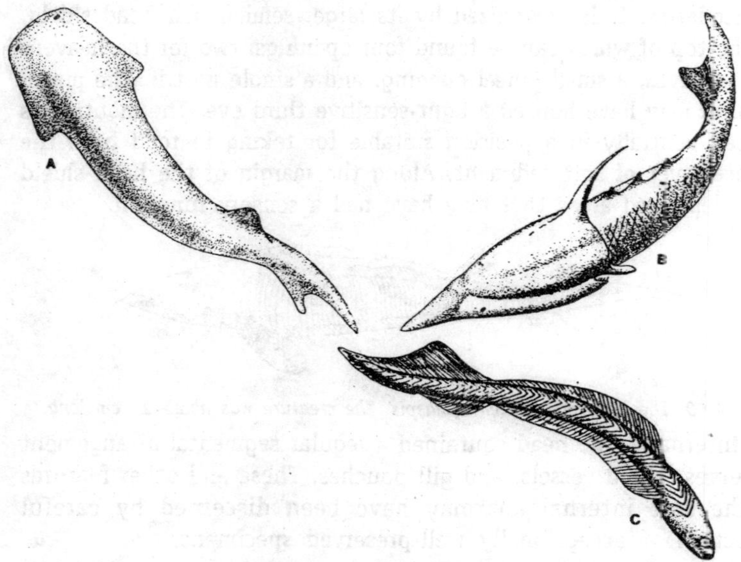

Fig. 4.39. The Early Paleozoic ostracoderms Thelodus (A), Pteraspis (B), and Jamoytius (C) drawn to the same scale.

the jawless fishes of the Ordovician and Silurian. Early Paleozoic agnaths are collectively termed *ostracoderms*. The name means "shell skin" and refers to the bony exterior that was a distinctive trait of many of these fishes. Ostracoderms composed a rather diverse group that included the unarmored forms *Thelodus* and *Jamoytius*, as well as such armored creatures as *Pteraspis*. The purpose of the bony armor that is the hallmark of many of the ostracoderms is still being debated. A widely held view is that the armor provided protection, although the fossil record shows few creatures of the same age and habitat that could have preyed on the ostracoderms. Nevertheless, it does not seem unreasonable that some of the smaller ostracoderms might have provided an occasional meal for a large cephalopod or eurypterid. Another theory stipulates that the dermal armor was not primarily for protection but rather was a device for storing seasonally available phosphorus. According to this idea, phosphates could be accumulated as calcium phosphate in the dermal layers during times of greater availability and utilized during periods when supply was deficient. This cache of phosphorus may have been vital to early vertebrates in order for them to maintain a suitable level of muscular activity.

Hemicyclaspis is one of the most widely known of the ostracoderms. It is recognized by its large, semicircular head shield, on the top of which can be found four openings: two for the heaven-directed eyes, a small pineal opening, and a single nostril. The pineal opening may have housed a light-sensitive third eye. The mouth was located ventrally in a position suitable for taking in food from the surface layer of soft sediment. Along the margin of the head shield were depressed areas that may have had a sensory function.

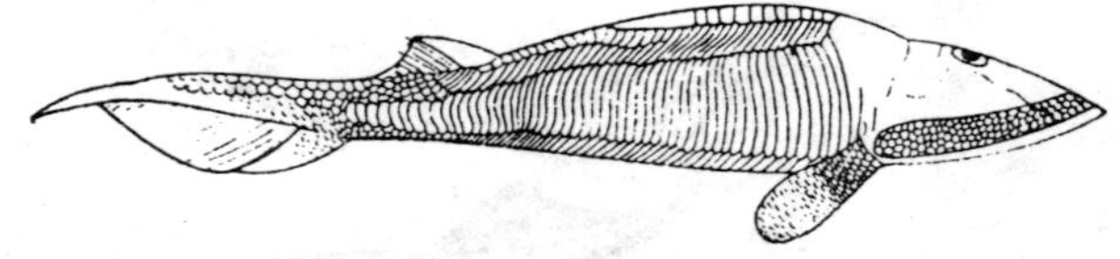

Fig. 4.40. The ostracoderm Hemicyclaspis. The creature was about 15 cm long.

Internally, the head contained a regular segmental arrangement of nerves, blood vessels, and gill pouches. These and other features of the soft internal anatomy have been discerned by careful dissection of exceptionally well-preserved specimens.

The ostracoderms continued into the Devonian but did not survive beyond that period. For the most part they were small, rather sluggish creatures restricted to mud-straining or filter-feeding modes of life. They were to be gradually replaced by fishes that had developed bone-supported, movable jaws. The evolution of the jaw was no small accomplishment, for it enormously expanded the adaptive range of the vertebrates Fishes with jaws were able to bite and grasp. These new abilities led to more varied and active was of life and to new sources of food not available to the agnathans.

The evolutionary development of the vertebrate jaw involved a remarkable transformation in which an older structure was modified

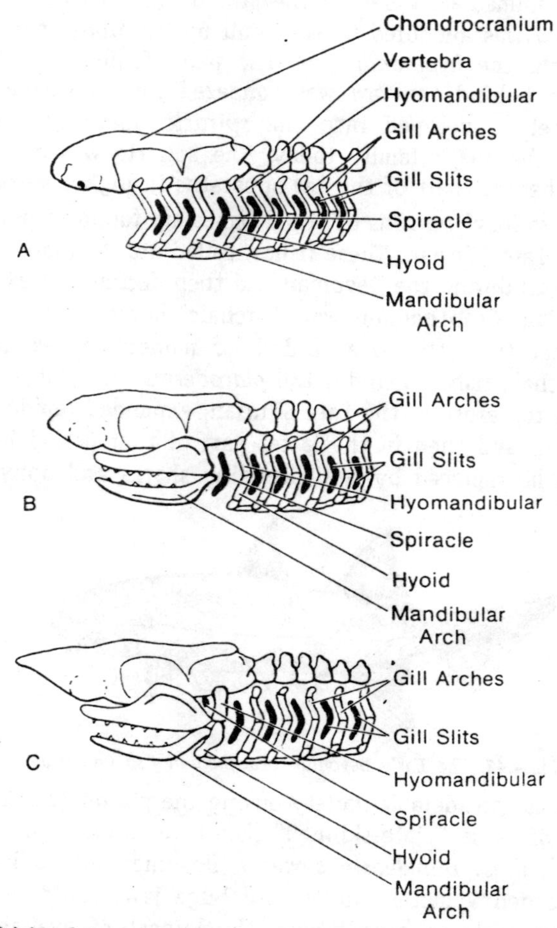

Fig. 4.41. Origin of jaws. A—Gill arches as they might exist in a primitive jawless fish. B—Early jawed fish with a complete gill slit behind jaws. C—A jawed fish (like a shark) with the first gill slit reduced to a spiracle.

to perform a role that was entirely different from its original function. Those older structures were supports called *gill arches* that were located between the *gill slits*. As the mouth extended posteriorly, the first two of these arches were sequentially eliminated. The third set of gill arches was gradually modified into jaws. Classic anatomic studies have clearly shown the similarities in jaw architecture and neurology to elements of the gill arches and the nerves that serve these supports. In the earliest fishes, the upper jaw was attached to the skull by ligaments only. A full gill slit occurred immediately behind the jaw. In more advanced fishes, the upper jaw was anchored to the skull by the upper part of the next gill arch, the *hyomandibular*. The pair of gill slits that had once existed behind the jaw was squeezed into a smaller space and ultimately developed into the *spiracle* openings found in members of the shark family today. The spiracle was a structure destined to become part of the ear apparatus in higher vertebrates.

The oldest fossil remains of jawed fishes are found in nonmarine rocks of the Late Silurian. These fishes, called *acanthodians*, became most numerous during the Devonian and then declined to extinction in the Permian. Acanthodians were "archaic" jawed fishes and were quite distinct from the great orders of modern fishes. Another group of archaic fishes includes the *placoderms*, or "plate-skinned" fishes. They too arose in the Late Silurian, expanded rapidly during the Devonian, and then in the latter part of that period began to decline and be replaced by the ascending sharks and bony fishes.

Fig. 4.42. The Early Devonian acanthodian fish Climatius.

There was considerable variety among the placoderms. The most formidable of these "plate-skinned" fishes was a carnivorous group called *arthrodires*. *Dunkleosteus* was a Devonian arthrodire whose length exceeded 9 meters and whose huge jaws could be opened exceptionally wide to engulf even the largest of available prey. Other placoderms, called *antiarchs*, had the heavily armored form

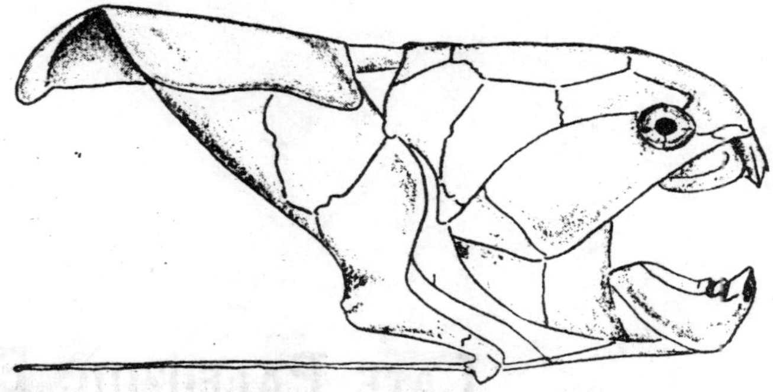

Fig. 4.43. Dunkleosteus, a highly predaceous Upper Devonian placoderm.

and mudgrubbing habits of their predecessors, the ostracoderms. The placoderms known as *rhenanids* anticipated the form and habits of living rays and skates.

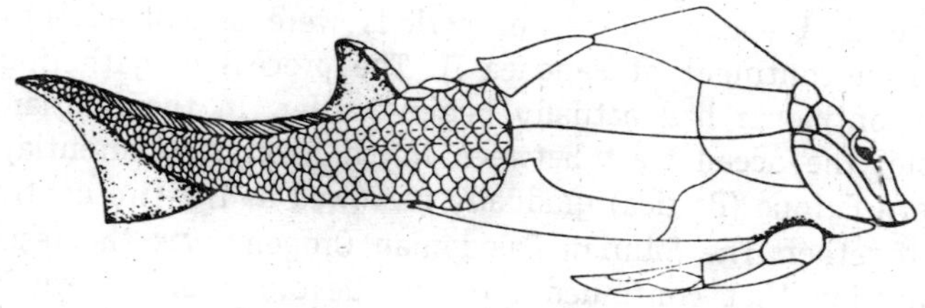

Fig. 4.44. The Devonian antiarch fish Pterichthyodes.

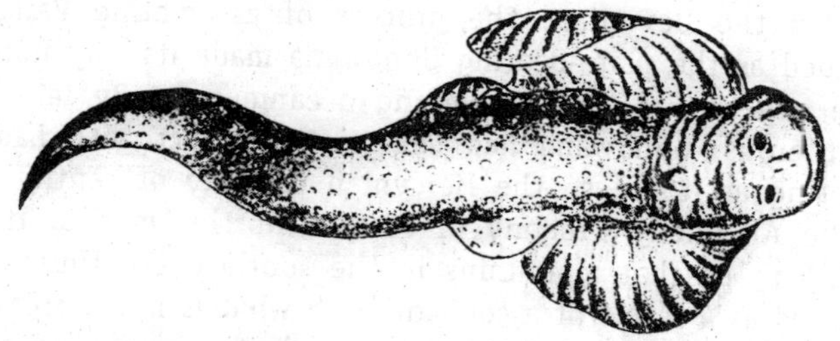

Fig. 4.45. The rhenanid skatelike placoderm Gemuendina.

Among the Late Silurian acanthodians and placoderms were the ancestors of the bony and cartilaginous fishes that were to dominate the marine realm in succeeding periods. The precise ancestor has not yet been discovered but will be keenly sought in the years ahead.

5

Late Paleozoic Era

The Devonian, Mississippian, Pennsylvanian, and Permian Periods compose the Late Paleozoic. This was a time when most of the separate land masses of earlier periods were assembled into the great supercontinent of Pangaea II. The process of gathering the errant continents had actually begun earlier. In the Silurian, for example, the ocean tract between North America (Laurentia) and northern Europe (Baltica) gradually narrowed as the two continents came together. The Silurian Caledonian Orogeny was the result of the collision that continued into the Devonian of eastern North America as the *Acadian Orogeny*. North America was now sutured to Europe, and the combined land mass has been named Laurussia.

After the Devonian, the process of assembling Pangaea II continued as the plate-bearing Gondwana made its way northward at the expense of the intervening oceanic area. By late in the Carboniferous, Gondwana had come into contact with Laurussia, generating as it did so the Hercynian Orogeny of central Europe and the Allegheny Orogeny of eastern North America. It seems startling to us that mountains in the southeastern United States originated as a result of a collision with what is now northwestern Africa, but this is indeed what occurred. The clustering to form Pangaea II was nearly complete by Late Permian. Great mountain ranges marked the suture zones of the once separated paleocontinents, and an enormous ocean (Panthalassa) spanned the globe across nearly 300° of longitude.

The Late Paleozoic was a time of diverse sedimentation, progress in organic evolution, and diverse climatic conditions. During this

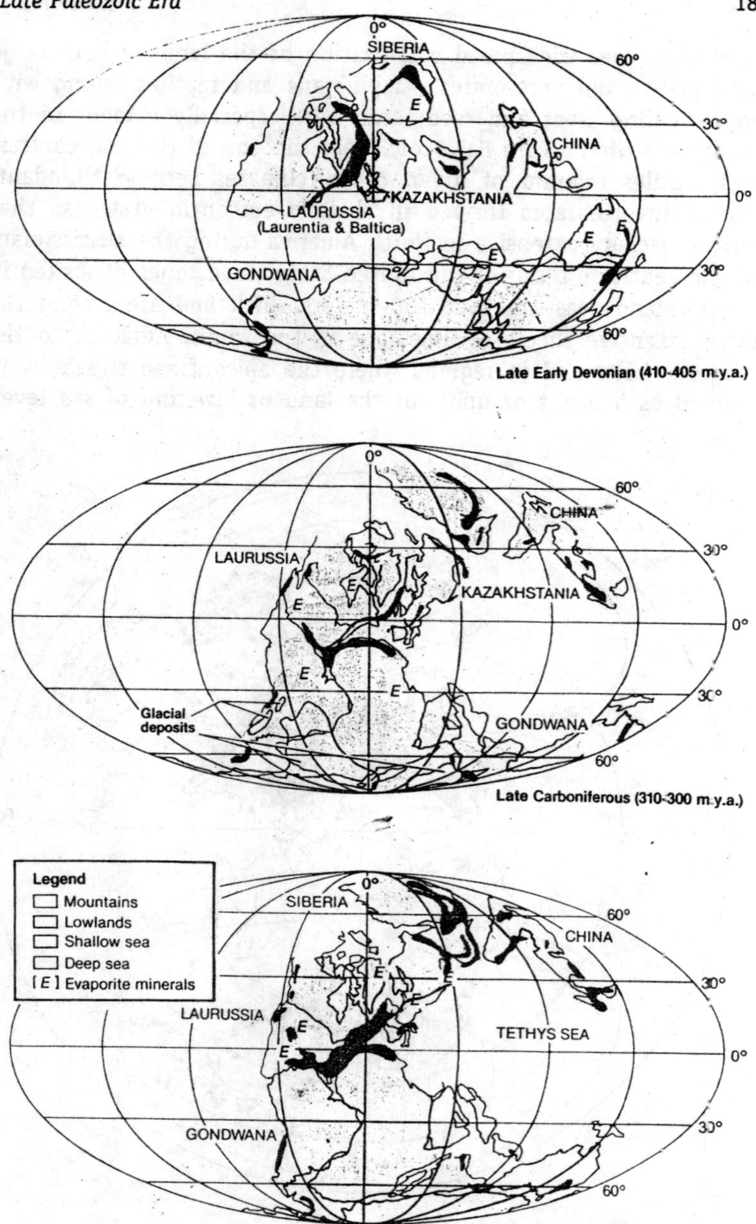

Fig. 5.1. Global paleogeography of the Late Paleozoic.

time there was widespread colonization of the land by both large land plants and vertebrates. Amphibians and reptiles, along with spore-bearing trees and seed ferns were especially evident in the landscapes of the Late Paleozoic. Near the end of the era, conifers and reptiles tolerant of dryer, cooler climates become abundant. Marine invertebrates thrived in shallow epicontinental seas that were especially extensive in North America during the Mississipian. In the central interior of the United States limestones deposited in these inland seas are over 700 meters thick and are among the most extensive sheets of limestone in the world. Adjacent to the orogenic belts and in regions where the epicontinental seas were drained as a result of uplift of the land or lowering of sea level,

Fig. 5.2. World geography about 250 million years ago during the Permian. The distribution of glacial tillites is represented by filled triangles. Coal is represented by filled circle and evaporites by the irregularly shaped filled area.

sediment-laden streams deposited their loads of sand and mud. These nonmarine sediments contain the fossil plants and animals that aid in inferring climatic conditions on the continents. The accumulation of abundant vegetation in swampy areas provided the raw materials from which the famous coal deposits of the Pennsylvania were formed. Now consolidated sands of desert dunes and layers of evaporites attest to arid conditions at many localities around the globe. Tillites and glacial striations on bedrock provide evidence of glaciation. This is not unexpected, for the more southerly parts of Gondwana were located near the poles. Evidence of Late Carboniferous or Permian ice sheets can be found on the now separated continents of South America, Africa, Australia, Antarctica, and India. Decades before geologists had an understanding of plate tectonics, the ancient tillite deposits on these continents were cited as evidence for continental drift.

Lands, Seas, and Mobile Belts

North America

During the Late Paleozoic, the North American portion of Pangaea II was nearly encircled by geosynclines. To the northeast lay the Northern Appalachian belt, which continued southward as the Southern Appalachian and Ouachita Geosynclines. The western edge of the continent was occupied by the western or Cordilleran Geosyncline. The encirclement was completed along the northern periphery of the continent by the Franklinian Geosyncline, which extended eastward for over 2000 km. At the center of the ring of geosynclines lay the stable cratonic region.

Sedimentation on the craton

The Kaskaskia Sequence. The final event of the Early Paleozoic on the North American craton had been the withdrawal of the inland seas. The Early Paleozoic formations were subjected to erosion, which exposed older and deeper formations in those locations that had experienced upraching. The gradual flooding of the old Early Paleozoic surface was the first event of Late Paleozoic cratonic history. This inland water body has been termed the *Kaskaskia Sea*. Except for minor regressions, it persisted over extensive areas of the craton well into the Mississippian Period, at which time the seas withdrew. Erosion of the layers of sediment deposited on the Kaskaskia Sea floor began, only to be interrupted by the advance of a second sea, the *Absaroka Sea*.

Along the eastern side of the North American craton, Kaskaskia sedimentation consists initially of clean quartz sands. The most famous of these sandy formations is the Oriskany Sandstone of New York and Pennsylvania. Because of its purity, Oriskany sandstone is extensively used in making glass.

Although quartz is by far the dominant mineral in the Oriskany, the formation also contains a very small percentage of other silicate minerals, which, like quartz, are resistant to weathering and erosion but are heavier than quartz. These so-called heavy minerals can be used to determine the kinds of parent rocks and source areas from which clastic sediments were derived. After study of the heavy minerals from scattered exposures of Oriskany sandstones, it became apparent that there were two distinctly different assemblages of heavy minerals. In the area south of New York, the formation contained exceedingly small amounts of only the most stable heavy minerals tourmaline, zircon, and rutile. The grains in these sandstones were exceptionally well rounded and worn, and less stable heavy minerals were lacking. This part of the Oriskany was considered to have been derived from older clastic sedimentary units to the east and north. In contrast, the Oriskany of the New York State area contains a heavy mineral assemblage of relatively unstable pyroxene, amphibole, biotite, and garnet. Garnet is ordinarily derived from metamorphic source rocks, whereas the others are regular components of a variety of igneous and metamorphic rocks. The grains are unaltered. One may infer, therefore, that the sands composing the Oriskany of the New York area may have had an igneous or metamorphic source area and were not recycled from older sandstones. The Oriskany is an example of one method utilized by geologists to infer the history of a rock unit.

The Oriskany Sandstone is the initial deposit of a transgressing sea, as is indicated by the way the sands overlap older strata. As this marine advance continued, limy sediments were deposited over the Oriskany, and corals began to build reef structures. In areas where water circulation was restricted, salt and gypsum were deposited. During the Middle and Late Devonian of the region east of the Mississippi Valley, carbonate sedimentation gave way to shales. The change to clastic deposition was a consequence of mountain building associated with the Acadian Orogeny in the Northern Appalachians. Highlands developed during this orogenesis were rapidly eroded, and clastics were transported westward as a vast apron of sediments that geologists refer to as the *Catskill Delta*. This great wedge of sediment is coarser and thicker near the source areas and merges into a thinner but extensive sequence of black shales in the eastcentral region of the United States. The Catskill Delta was located in the equatorial zone. Hence it is not surprising that the sediments are deeply weathered and stained red by iron oxides. In these tropical stream and lake deposits, the first abundant remains of truly large land plants are found.

In the far west part of the craton, Middle and Late Devonian rocks are largely limestones, although there are shales as well. In a depressed area known as the Williston Basin, extensive reefs were developed. Arid conditions and restricted circulation in reef-enclosed basins resulted in the deposition of impressive thicknesses of gypsum and salt. Some of Canada's largest oil fields derive their product from permeable reefs into which oil migrated and became trapped by enclosing layers of impermeable shale.

Late Devonian and early Mississippian seas deposited a remarkably uniform black of blanket shales that extended from the eastern geosyncline to the Mississippi Valley. Although there are many local names for this sediment, the most widely used is *Chattanooga Shale*. The clayey sediment had as its source the continuing erosion of high lands to the east that had periodically uplifted during the Late Devonian and Mississippian. In places, the Chattanooga consists largely of fine-grained calcite and compressed carbonized plant material. Gradually, as the highland source areas were reduced, the quantity of clastic materials decreased, and carbonates became the most abundant and widespread deposit. Cherty limestones, limestones composed entirely of the remains of billions of crinoids and wave-worn debris of other invertebrates, and limestones formed from myriads of tiny calcium carbonate spheres (oölites) formed massive beds and covered the entire central and western regions of the craton. Oölites are spherical grains formed by the precipitation of calcium carbonate around a nucleus. Oölites are known to occur today in shallow coastal areas with high turbulence, such as breaker zones along shorelines. The seaway in which these carbonates were deposited was the most extensive North America had experienced since Ordovician time and was the last of the great Paleozoic floodings of the North America craton.

In Late Mississippian time, as the Kaskaskia Seas began their final withdrawal, large quantities of clastics and thin, laterally nonpersistent limestones were deposited. These rocks are reservoirs for petroleum in Illinois and therefore have been studied extensively. Detailed maps of thicknesses of some of the sandstone units show that they are thickest along branching, sinuous trends that suggest old stream valleys developed on the former sea floor. Studies of grain size, cross-bedding. Studies of grain size, cross-bedding, and current-produced sedimentary structures strongly indicate that the clastics were derived from the northern Appalachians and were transported southwestward across the central interior.

The Absaroka Sequence. When the Kaskaskia Seas had finally left the craton, at the end of Mississippian time, the exposed terrain was subjected to erosion that resulted in one of the most widespread regional unconformities in the world. Not only was erosion areally extensive, but also over arches and domes it beveled away entire systems of older rocks. The unconformity provides a criterion for separating those strata equivalent to the Carboniferous of Europe into the Mississippian and Pennsylvanian systems. It is appropriate to use these two names in eastern and central North America, not only because of the unconformity but also because the rocks above the erosional hiatus differ markedly from those below. Indeed, the overlying Pennsylvanian strata are a consequence of quite different tectonic circumstances.

It was not until near the beginning of Middle Pennsylvanian time that the seas were able to encroach onto the long exposed surface of the craton. The deposits of this seaway are those of the Absaroka Sequence. In general, the Pennsylvanian section of rocks near the eastern geosyncline are thicker, and virtually all are continental sandstones, shales, and coal beds. This eastern section of Pennsylvanian rocks gradually thins away from the Appalachian belt and changes from predominantly terrestrial to about half marine rocks and half nonmarine rocks. Still farther west, the Pennsylvanian outcrops are largely marine limestones, sandstones and shales.

One of the most notable aspects of Pennsylvanian sedimentation in the middle and eastern states is the repetitive alternation of marine and nonmarine strata. A group of strata deposited during one alternation of such a cycle is called a *cyclothem*. A typical cyclothem in the Pennsylvanian of Illinois frequently contains 10 units. Units one through five are continental deposits, the uppermost of which is a coal bed. The strata deposited above the coal bed represent an advance of the sea over an old forested

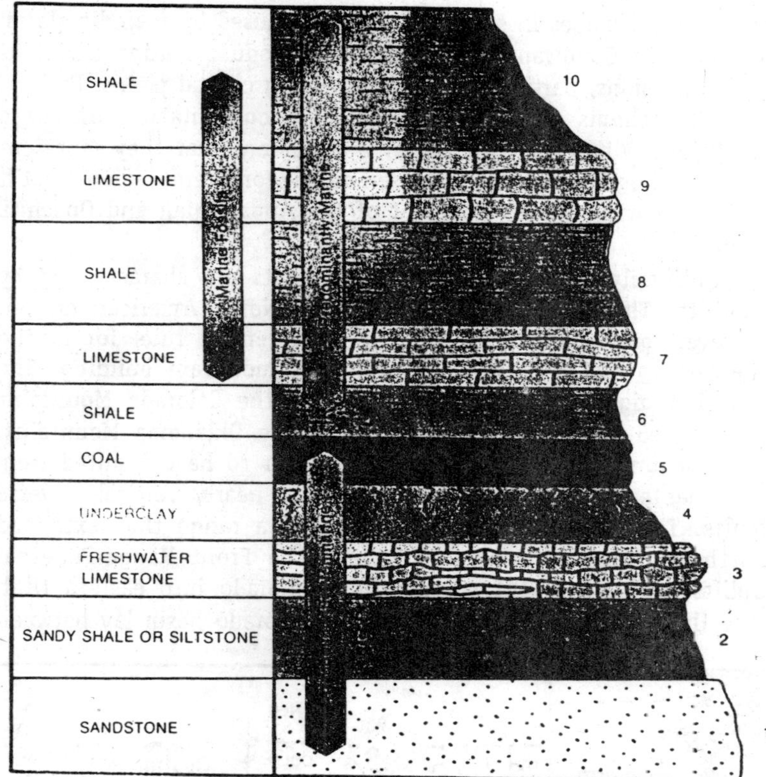

Fig. 5.3. A coal-bearing cyclothem showing the ideal sequence of layers. Many cyclothems do not contain all 10 units, as in this illustration at an ideal sequence. Some units may not have been deposited because the changes from marine to nonmarine conditions were abrupt, and units may also have been remove by erosion following marine regressions.

area. In Missouri and Kansas, at least 50 changes of this kind are recognized within a section about 750 meters thick. Some extend across thousands of kilometers. It is apparent that the advances and retreats of the seas were frequent and widespread. What caused these oscillations? One explanation involves periodic regional subsidence of the land to a level slightly below sea level so that marginal seas could spill onto the level swampy lowlands. A short time later, subsidence might cease and sediments be built up above sea level to extend the shoreline seaward and re-establish continental conditions. Alternatively, the re-establishment of dry lands may have resulted from temporary regional uplifts. Finally, worldwide or eustatic changes in sea level, possibly caused by periodic glacial advances in Gondwanaland, might also produce marine invasions and regressions, particularly along low-lying coastal plains. Perhaps the cyclothems were caused by some combination of these conditions. Whatever their cause, it is evident that they represent short-term oscillations superimposed on the long-term regional uplift associated with the development of the Appalachian and Ouachita Mountains.

Ordinarily, cratonic areas of continents are characterized by stability. The southwestern part of the North American craton, however, provides an exception to this general rule, for during the Pennsylvanian this was a region of mountain building. The resulting highlands are generally termed the Colorado Mountains (also called the ancestral Rockies) and the Oklahoma Mountains. These mountains and related uplifts appear to have resulted from movements of crustal blocks along large, nearly vertical, reverse faults. The Colorado Mountains included a range that extended north-south across central Colorado (the Front Range-Pedernal Uplifts) and a segment curving from Colorado into eastern Utah (the Uncompahgre Uplift). The Central Colorado Basin lay between

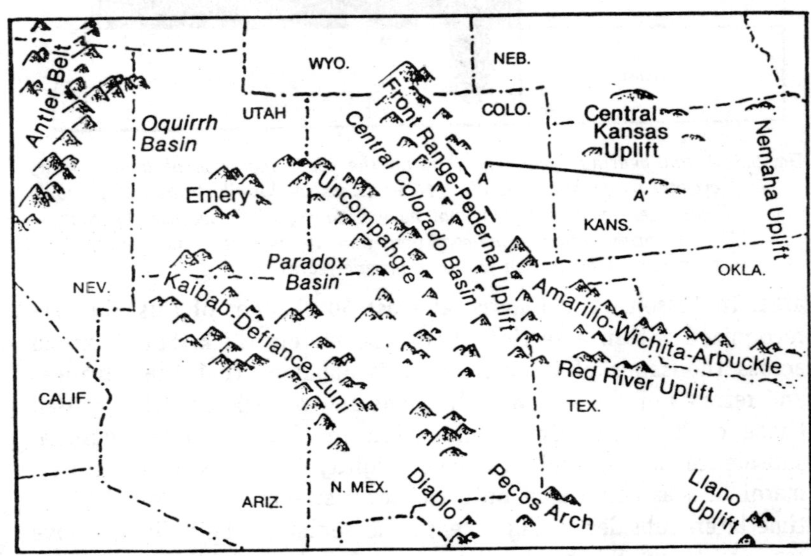

Fig. 5.4. Location of the principal highland areas of the southwestern part of the craton during Pennsylvanian time.

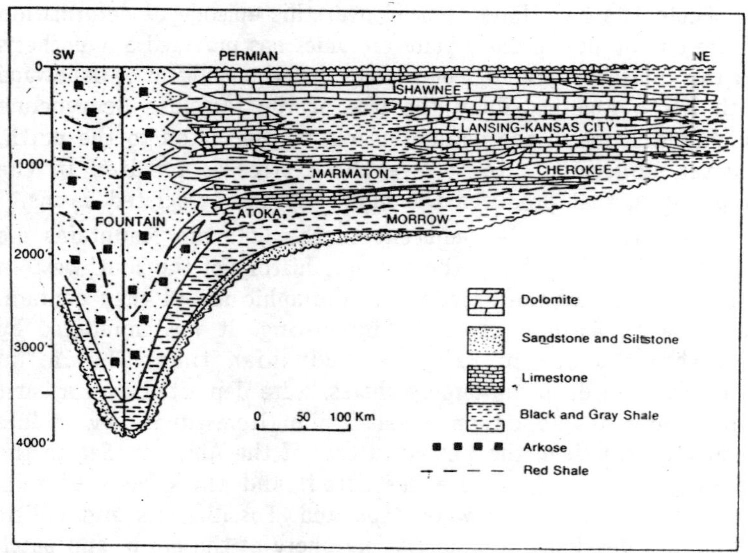

Fig. 5.5. Pennsylvanian cross-section across eastern Colorado and western Kansas showing great accumulation of coarse arkosic sandstones east of the Colorado Mountains. Line of section is along A–A´ in Figure 5.4.

these highland areas. A separate range, the Zuni–Fort Defiance Uplift, extended across northeastern Arizona on the southwestern perimeter of the Paradox Basin. To the east of the Colorado Mountains lay the southeastward-trending Oklahoma Mountains. Eroded stumps of this once ruged range form today's greatly reduced Arbuckle and Wichita Mountains. Remains of the Amarillo Mountains are now buried beneath younger rocks and are known principally from exploratory drilling for oil.

Judging from the tremendous volume of sediments eroded from the Colorado Mountains, it is likely that they attained heights in excess of 1000 meters. They also were probably subjected to repeated episodes of uplift, continuing in some places into the Permian. Erosion of these highlands eventually exposed their Precambrian igneous and metamorphic cores. As erosion and weathering continued, great wedge-shaped deposits of red arkosic sediments were spread onto adjacent and intervening basins. A small part of this massive accumulation of clastic sediment is dramatically exposed in the Red Rocks Amphitheatre just a few miles southwest of Denver and the "flat irons" near Boulder, Colorado.

Geologists have long puzzled over this episode of deformation of the craton, but recently plate tectonics has provided a hypothesis for this unusual event. It seems likely that the collision of Gondwana with North America along the site of the Ouachita Geosyncline generated stress in the bordering area of the craton to the north. Crustal adjustments to relieve these stresses resulted in the deformation that produced the highlands and associated basins.

The basins that lay adjacent to the Colorado Mountains are important in working out the geologic history of the southwestern craton because of the excellent stratigraphic record they contain. The Paradox Basin is especially interesting. It was inundated by the Absaroka Sea in Early Pennsylvanian time. The initial Pennsylvanian deposits, largely shales, were deposited over a karst topography developed on Mississippian limestones. By Middle Pennsylvanian time, the western access of the Absaroka Sea to the Paradox Basin had become restricted, and thick beds of salt, gypsum, and anhydrite were deposited. Fossiliferous and oölitic limestones developed around the periphery of the basin, and patch reefs grew abundantly along the western side. The association of porous reefs and lagoonal deposits resulted in several sites suitable for the later entrapment of petroleum.

Near the end of the Pennsylvanian, the Paradox Basin was filled to above seal level by arkosic sediments shed from the recently uplifted Uncompahgre highlands. Also at this time, the Absaroka Sea, which had begun its transgression at the beginning of the Pennsylvanian, began a slow and irregular regression near the end of the same period. The withdrawal was still incomplete in Early Permian time, so that marine sediments continued to be deposited in a rather narrow zone from Nebraska through western Texas. Fossiliferous limestones characterized these inland seas, although near highlands in Colorado, Texas, and Oklahoma, coarse clastics accumulated to sufficient thicknesses to bury surrounding uplands. The deposits along what was the eastern edge of the seaway have been eroded away, but at several places along the western sides, one can observe the change from richly fossiliferous beds below to barren shales, red beds, and evaporites above. The thick and extensive salt beds of Kansas provide testimony to the gradual restriction and evaporation of Permian seas in the central United States. The "last stand" for Permian marine conditions occurred in the western part of Texas and southeastern New Mexico, where a remarkable sequence of interrelated lagoon, reef, and open-basin sediments were deposited. In this region, several irregularly subsiding basins developed between shallowly submerged platforms. Dark-coloured limestones, shales, and sandstones were deposited in the deep basins, whereas massive reefs formed along the basin edges. Behind the reefs, in what must have been the shallow waters of extensive lagoons, the deposits are thin limestones evaporites, and red beds. Late in the Permian, the connections of these basins to the south became so severely restricted that the waters gradually evaporated, leaving behind great thicknesses of gypsum and salt.

Much paleoenvironmental information has been gleaned from a study of the western Texas Permian rocks. From the lack of medium and coarse-grained clastics, one may assume that surrounding regions were low-lying. The gypsum and salt suggest a warm, dry climate in which basins were periodically replenished with sea water and experienced relatively rapid evaporation. Careful mapping of the rock units has helped to establish an estimated depth of about 500 meters for the deeper basins and only a few meters for the intrabasinal platforms. The basin deposits are dark in colour and rich in organic carbon as a likely consequence of accumulation under stagnant, oxygenpoor conditions. Perhaps upwelling of these deeper waters may have provided a bonanza of nutrients on which phytoplankton and reef-forming algae thrived. Along with the algae, the reefs contain the skeletal remains of over 250 species of marine invertebrates. Today, because of their relatively greater resistance to erosion, these ancient reefs form the steep EI Capitan promontory in the Guadalupe Mountains of Texas and New Mexico. Here one can examine the forereef composed of broken reefs debris that formed a sort of talus deposit caused by the pounding of waves along the southeast side.

The Eastern Geosynclines

During the latter half of the Paleozoic Era, the great. Appalachian and Ouachita Geosynclines experienced their culminating and most intense episodes of orogenesis. The crumpling of these geosynclinal tracts was a consequence of the reassembly of the formerly separated continents into Pangaea II. The northern part of the Appalachian-Caledonian belt had taken the shock of a collision with Europe during the Silurian and Devonian. That collision was the cause of the Caledonian Orogeny in Europe and the Acadian Orogeny in northeastern North America. As stated already the southern half of the Appalachian Geosyncline and the Ouachita Geosyncline as well were crushed into mountainous tracts when they were struck by the northwestern bulge of Africa during the Late Carboniferous. This great encounter between Laurussia and Gondwana was the cause of the Allegheny Orogeny. As noted earlier, the collision from the south not only raised the old geosynclinal tracts but also transmitted stresses into the interior, causing deep-seated deformations such as those that produced the Colorado and Oklahoma Mountains.

The effects of the Devonian deformation, called the Acadian Orogeny, are clearly seen in a belt from Newfoundland into West Virginia. Here one finds thick, folded sequences of eugeosynclinal clastics interspersed with rhyolitic volcanic rocks and granitic intrusions. The intensity of the compression that affected these rocks is reflected in their metamorphic minerals, which indicate that mineralization occurred at temperatures exceeding 500°C and pressures equivalent to burial under 15,000 meters of rock. The overall result of the Acadian Orogeny was to demolish forever the Appalachian eugeosyncline as a marine depositional trough and establish in its place mountainous areas in which erosion was the prevalent geologic process. Here and there in isolated basins among the mountains, Devonian nonmarine sediments were deposited. However, the greatest volume of erosional detritus was spread outward from the highlands as a great wedge of terrigenous sediment called the Catskill Delta. The pattern of sedimentation seen in the Taconic Orogeny with its Queenston Delta was repeated, except that the Catskill deposits came at a time when land plants were abundant and were able to provide a green mantle for the alluvial plains and hills. The Catskill sediments are dominated by sandstones, and shales in which the contained iron is deeply oxidized. The oxide takes the mineral form of hematite and causes the reddish or brownish colouration characteristic of Catskill "red beds." Because they are so deeply oxidized, red beds imply subaerial deposition. However, this inference must be confirmed by other evidence of continental deposition, because the colour may be secondarily derived from reddish sediments in the source area. In the case of the Catskill red beds, there are numerous traces of rootlets and shrublike vegetation. Toward the west, the red beds give way to gray sandstones and shales that contain fossil tree stumps. Here, the grayish colouration results from the ferrous

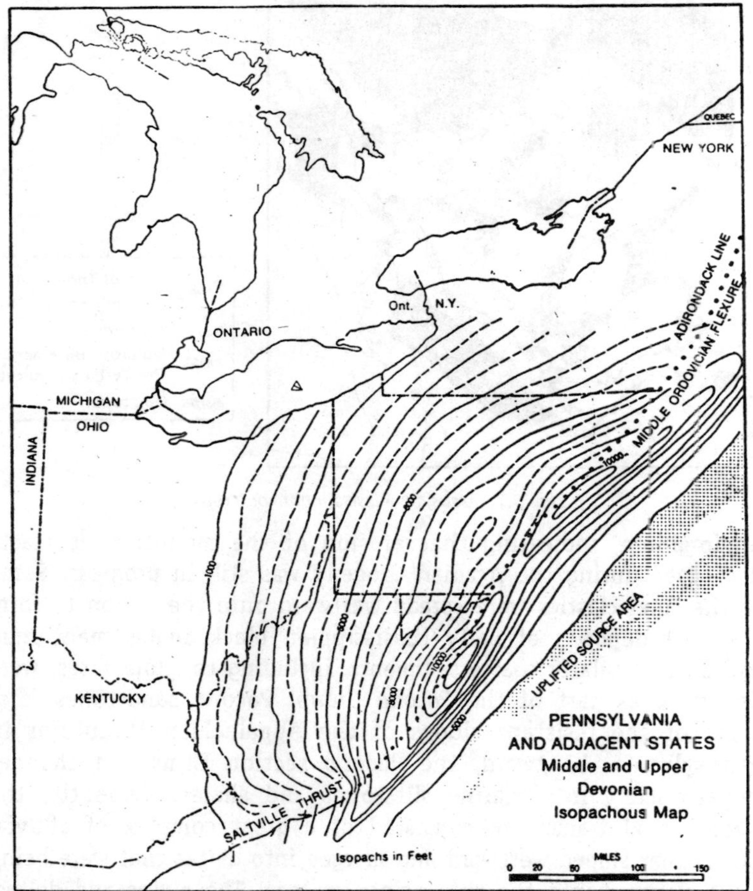

Fig. 5.6. Isopach (thickness) map of Middle and Upper Devonian strata in West Virginia, Pennsylvania, and parts of adjacent states.

variety of iron oxide and may imply deposition in frequently swampy or marshy environments where there is a relative deficiency of oxygen.

On the European side of the adjoined continents, the blanket of coarse debris derived from the Caledonian highlands spread southwestward, creating a vast land are named—for its most famous formation—*The Old Red Sandstone*.

Mississippian strata crop out in the Appalachian region from Pennsylvania to Alabama. Nonmarine shales and sandstones

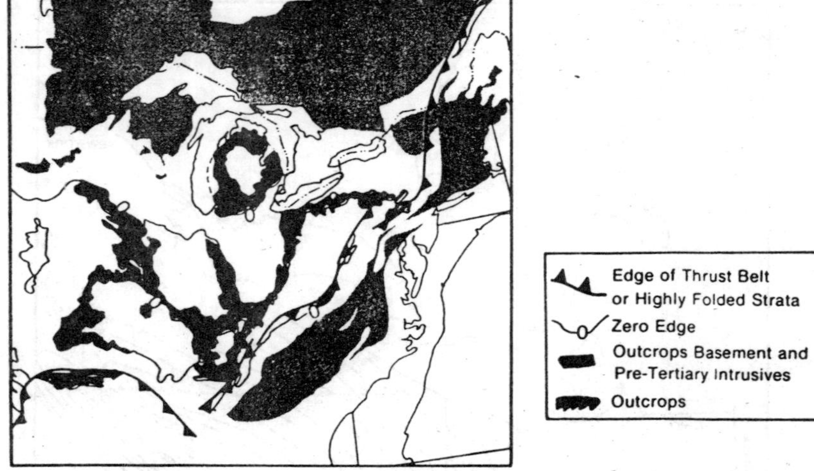

Fig. 5.7. Areas of Mississippian outcrops.

predominate, indicating that erosion of the mountainous tracts developed during the Acadian Orogeny was still in progress. Some of the finer clastics were spread westward onto the craton to form the vast deposits of Early Mississippian black shales mentioned earlier. Particularly coarse sediments, including conglomerates, were deposited as part of the *Pocono Group*. Pocono sandstones form some of the resistant ridges of the Appalachian Mountains in Pennsylvania. Westward, the Pocono section thins and changes imperceptibly into marine siltstones and shales. Evidently, the depositional framework consisted of a great complex of alluvial plains that sloped westward and merged into deltas that were being built outward into the epicontinental seas. The plains and deltas, standing only slightly above sea level, were backed by the rising mountains of the Appalachian fold belt. A large part of the coarser clastics had settled before reaching the southern portions of the geosyncline, and marine limestones are the most prevalent rocks.

Pennsylvanian rocks of the Appalachians are characterized by cross-bedded sandstones and gray shales that were deposited by rivers or within lakes and swamps. Coal seams are, of course, prevalent in the Pennsylvanian System of the eastern United States and reflect the luxuriant growths of mangrove-like forests that clothed the lands. This was an ideal environment for coal formation. Vegetation that accumulated in the poorly drained swampy areas was frequently inundated and killed off. Immersed in water or covered with muck, the dead plant material was protected from rapid oxidation; however, it was attacked by anaerobic bacteria. These organisms broke down the plant tissues, extracted the oxygen, and released hydrogen. What remained was a fibrous sludge with a high content of carbon. Later, such peatlike layers were covered with additional sediments—usually siltstones and shales—and then compressed and slowly converted to coal.

The culminating deformational event in the southern Appalachians has been termed the *Allegheny Orogeny*. This great episode of mountain building probably began during Pennsylvanian

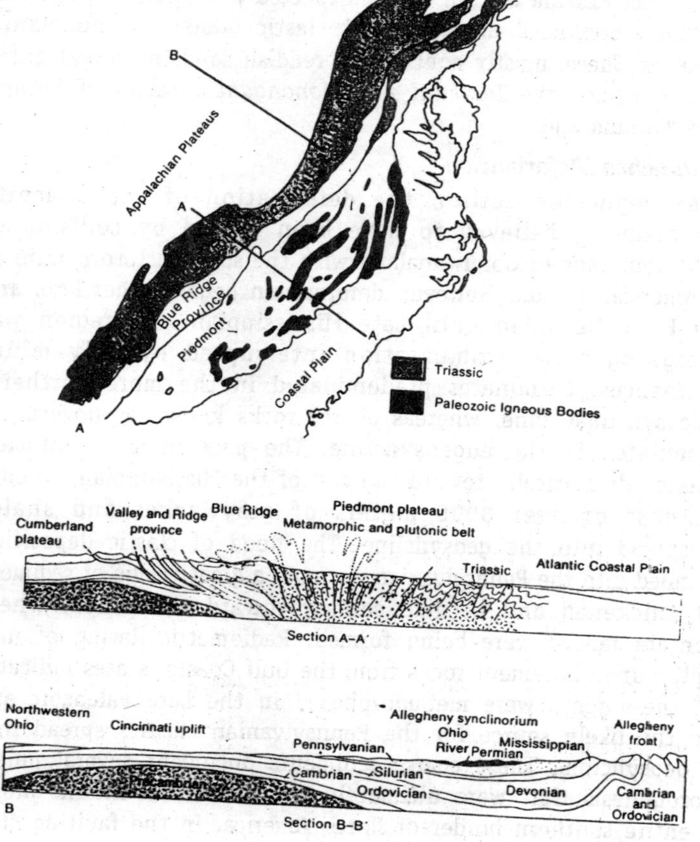

Fig. 5.8. A—Map of physiographic provinces of the Appalachian region. B—Two cross-sections showing the structural relations of the Appalachian Mountains.

time and continued throughout the Permian and into the Triassic. It affected a belt that extended for over 1600 km from southern New York to central Alabama. The results of the orogeny were profound and included Permian compression of geosynclinal strata as well as the bordering tract of the craton. The great folds now visible in the Valley and Ridge Province were developed during this orogeny. Less visible at the surface but no less impressive are enormous thrust faults formed along the east side of the southern thrust Appalachians. The folds are asymmetrically overturned toward the northwest, and the fault surfaces are inclined southeastward, suggesting that the entire miogeosyncline was moved forceably against, the central craton. Not unexpectedly, erosion of the rising mountains produced another great clastic blanket of nonmarine sediments. These mostly continental reddish sandstones and gritty shales compose the *Dunkard* and *Monongahela* series of Permo-Pennsylvanian age.

The Ouachita Deformation

As suggested earlier, the deformation of the Ouachita Geosyncline is believed to have been caused by collision of the African edge of Gondwanaland with the southeastern margin of the American craton. However, deformation began rather late, and from Early Devonian until Late Mississippian, the region was undergoing slow sedimentation interrupted by only minor disturbances. Carbonates predominated in the more northerly miogeosynclinal zone, whereas cherty rocks known as *novaculites* accumulated in the eugeosyncline. The pace of sedimentation increased dramatically toward the end of the Mississippian, when a thickness of over 8000 meters of graywackes and shales was spread into the geosyncline. The flood of clastic deposition continued into the Pennsylvanian, forming a great wedge of sediment that thickened and became coarser toward the south, where mountain ranges were being formed. Radiometric dating of now deeply buried basement rocks from the Gulf Coastal states indicates that these rocks were metamorphosed in the Late Paleozoic and were the likely source for the Pennsylvanian clastics spread into the geosyncline. These coarse sediments document several pulses of orogenesis that were ultimately to produce mountains along the entire southern border of North America. In the faulting that accompanied the intense folding, eugeosynclinal rocks were thrust northward onto the miogeosyncline. By Permian time, stability returned, and strata of this final Paleozoic period are relatively undisturbed.

Since the time of active mountain building, erosion has leveled most of the highlands, leaving only the Ouachita Mountains of Arkansas and Oklahoma and the Marathon Mountains of southwest Texas as remnants of once lofty ranges. Although the Ouachita Geosyncline has been traced over a distance of nearly 2000 km, only about 400 km are exposed. Thus, the actual configuration of geosyncline has been determined by the examination of millions of well samples and other data obtained during drilling activities associated with petroleum exploration.

The Late Paleozoic history of the western or Cordilleran Geosyncline was almost as lively as that of the Appalachian. In the far west, patches of highly deformed and intruded Late Paleozoic cherts and volcanics and thick sections of graywacke attest to deformation that is thought to have been caused by the collision of an eastward-moving island arc against the western border of North America. As a result of this encounter, an elongate area of uplift and instability known as the Antler Orogenic Belt developed along the boundary between the eugeosyncline and miogeosyncline. Tectonic activity along this belt resulted in highlands that provided source materials for over 1000 meters of conglomerates and sandstones spread across Nevada and Idaho. The Antler Orogeny was accompanied by thrust-faulting on a colossal scale, with eugeosynclinal sequences moved as much as 80 km eastward over miogeosynclinal sediments of the former continental shelf.

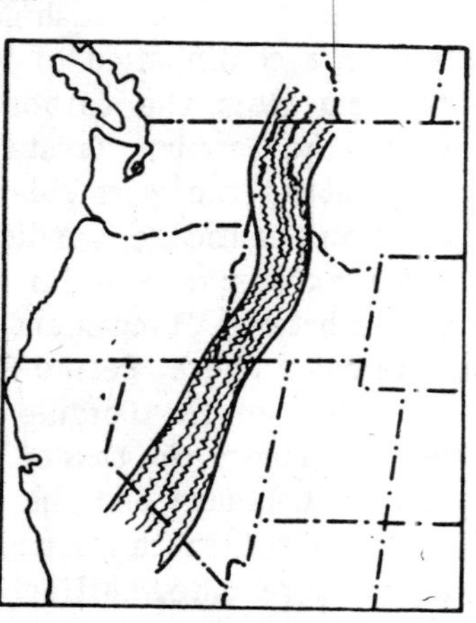

Fig. 5.9. Location of the Antler Orogenic belt.

The Antler Orogeny, which had begun in the Late Devonian, continued actively into the Mississippian and Pennsylvanian. An upland tract known as the Manhattan Geanticline shed sediments into both the miogeosyncline on the east and the eugeosyncline on the west. Especially thick deposits of Pennsylvanian and Permian miogeosynclinal sediments accumulated in the area now occupied by the Wasatch and Oquirrh Mountains. In the latter area, the Oquirrh Formation is over 9000 meters thick.

Mississippian and Pennsylvanian eugeosynclinal deposits west of the Manhattan Geanticline include a great volume of coarse clastics and volcanics. Over 2000 meters of sandstones, shales, lavas, and ash beds are found in the Klamath Mountains of northern California. Volcanic rocks in western Idaho and British Columbia attest to a continuation of vigorous volcanism from the Carboniferous through the Permian. Crustal deformation along the west side of the Manhattan Geanticline is indicated by areally extensive angular unconformities between Permian and Triassic sequences. There Permo-Triassic disturbances of the Cordilleran Belt have been named the *Cassiar Orogeny* in British Columbia and the *Sonoma Orogeny* in the southwestern United States. Like the earlier Antler Orogeny, the Sonoma event was probably caused by the collision of an eastward-moving island arc against the North American continental margin in west-central Nevada. Oceanic rocks and remnants of the arc were thrust onto the edge of the continent and became part of North America.

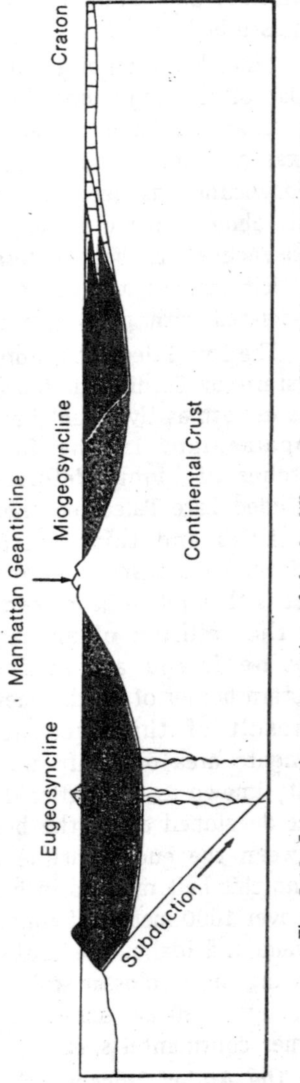

Fig. 5.10. Interpretive cross-section of the Cordilleran Geosyncline during Mississippian time.

Permian miogeosynclinal conditions were quieter than those in the eugeosyncline to the west. Quartz sandstones and limestones are the predominant deposits. One of the better known of these Permian miogeosynclinal rock units is the *Phosphoria Formation*, which was deposited in a shallow sea that covered Wyoming and eastern Idaho. As indicated by its name, the formation includes phosphatic limestone and phosphorite deposits. The *phosphorite*, a dark gray, concretionary variety of calcium phosphate, is mined for the manufacture of fertilizers and other chemical products. The unusual concentration of phosphates may have resulted from upwelling of phosphorus rich sea water from deeper parts of the basin. Once within the area of phosphoria deposition, microorganisms may have caused the precipitation of the phosphate salts.

The Northern Geosyncline

The Late Paleozoic record of the Northern of Franklinian Geosyncline is initiated by Devonian shales and limestones. Coral reefs are abundantly preserved in the Devonian System. The existence of these reefs is not surprising, for the region lay at about 15° north latitude during the Devonian and temperatures were much warmer than today. Following the deposition of the Devonian marine rocks, a highland area was developed along the northern border of the geosyncline. Coal-bearing deltaic strata were deposited in the adjacent lowland. These and older deposits were folded in Early Mississippian time, then eroded, and ultimately covered by a series of limestones and evaporites.

Europe During the Late Paleozoic

During the Late Paleozoic, Europe was bordered by the Uralian Geosyncline on the east and the Hercynian on the south. Along the northern margin of Europe, the Caledonian Orogeny had created the vast land area of the Old Red Continent. Uplands provided a source for clastic sediments, which were swept out into numerous basins to accumulate to thicknesses exceeding 10,000 meters. Judging from the many interlayers of ash and lava, volcanic activity was frequent on the Old Red Continent. Fossil fishes and sedimentary evidence indicate Europe's climate was tropical and possibly semiarid at the time.

South of the Caledonian Mountains, the continental deposits gradually thin and grade into marine shales and limestones of the Hercynian Geosyncline. For a time, quiet prevailed in the Hercynian Geosyncline. However, in Late Devonian and Early Mississippian, the belt was intensely folded, metamorphosed, and intruded by granites as ancestral Europe and Gondwanaland collided. The event has been named the *Hercynian Orogeny*, and its result was a great range of mountains across southern Europe. For the most part, the eroded stumps of these ranges are now buried, but here and there patches of the covering younger rocks have been eroded away, revealing the intensely folded, faulted, and intruded older rocks that lie beneath.

From the Hercynian uplands, gravel, sand, and mud were carried down into basins and coastal environments. The clastic deposits were coastal environments. The clastic deposits were quickly clothed in dense tropical forests. Burial and slow alteration of vegetative debris from these forests provided the material for coal formation in the great European coal basins. Plant fossils found in these coal seams are of the tropical kind and differ from the more temperate Gondwana flora of the Southern Hemisphere.

Although the greatest amount of Hercynian deformation occurred toward the end of the Early Carboniferous (Mississippian) in Europe, spasms of unrest continued in both the Late Carboniferous and the Permian. These later episodes of folding correlate with similar deformations in the Southern Appalachian and Ouachita orogenic belts of North America.

No less important than the Hercynian Geosyncline in Europe was the lengthy Uralian Geosyncline, which extended along a belt now occupied by the Ural Mountains. The geosyncline was already in existence at the beginning of the Paleozoic Era. Interestingly, there are discrepancies between the polar wandering curves for Paleozoic rocks from the cratons on either side of the Urals. The discrepancies indicate that the Russian and Siberian platforms were widely separated during Early Paleozoic time and that they began to converge in the middle part of the era and ultimately collided by the end of the Paleozoic. Indeed, the eugeosynclinal tract of the Urals consists largely of oceanic material scraped off against the edge of the converging plates. The collision resulted in the formation of a great mountain system along the entire Uralian orogenic belt and unified ancestral Siberia with eastern Europe.

In central Europe and parts of Russia, there were temporary Late Permian marine incursions that precipitated evaporites. The famous German potassium salts are a product of one of these inland seas, which has been named the *Zechstein Sea*. Apparently, aridity was as characteristic of western Europe during the Permian as it was of Texas and New Mexico.

Gondwana During the Late Paleozoic

During the Late Paleozoic, the great land mass of Gondwana remained fairly intact. It moved across the south pole and more fully entered the side of the earth on which Laurussia was located.

In its northward migration, Gondwana caused the closure of the ocean that separated it from Laurussia, causing the Hercynian Orogeny of Europe and the Allegheny Orogeny of North America. Orogenic activity associated with subduction zones was also evident in the Late Paleozoic history of Gondwana, particularly along the Andean Geosyncline of South America and the Tasman Geosyncline of eastern Australia.

The most dramatic paleoclimatologic event of Gondwana's Late Paleozoic history was the growth of vast continental glaciers. Extensive layers of tillite and the scour marks of glaciers have been found at hundreds of locations in South America, South Africa, Antarctica, and India. There are indications of at least four and possibly more glacial advances, suggesting a pattern of cyclic glaciation not unlike that experienced by North America and Europe less than 100,000 years ago. The orientation of striations chiseled into bedrock by the moving ice suggests the glaciers moved northward from centers of accumulation in southwestern Africa and eastern Antarctica. During the warmer interglacial stages and in outlying less frigid areas, *Glossopteris* and other plants tolerant of the cool, damp climates grew in profusion and provided the materials for thick seams of coal.

In time the ice receded, and Gondwana's vast cratonic areas became sites for deposition of Permian nonmarine red beds and shales. As we shall see in the next section, some of these sediments contain the fossil remains of the ancestors of the earth's first mammals.

Life of the Late Paleozoic

One of the lessons learned from the fossil record is that once a major taxonomic category of organisms has evolved and proliferated, that phylum tends to persist. Of course, members of the phylum diversify according to evolutionary factors that act upon them. For this reason, life of the Late Paleozoic represents a continuation of the evolutionary trends initiated in the Early Paleozoic. There are some important innovations, however. During the Early Paleozoic, most life appears to have been marine, whereas early in the Later Paleozoic, plants and animals had begun to proliferate upon the continents. Great forests appeared and changed the appearance of the landscape. From the fishes, which made their debut in the Early Paleozoic, amphibians and reptiles evolved. For the first time ever, vertebrates walked across the lands, and insects began to diversify and populate the continents. In the sea, continuations rather than new appearances were the rule. For example, trilobites persisted into the Permian, although their abundance and diversity were markedly decreased. Corals and bryozoans expanded and replaced stromatoporoids as the principal reef-forming organisms. Graptolites and other once flourishing groups suffered extinction.

Plants

The first unquestioned occurrences of vascular land plants are in late Ordovician rocks. These early invaders of the terrestrial environment were the *psilopsids* described in the previous chapter. The psilopsids preceded in evolution a variety of higher plants, many of which were lofty, well-rooted leafy trees that grew in Devonian forests. One of these forests has left an ample fossil record in the vicinity of Gilboa, New York. The Gilboa Forest contained trees over 7 meters tall. However, impressive as they were, the Devonian forests were dwarfed by their Carboniferous descendants.

When one surveys the entire history of vascular plants, three major advances become apparent. Each involved the development of increasingly more effective reproductive systems. The first advance led to seedless, spore-bearing plants, such as those that were ubiquitous in the great coalforming swamps of the Carboniferous. The second saw the evolution of seed-producing, pollinating, but nonflowering plants ("gymnosperms"). This also was a Late Paleozoic event. The evolution of plants with both seeds and flowers ("angiosperms") came late in the era that followed the Paleozoic.

Among the moisture-loving plants of the Carboniferous were the so-called scale trees, or *lycopsids*. Today, they are represented by their smaller survivors, the club mosses, of which the ground pine *Lycopodium* is a member. Small size was not particularly characteristic of the Late Paleozoic lycopsids. The forked branches of *Lepidodendron* reached 30 meters into the sky. The elongate leaves of the scale trees emerged directly from the trunks and branches. After being released, they left a regular pattern of leaf scars. In *Lepidodendron*, the scars are arranged in diagonal spirals, whereas species of *Sigillaria* have leaf scars in vertical rows.

Another dominant group of plants that grew side by side with the Carboniferous lycopsids were the *sphenopsids*. Living sphenopsids include the scouring rushes and horsetails. Fossil

Fig. 5.11. Fossilized bark of the Carboniferous tree Lepidodendron.

sphenopsids, such as *Calamites* and *Annularia*, possessed slender, unbranching, longitudinally ribbed stems with a thick core of pith and rings of leaves at each transverse joint. At the top, a cone bore the spores that would be scattered in the wind.

True ferns were also present in the coal forests. Many were tail enough to be classified as trees. Like the lycopsids and sphenopsids, they reproduced by means of spores carried in regular patterns on the undersides of the leaves.

Seed plants also made their debut during the Late Paleozoic. They probably arose from Devonian fernlike plants. These plants, appropriately dubbed "seed ferns," had fernlike leaves but unlike true ferns reproduced by means of seeds.

One of the most widely known seed fern groups was *Glossopteris*, which was widespread in the southern hemisphere during the Carboniferous and Permian. Most species of *Glossopteris* were plants

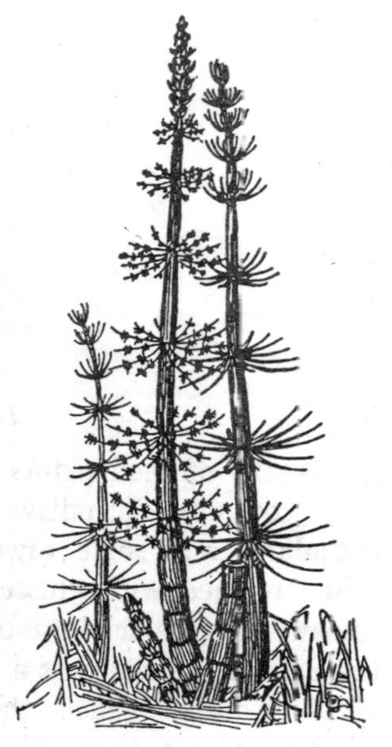

Fig. 5.12. Calamites.

with thick, tongue-shaped leaved. Because of certain anatomic traits, and of their association with glacial deposits, *Glossopteris* and associated plants are thought to have been adapted to cool climates.

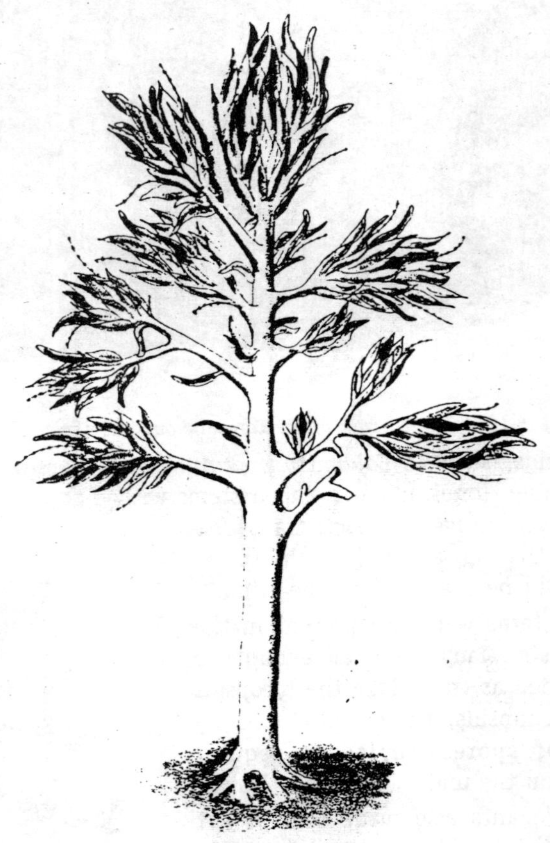

Fig. 5.13. Cordaites.

Fossils of seed plants are present in the northern hemisphere also. Cordaites (primitive members of the conifer lineage) were abundant, and some towered 50 meters). Their branching limbs were crowned with clusters of large, straplike leaves. These and somewhat more modern cone-bearing plants spread widely during the Permian, perhaps as a consequence of dryer climatic conditions. The first ginkgoes made their appearance during the Permian. Today, only a single species, *Ginkgo biloba*, remains as a survivor of this once flourishing group.

Continental Invertebrates

Because of the greater hazards of postmortem destruction of continental invertebrates, their fossil record is not as complete as that for marine invertebrates. Nevertheless, logic dictates that many invertebrate animals populated the soil, rivers, and lakes of the Late Paleozoic. The oldest known insects are wingless species found in Devonian rocks. Carboniferous strata contain a more complete but still inadequate insect record. Included were giant dragonflies with wingspans of over 70 cm. Cockroaches that reached lengths of 10 cm creeped about among the rotting vegetation. Eurypterids, although not common, persisted throughout the Late Paleozoic. Several species of Devonian and Carboniferous strata in North America, Germany, and Great Britain. In Nova Scotia, an interesting collection of land snails has been collected from hollows within the preserved stumps of trees. Scorpions and centipedes are found in abundance in a concretionary Pennsylvanian siltstone exposed along Mazon Creek in northern IIlinois.

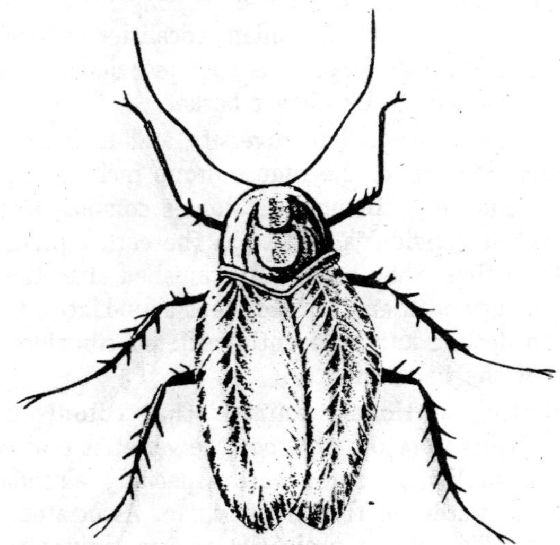

Fig. 5.14. Reconstruction of a primitive Pennsylvanian cockroach.

Of all the land invertebrates, it is apparent that the arthropods and gastropods have been the most persistently successful. Many paleontologists believe this may be because of attributes already evolved by their aquatic ancestors. As protection against desiccation, arthropods had evolved relatively impervious exoskeletons. Snails derived similar benefits from their shells. Both groups included very active animals with sufficient mobility to seek out food aggressively.

Marine Invertebrates

Most of the familiar groups of Early Paleozoic marine invertebrates continued into the Late Paleozoic, although some groups diminished in number and variety, whereas others expanded and diversified. The foraminifers, which had a modest start in the Early Paleozoic, experienced their first major expansion in the Carboniferous. One reflection of their numerical increase is seen in the Mississippian Salem Limestone, which locally consists almost entirely of specimens of *Endothyra*. The *fusulinids*—spindle-shaped descendants of endothyrids—proliferated during the Pennsylvanian and Permian. As a group, fusulinids looked superficially similar, but individual species evolved complex and distinctive internal structures that permit their use as index fossils in many parts of the world.

The most striking innovation among *sponges* was the expansion of siliceous forms during the Devonian. Localities in New York yield great numbers of fossilized siliceous sponges, many of which were similar to the modern Venus flower basket.

Rugose corals continued to diversify and increase during the Late Paleozoic. In general, the rugose group included both solitary "horn corals" and large compound colonies composed of hundreds of closely packed individuals. Although the earlier tabulate species continued for a time, they had nearly vanished after the Devonian. Following their Devonian and Carboniferous abundances, the rugosans also began to decline and apparently suffered complete extinction by the end of the Paleozoic.

Among other stationary animals that colonized the Late Paleozoic sea floors were the *Bryozoa*. The varieties that constructed lacy, delicate, fanlike colonies were especially abundant in the shallow tropical waters of the Mississippian. Associated with these so-called *fenestellid* colonies were the bizarre corkscrew bryozoans known by the generic name *Archimedes*.

Brachiopods were present by the millions throughout the Paleozoic, but they may have reached their zenith in number and variety during the Devonian. The spiriferid brachiopods, so named because they possessed internal helicoid spirals to support the lophophores, were especially abundant during the Devonian. During

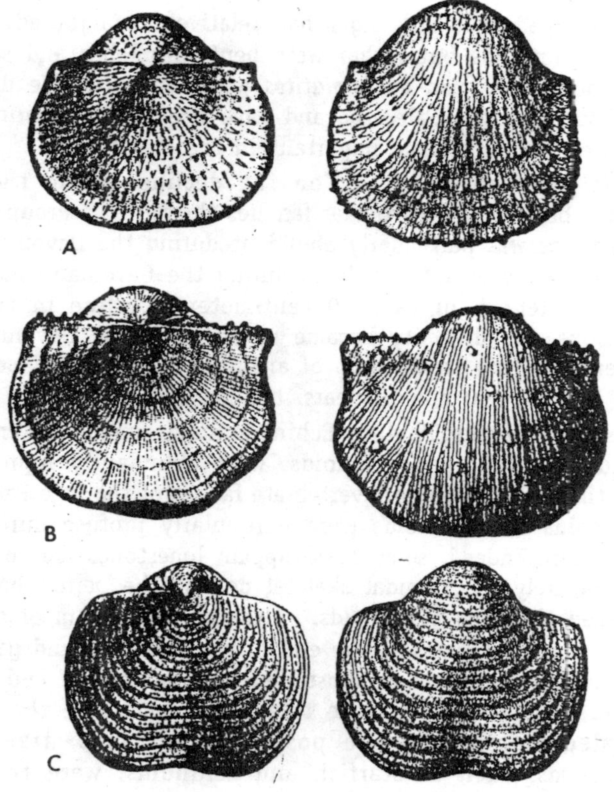

Fig. 5.15. The Pennsylvanian productid brachiopods Juresania (A), Linoproductus (B), and Echinoconchus (C).

the Carboniferous and Permian, the productid brachiopods, which has distinctly spinose and inflated convex ventral valves, became the most abundant group. Indeed, among invertebrate paleontologists, the Late Paleozoic is frequently proclaimed the "Age or Productids."

Although the greatest advances among the *mollusc* clan came in eras that followed the Paleozoic, they were nevertheless persistently present throughout the Late Paleozoic, *Pelecypods* flourished in the frequently sandy and muddy sea bottoms of the time. Air-breathing *gastropods* made their first appearance, whereas the older, aquatic group persisted. However, the single most important molluscan event was the debut of that important cephalopod group known as *ammonoids*. In this group, sutures were developed that no longer had relatively straight edges like those of nautiloids but rather were bent into lobes and saddles. The ammonoids known an *goniatites* persisted in marine deposits throughout the Late Paleozoic and gave rise to the *ceratites* and *ammonites*, which became important in the Mesozoic.

Trilobites were generally on the decline during the Late Paleozoic, but, locally, particular families thrived. The group known as proparians was particularly abundant during the Devonian. The largest trilobites ever found lived during the Devonian, one giant reaching a length of over 70 centimeters. Decline in trilobite populations and diversity became severe after the Devonian. By Late Permian, this great group of arthropods, which had persisted over a span of 350 million years, became extinct.

Members of the Phylum Echinodermata, including crinoids, blastoids, starfish, and echinoids, were abundantly represented among the shallow marine invertebrate faunas of the Late Paleozoic. The crinoids and blastoids were particularly profuse during the Mississippian. Indeed, some Mississippian limestones are composed almost entirely of crinoidal skeletal debris. The period has been nicknamed the "Age of Crinoids." Although one group of crinoids survived the hard times at the end of the Paleozoic and gave rise to those still living today, most died out before the end of the Permian. The blastoids became totally extinct before the end of the Paleozoic. During the post-Paleozoic eras, free-living echinoderms, such as starfish and echinoids, were the most ubiquitous representatives of the phylum.

The Vertebrates

The Late Paleozoic was a particularly eventful time span for the vertebrates. This was the interval during which animals evolved the limbs and other features that permitted them to invade the land. It was the time that two great classes of vertebrates arose, the amphibians and the reptiles.

Fishes

In the previous chapter, the evolutionary progression from the jawless ostracoderms to the first fishes with movable lower jaws was described. Those earliest jawed fishes, the acanthodians, first appear in nonmarine rocks of the Late Silurian. They increased in numbers during the Devonian and then slowly declined toward extinction in the Permian. Placoderms, another group of archaic jawed fishes, had a similar duration on earth. Among the

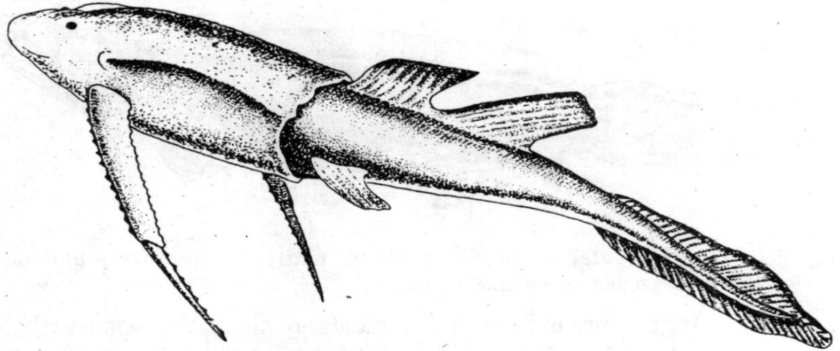

Fig. 5.16. The Devonian placoderm Bothriolepis.

placoderms were the savage predators *Dinichthyes*. These placoderms, some of which were over 9 meters in length, had huge jaws lined with scissors-sharp bone. Other placoderms were less formidable. *Bothriolepis*, for example, seems to have been well adapted for gleaning food from the floors of lakes or streams, in somewhat the same manner as some ostracoderms had done in an earlier age.

During the Devonian there was a veritable piscine explosion as fishes began their dominance of the oceans as well as streams and lakes. Two important categories of fishes, the cartilaginous chondrichthyans and the bony osteichthyans, made their debut in the Devoniar. Today the cartilaginous fishes are represented by sharks, rays, and skates. Among the better known of the Late Paleozoic sharks were species of *Cladoselache*. Remains of this shark are frequently encountered in the Devonian shales that outcrop on the south shore of Lake Erie. During the Late Carboniferous, a group represented by *Xenacanthus* managed to penetrate the freshwater environment. A third group of Paleozoic cartilaginous fishes were the bradyodonts. These had flattened bodies like modern

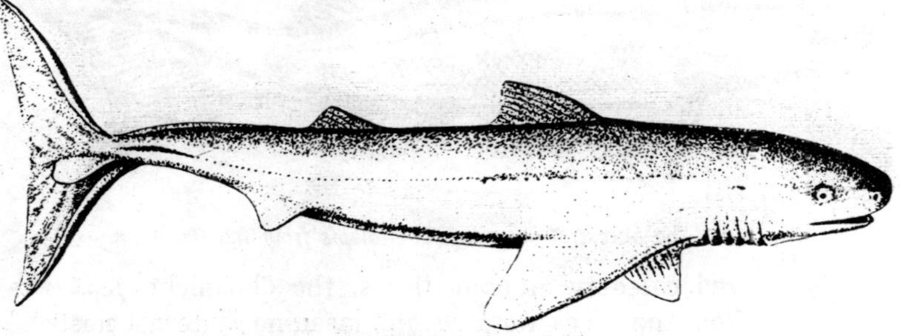

Fig. 5.17. Cladoselache, a Devonian shark.

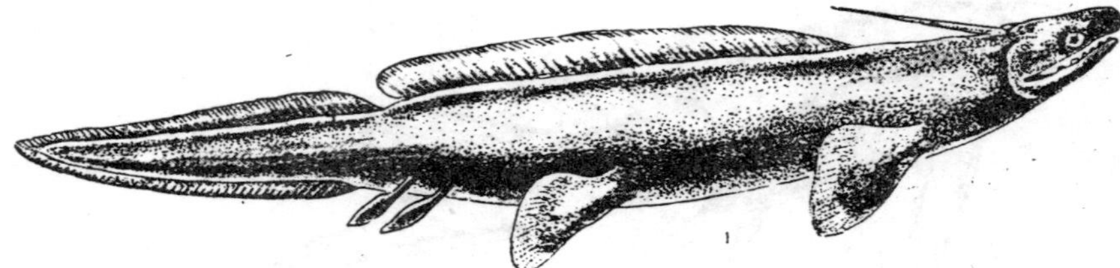

Fig. 5.18. The freshwater shark Xenacanthus, whose remains are found in Pennsylvanian nonmarine strata.

rays and blunt, rounded teeth for crushing shellfish. Apparently, modern sharks arose from cladoselachian ancestors but retained their archaic traits until the Jurassic.

Because of the role of bony fishes in the evolution of tetrapods (four-legged animals) and because they are the most numerous, varied, and successful of all aquatic vertebrates, their evolution is of particular importance. Bony fishes may be divided into two categories, namely the familiar "ray-fin," or *Actinopterygii*, and the "lobe-fins," or *Choanichthyes*.

As implied by their name, ray-fin fishes lack a muscular base to their paired fins, which are thin structures supported by radiating bony rays. Unlike the Choanichthyes, they do not possess paired nasal passages that open into the throat. The ray-fins began their evolution in Devonian lakes and streams and quickly expanded into the marine realm. They became the dominant fishes of the modern world. The more primitive Devonian bony fishes are well represented by the genus *Cheirolepis*. From such fishes as these evolved the more advanced bony fishes during the Mesozoic and Cenozoic.

Fig. 5.19. The ancestral bony fish Cheirolepis from the Devonian.

The second category of bony fishes, the Choanichthyes, take their name from the Greek word *choana* meaning "internal nostril." Choanichthyids have a pair of openings in the roof of the mouth that lead to clearly visible external nostrils. Such fish were able to rise to the surface and take in air through their nostrils. Choanichthyans also possessed sturdy muscular fins and lungs. Lungs and fins do not seem to go together in modern fishes, but in Late Paleozoic fishes the combination was not uncommon. Studies of living choanichthyans indicate that lungs probably began their evolution as saclike bodies developed on the ventral side of the esophagus and then became enlarged and improved for the extraction of oxygen. (In modern fishes, the lung has been converted to a swim bladder, which aids in hydrostatic balance.)

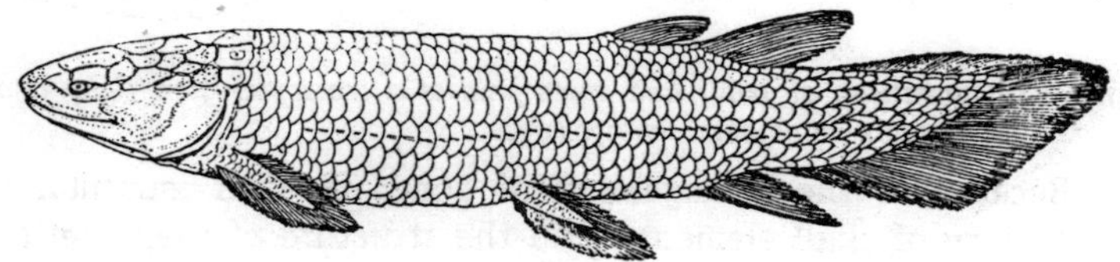

Fig. 5.20. Dipterus, a Devonian lungfish.

Two major groups of lungfishes lived during the Devonian. They are designated the *Dipnoi* and *Crossopterygii*. The Dipnoi, represented in the Devonian by *Dipterus*, were not on the evolutionary track that was to lead to tetrapods. They are, nevertheless, an interesting group, which includes the living freshwater lungfish of Australia, Africa, and South America. Their restricted presence south of the equator suggests that Gondwanaland was the probable center of dispersal for dipnoans. Dipnoi means "double breather." The name was suggested by the observation that living species are able to breathe by means of lungs during dry seasons. At such difficult times, they burrow into the mud before the water is gone. When the lake or stream is dry, they survive by using their accessory lungs, and when the waters return they switch to gill respiration.

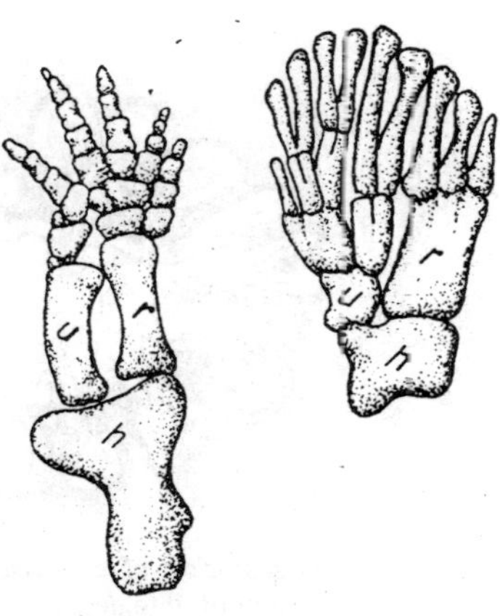

Fig. 5.21. Comparison of the limb bones of a crossopterygian fish (upper right) and an early amphibian.

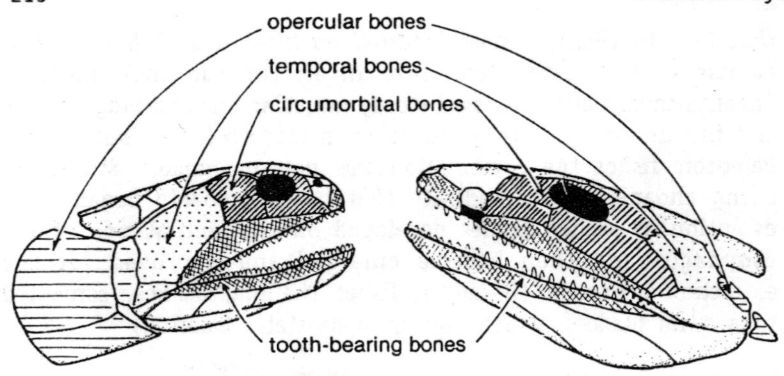

Fig. 5.22. Comparison of skulls and lower jaws of a crossopterygian (left) and the Devonian amphibian Ichthyostega.

Because of the arrangement of bones in their muscular fins, the pattern of skull elements, and the structure of their teeth, the fossil Crossopterygii are considered the ancestors of the amphibians. A rather advanced fish that exemplifies Devonian crossopterygians is *Eusthenopteron*. In this genus, the paired fins were short and muscular. Internally, a single basal limb bone, the humerus (or femur for the pelvic structure), articulated with the girdle bones and was followed by two bones, the ulna and radius, for the pectoral fins, and the tibia and fibula for the posterior fins.

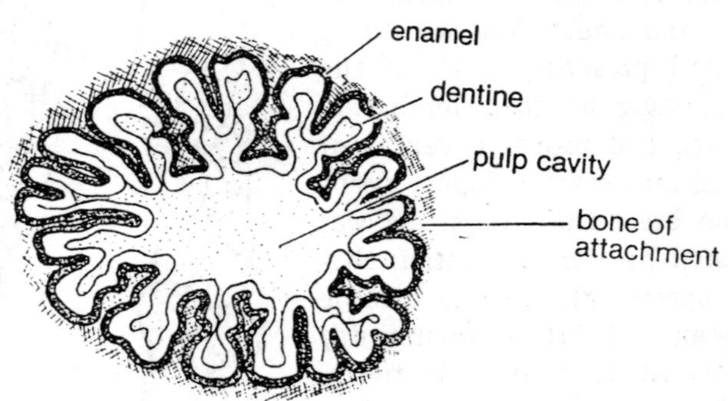

Fig. 5.23. Cross section of a crossopterygian tooth clearly exhibits the distinctive pattern of infolded enamel. This same characteristic is a characteristic ofthe teeth of the amphibian descendents of crossopterygian fishes.

Of course, the robust skeleton and sturdy limbs of the crossopterygians did not evolve because fishes had miraculously taken on a desire for life on land. These were adaptations for

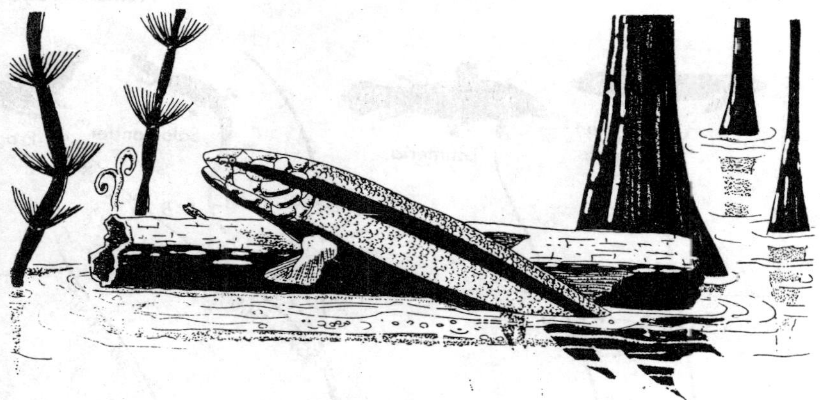

Fig. 5.24. Eusthenopteron, a Devonian crossopterygian fish that had evolved in the direction of early amphibians.

moving to water and remaining in water during periods of drought. The air-gulping fishes found land a hostile environment and occasionally dragged themselves onto the land only as a means of reaching another pond or fresher body of water.

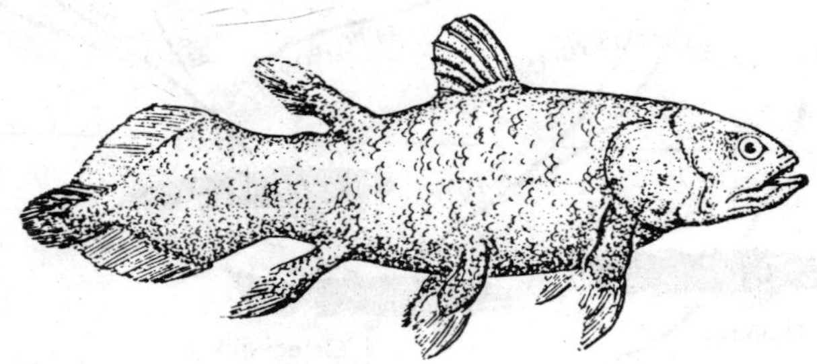

Fig. 5.25. Latimeria, a surviving coelacanth living in the ocean near Madagascar and southeast Africa.

During the Devonian, two distinct branches of crossopterygians had evolved. One of these led ultimately to amphibians, and the other led to salt water fishes called coelacanths. Coelacanths were thought to have undergone extinction during the late Cretaceous, but their survival down to the present day has been documented by several catches of the coelacanth *Latimeria* near Madagascar.

Amphibians

It required tens of millions of generations to convert the crossopterygian fishes into animals that could live comfortably on solid ground. Even so, the conversion was not complete, for

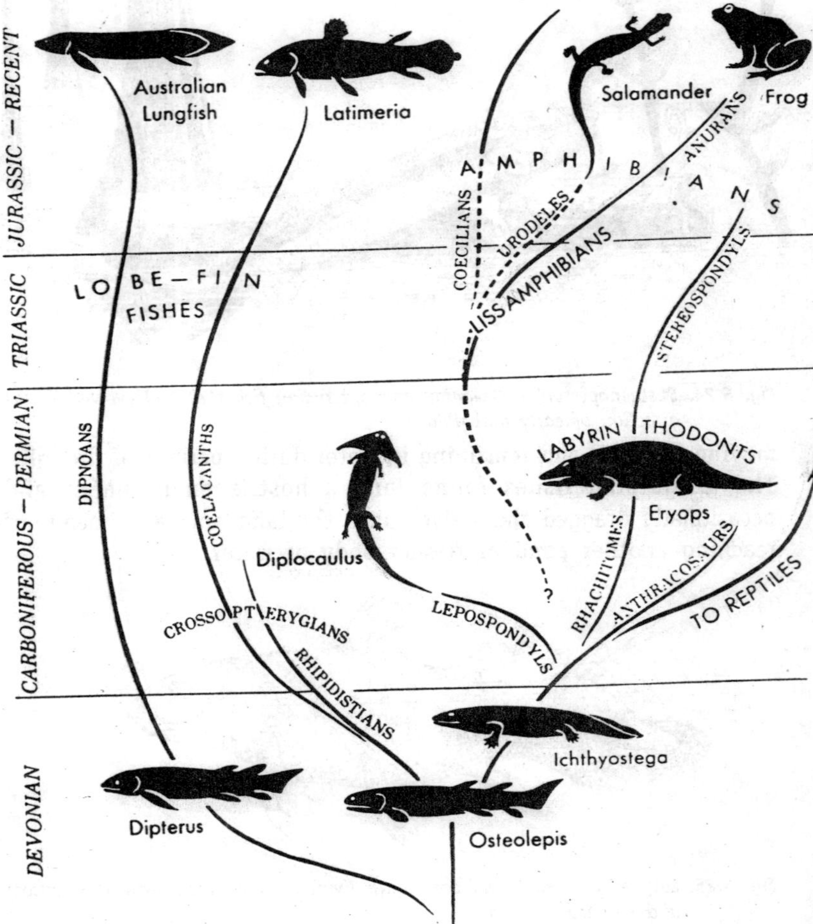

Fig. 5.26. The evolution of amphibians and lobe-in fishes.

amphibians continued to return to water to lay their fishlike naked eggs. From these eggs came fishlike larvae, which, like fish, used gills for respiration.

A number of changes accompanied the shift to land dwelling. A three-chambered heart developed to route the blood more efficiently to and from the now more efficient lungs. The limb and girdle bones were modified to overcome the constant drag of gravity and better hold the body above the ground. The spinal column, a simple structure in fishes, was transformed into a sturdy but flexible bridge of interlocking elements. In order to improve hearing in gaseous rather than fluid surroundings, the old hyomandibular bone, used in fishes to prop the braincase and upper jaw together, was pressed into service as an ear ossicle—the *stapes*. The fish spiracle (a vestigial gillslit) became the amphibian eustachian tube and middle ear. To complete the auditory apparatus, a tympanic membrane ("ear drum") was developed across a prominent notch in the rear part of the amphibian skull.

The fossil record for amphibians begins with a group called *ichthyostegids*. As suggested by their name, these creatures retained many features of their piscine ancestors. The amphibians that followed the ichthyostegids fall mostly within a group collectively termed the *labyrinthodonts*. The labyrinthic wrinkling and folding of tooth enamel provided the inspiration for the labyrinthodont name. During the Carboniferous, large numbers of labyrinthodonts wallowed in swamps and streams, eating insects, fish, and one another. A labyrinthodont that exemplifies the culmination of their lineage is *Eryops*. Eryops was a bulky, inelegant creature with the flattish skull so typical of labyrinthodonts and bony nodules in the skin for protection against some of its more vicious contemporaries. The labyrinthodonts declined during the Permian, and only a relatively few survived into the Triassic.

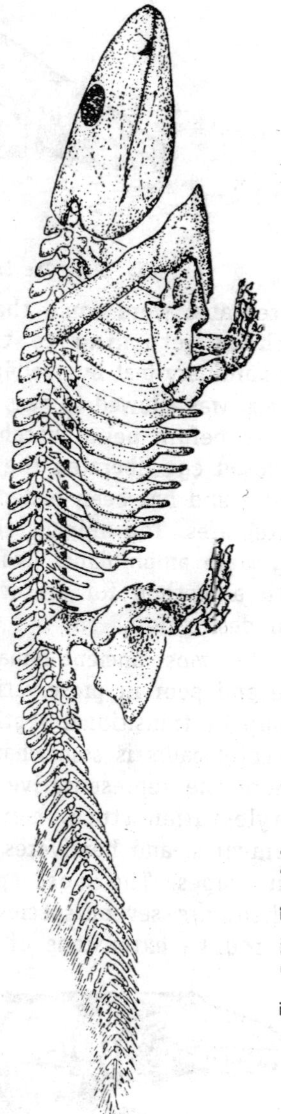

Fig. 5.27. The skeleton of Ichthyostega still retains the fishlike form of its crossopterygian ancestors.

Reptiles

The evolutionary advance from fish to amphibians was no trivial biological achievement, yet the Late Paleozoic was the time of another equally significant event. Among the evolving land

Fig. 5.28. The large Permian amphibian Eryops.

vertebrates were some that had developed a way to reproduce without returning to the water. They accomplished this extraordinary feat by evolving enclosed eggs in which the embryonic animal was allowed to pass through larval and other developmental stages before being hatched in an essentially adult form. This enclosed egg liberated the tetrapods from their reliance on water bodies and has been hailed as a major milestone in the history of vertebrates. The animals that first evolved the so-called amniotic egg were amphibians. Evolutionary processes had provided them with a method for protecting developing young from predation and desiccation.

The most ancient remains of reptiles are Late Carboniferous in age and poor in preservation. The fossils are representatives of a group of transitional reptiles called *cotylosaurs*. The preservation of cotylosaurs is somewhat better in Permian sediments of Texas, where the representative form *Seymouria* was discovered. From cotylosaurian stock, more advanced, large and small reptiles, carnivores, and herbivores evolved and came to dominate Permian landscapes. The most spectacular of these reptiles were the *pelycosaurs*, several species of which sported erect "sails" supported by rodlike extensions of the vertebrae. The pelycosaurs were a

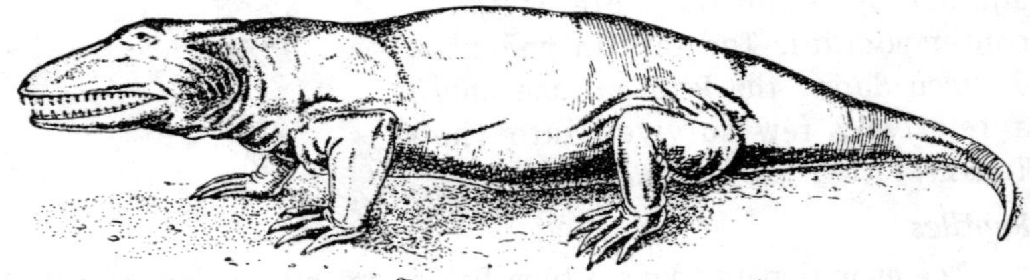

Fig. 5.29. The primitive reptile Seymouria, sometimes called the "stem reptile" because it possessed skeletal characteristics transitional between amphibians and reptiles.

varied group. Some, like *Edaphosaurus*, were plant eaters, whereas others, as indicated by their great jaws and sharp teeth, ate flesh. *Dimetrodon* was one such predator. There have been many attempts to explain the function of the pelycosaurian sail. The most reasonable explanation is that sail-fin acted in temperature regulation by serving sometimes as a collector of solar heat and at other times as a radiator giving off surplus heat. Because pelycosaurs are considered the group from which mammal-like reptiles arose, it is especially interesting that they may have been attempting some sort of body temperature control.

To humans, the mammal-like reptiles are among the most fascinating of fossil vertebrates. These creatures, collectively known as *therapsids*, were widely dispersed during Permian and Triassic time. There is a fabulous record of these reptiles and their contemporaries in the Karoo Basin of South Africa and more modest discoveries in Russia, South America, Australia, India, and Antarctica. Therapsids were predominantly small to moderate-sized animals that displayed at least the beginnings of several mammalian skeletal traits. There were fewer bones in the skull than in reptile skulls generally, and there was a mammal-like enlargement of the lower jaw bone (dentary) at the expense of more posterior elements of the jaw. A double ball and socket articulation had evolved between the skull and neck. Teeth showed a primitive but distinct differentiation into incisors, canines, and cheek teeth. The limbs were swung more directly into vertical alignment beneath the body, and the ribs were reduced in the neck and lumbar region for greater over flexibility. These features are well developed in the Permian therapsid *Cynognathus*. The therapsids continued from the Permian through the Triassic but became extinct early in the Jurassic. However, before they died out completely, they gave origin to the early mammals.

Extinctions

The fossil record indicates that there has been a persistent increase in the numbers of different species through geologic time. It also shows that this trend has been upset or even reversed at times as a result of environmental adversity. Extinctions of large numbers of previously successful animals and plants may have resulted from their inability to adapt to changing conditions. Early geologists were able to identify such times of biologic crisis and used extinctions to define further the termination of the Paleozoic and Mesozoic Eras.

There were drastic geographic changes as the Paleozoic drew to a close. Pangaea was experiencing mountain building and regional warping, there was violent volcanism in several regions of the earth, and the seas had withdrawn from the extensive tracts they once covered. Climates became more severe, and a protracted glacial age occurred in the south. There was evidence of marked seasonal changes and aridity in many areas of the globe. As one or another group of animals gradually diminished in response to these changes, they contributed to the demise of others farther along in the food chain. Then as now, the balance in the biologic environment could be easily upset, causing far-reaching waves of extinction.

Among the invertebrate groups that became extinct near the end of the Paleozoic were the trilobites, eurypterids, many families of bryozoans, and blastoids. Entire families of other invertebrate taxa were lost as well. The archaic jawed fishes (acanthodians) became extinct during the Permian, whereas amphibians, cotylosaurs, and several groups of plants declined conspicuously. The misfortunes of the invertebrates may have been associated with the drastic reduction of the inland seas, whereas the amphibians may have suffered in the competition with evolving reptiles. Among those reptiles were a few hardy groups that were able to meet the new challenges and become the founders of the reptilian communities that were to dominate the Mesozoic terrestrial scene.

Climates of The Late Paleozoic

The main climatic zones of the Late Paleozoic tended to parallel latitudinal lines just as they do today. Of course, the continents were located differently, so that the north pole was in South Africa and the north pole was over open ocean. The Late Paleozoic equator trended northeastward across Canada and southward across Europe. Coal beds, evaporites, coral reefs, and dune-deposited red beds developed within 40° of the paleoequator. Today, we cannot help but be startled to find the fossils of tropical amphibians and plants at locations within the arctic circle. As already noted, cooler climates prevailed in Gondwanaland because of its Late Paleozoic proximity to the south pole, and by Permian time the southern continents were in the grip of a major ice age. To the north, the apparent aridity of large regions was influential in the evolution and extinction of terrestrial animals and plants.

Mineral Products of The Late Paleozoic

Although a great variety of mineral deposits were formed during the Late Paleozoic, the fossil fuels (coal, oil, and gas) are particularly significant. Coal occurs in all post-Devonian systems. In northern hemisphere continents, it is particularly characteristic of the Late Carboniferous. Thick deposits of Pennsylvanian coal occur in the Appalachians, the Illinois Basin, and the industrial heartland of Europe. Single sequences of Pennsylvanian strata may include several coal beds, as in West Virginia, where 117 different layers have been named. In the so-called anthracite district of western Pennsylvania, orogenic compression has partially metamorphosed coal into an exceptionally high-carbon, low-volatile variety that is prized for its industrial uses. Permian coal seams are found in China, Russia, India, South Africa, and Australia. The coal industry in North America, which has deteriorated for the last few decades, is currently experiencing an upsurge as a consequence of decreasing supplies of petroleum and natural gas.

Commercial quantities of oil and gas are frequently found in Late Paleozoic strata. Devonian reefs within the Williston Basin of Alberta and Montana have been exceptionally productive reservoir rocks for petroleum. Devonian petroleum has also been produced in the Appalachians. Indeed, in 1859, the first American oil well was dug into a Devonian sandstone. (Oil was struck at a depth of only 20 meters). Carboniferous formations of the Rocky Mountains, midcontinent, and Appalachians also contain oil reservoirs. However, wells drilled into reefs and sandstones of the Permian Basin of western Texas have yielded the greatest amounts of oil from Late Paleozoic formations of the western United States. Oil trapped in Late Paleozoic strata beneath the North Sea is now being actively utilized to supply the needs of Europe.

Arid, warm climatic conditions, particularly in northern continents during the Late Paleozoic, provided a suitable environment for deposition of sodium and potassium salts. In Late Permian time, enormous amounts of phosphates were also deposited as part of the Phosphoria Formation, which is exposed in Montana, Idaho, Utah, and Wyoming. The Phosphoria shales are extensively quarried for phosphate, which is sold as an important plant food.

Mountain building, with its attendant intrusive and volcanic igneous activity, is nearly always accompanied by the emplacement of metallic ores. The Hercynian and Allegheny Orogenies of the Late Paleozoic generated ores of tin, copper, silver, gold zinc, lead, and platinum. Deposits of all the precious metals, as well as copper, zinc, and lead, are found in the Urals of Russia and in China, Japan, Burma, and Malaya. Tin, Tungsten, bismuth, and gold are mined in Australia and New Zealand. It is self-evident that Late Paleozoic rock sequences, with their stores of both metallic and nonmetallic economic minerals and their content of fossil fuels, are vitally important to the welfare of modern civilizations.

6

MESOZOIC ERA

The extinction of Paleozoic animals and plants that occurred at the close of the Permian Period did not go unnoticed by the founders of geology. The fossil evidence for this time of crisis provided a natural boundary that was used to mark the end of the Paleozoic sequence. Overlying younger strata contained different and generally more progressive organisms, yet not as modern as those that live today. Thus, it seemed appropriate to formulate a middle chapter in the history of life. They named that chapter the Mesozoic Era. In its turn, the Mesozoic also ended in a biologic crisis that is used to mark its separation from the youngest era of all—the Cenozoic.

The Mesozoic Era lasted an estimated 160 million years, ending approximately 65 million years ago. The three periods of the Mesozoic are unequal in duration. The Triassic lasted about 30 million years, the Jurassic 55 million years, and the Cretaceous about 75 million years.

During the lengthy span of the Mesozoic, the earth witnessed many changes. New families of plants and animals evolved and experienced often spectacular radiations. It was the era in which two new vertebrate classes, the birds and the mammals, first appeared. It was also a time in which the supercontinent Pangaea. II, formed in the previous era by the joining of errant ancestral continents, was gradually dismembered. As the continents moved slowly apart, the oceanic rifts between them deepended and broadened. A process of fragmentation and drift had begun that would ultimately lead to the present physical geography of the planet.

The Dismemberment of Pangaea II

Any discussion of the history of the Mesozoic Era must have Pangaea II as its starting point. The same forces that had drawn the plate segments together to form this great supercontinent were still in operation at the beginning of the Mesozoic; however, they now moved in different directions. In general, the dismemberment of Pangaea II seems to have occurred in four stages. The first stage began in the Triassic, with rifting and volcanism along tensional fault systems. Events such as these often result when a region of the lithosphere is subjected to tensional stresses. In this instance, the tension very likely was associated with the separation of North America and Gondwana. As the rifting progressed, Mexico was decoupled from South America, and the eastern border of North America parted from the Moroccan bulge of Africa. Oceanic basalts were added to the sea floor of the newly formed and gradually widening Atlantic Ocean. The largely tensional geologic structures, as well as the volcanic and clastic sedimentary rocks in the Triassic System of both the eastern United States and Morocco exhibit many striking similarities. These now widely separated regions were on

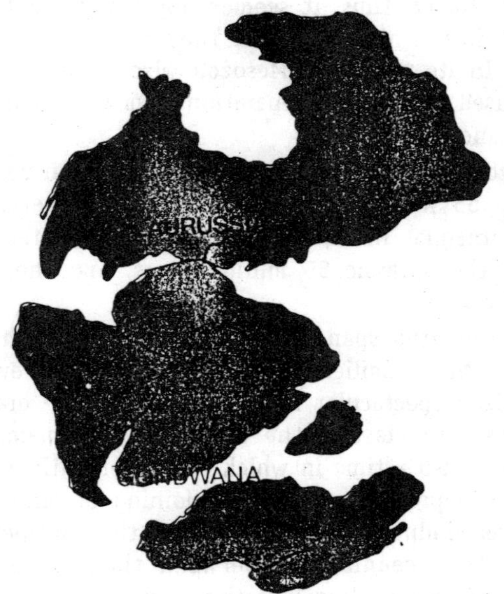

Fig. 6.1. Conceptual paleogeographic map of the Triassic world about 180 million years ago, when the breakup of Pangaea II was beginning.

opposite sides of the axis of spreading and were affected by similar forces and events. Temporarily, at least, Laurussia north of the Maritime Provinces remained intact. In addition, South America seems to have maintained its hold on Africa.

The second stage saw the beginning of the breakup of Gondwanaland. A great rift separated Antarctica from the southern ends of South America and Africa. This rift developed a branch that extended eastward from South Africa along what is now the eastern side of India. Great volumes of basaltic lavas poured from volcanoes located along these great rifts while the separated Gondwanaland segments began to move slowly northward, turning gently counterclockwise as they did so.

In stage three of the breakup of Pangaea II, the Atlantic rift began to extend itself northward, and clockwise rotation of Eurasia tended to close the eastern end of the Tethys Sea, a forerunner of the Mediterranean. By the end of Jurassic time, an incipient breach began to split South America away from Africa. The cleft apparently worked its way up from the south, creating a long seaway somewhat reminiscent of the present Red Sea. Australia and Antarctica remained intact, but India had moved well along on its long voyage to Laurasia. By Late Cretaceous, some 70 million years ago, South America had completely separated from Africa, and Greenland began to separate from Africa, and Greenland began to separate from Europe. However, northeasternmost North America still clung to Greenland and northern Europe.

The fourth stage in the dismemberment of Pangaea II was not a Mesozoic event. It occurred early in the Cenozoic, during which time the North Atlantic rift slowly penetrated northward until eventually the two great Laurasian continents were completely separated. Also during this final era of earth history, Antarctica and Australia parted company.

The Mesozoic History of North America

To the East and South

At the beginning of the Mesozoic, conditions in the eastern part of North America were much the same as they had been near the end of the Paleozoic. The rugged Appalachian ranges that had been raised during the Allegheny Orogeny were undergoing vigorous erosion. The coarse clastics derived from the uplands filled intermontane basins and other low areas during the Early and Middle

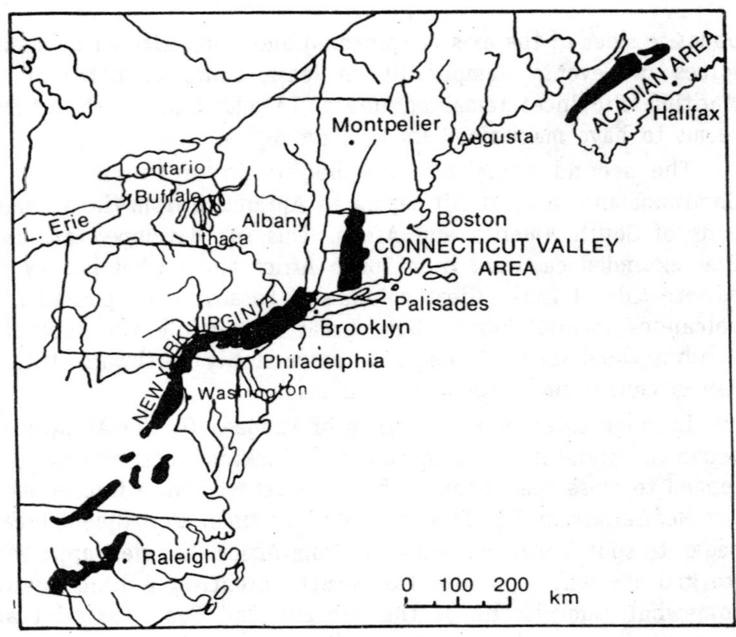

Fig. 6.2. Outcrop area of Triassic rocks in eastern North America.

Triassic. Then, during the Late Triassic, North America began to move away from Gondwanaland. As a possible result of the crustal stretching that accompanied the separation, as well as of gravitation adjustments associated with erosion of the mountains, a series of fault-bounded troughs developed along the eastern coast from Nova Scotia to North Carolina. The downfaulted blocks received great quantities of poorly sorted, arkosic, red sandstones and shales these rocks are known as the *Newark Series*. The poor sorting of the coarser Newark clastics and their high content of relatively unweathered feldspar grains indicate transportation and deposition by streams flowing swiftly down from the granite highlands that bordered the fault basins. Here and there, drainage in the basins became impounded, and lakes formed. The lake deposits contain the remains of freshwater crustaceans and fish. Ripple marks, mud cracks, raindrop impressions, and even the footprints of early members of the dinosaur line are frequently found on the lithified surfaces of Newark sediments. Lavas, steaming up along steeply dipping fault surfaces, flowed out upon the recently deposited sediment, and volcanoes spewed ash onto the surrounding hills and plains.

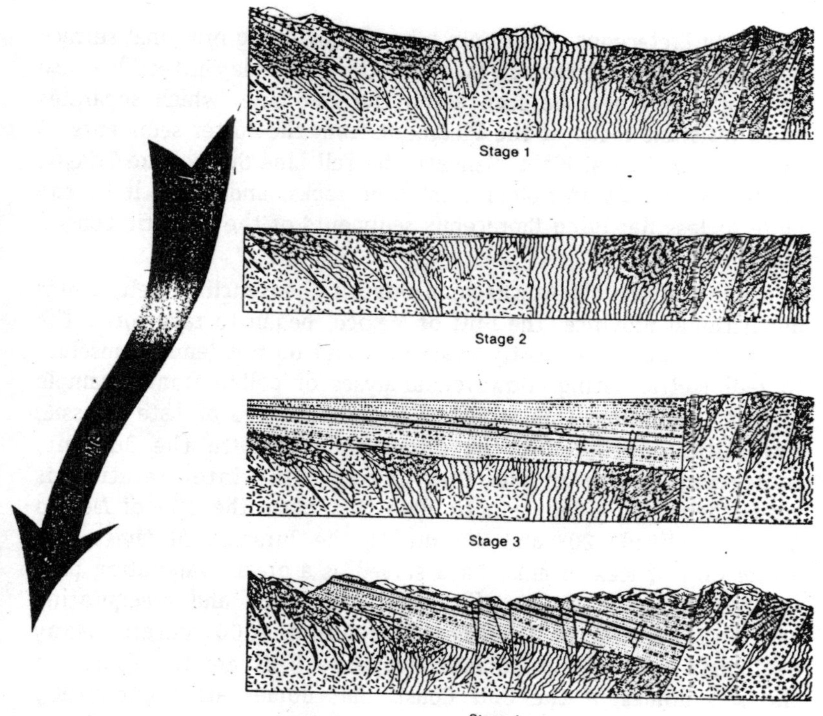

Fig. 6.3. Four stages in the Triassic history of the Connecticut Valley. (Stage 1) Erosion of the complex structures developed during the Allegheny Orogeny. (Stage 2) Mountains have been eroded to a low plain, and Triassic sedimentation has begun. (Stage 3) Newark sediments and basaltic sills, flows and dikes accumulate in troughs resulting from faulting. (Stage 4) In Late Triassic the area is broken into a complex of normal faults as part of Palisades Orogeny.

Three particularly extensive lava flows and an imposing sill are included within the Newark Group in the New Jersey-New York area. The exposed edge of a vertically jointed sill forms the Palisades of the Hudson River. Radiometric dates obtained for the Palisades basalt indicate that it solidified about 200 million years ago. Although most of the Newark beds are Triassic, recent spore and pollen studies suggest that Newark sediments in some areas accumulated during the early Jurassic as well.

By Late Triassic, the fault block mountains that had been produced as a result of rifting of Laurusia had been severely reduced by erosion. The topography was further reduced during the Jurassic and Early Cretaceous, until only a broad, low-lying erosional surface remained. That old surface is called the Fall Line Surface because its profile can best be seen along the Fall Line, which separates more resistant rocks of the Piedmont from the softer sediments of the Atlantic Coastal Plain. Beneath the Fall Line Surface are Triassic faulted sediments and older crystalline rocks, and above it lie the more or less flat-lying Cretaceous sediments of the Atlantic Coastal Plain.

South of the old Appalachian-Ouachita geosynclinal belt, a new depositional province, the Gulf of Mexico, began to take form. The initial deposits were mostly evaporites and do not lend themselves to radiometric dating. However, analyses of pollen from a sample of salt taken from this sequence suggest an age of Late Triassic. Evaporites continued to be deposited well into the Jurassic, indicating aridity in the Gulf region. This interpretation is strengthened by paleomagnetic data that place the Gulf of Mexico between latitude 20° and 30° during the Jurassic. At times, the entire Gulf of Mexico must have served as a great evaporating pan, concentrating the waters of the Atlantic Ocean and precipitating salt and gypsum to thickness exceeding 1000 meters. Many geologists believe that Jurassic evaporite beds are the source of the salt domes of the Gulf Coast. Salt domes are economically important structures associated with the entrapment of petroleum. When compressed by a heavy load of overlying strata, salt tends to flow plastically. As it moves upward in the direction of lesser pressures, the salt bends and faults overlying strata. The permeable, often faulted beds that slope downward away from the salt, or are arched over the dome, afford excellent sites for petroleum accumulation.

Evaporative conditions abated later in the Jurassic, and several hundred meters of normal marine limestones, limy muds, shales, and sandstones accumulated in the alternately transgressing and regressing seas of the Gulf embayment. These rocks and the evaporites beneath them are not exposed at the surface. They are deeply buried beneath a thick cover of Cretaceous and Cenozoic sediments. Were it not for the drilling activities of oil companies, the nature of these rocks could only be inferred from geophysical data.

The Cretaceous was a time of great marine inundations of land masses, and the Atlantic and Gulf coastal regions received their

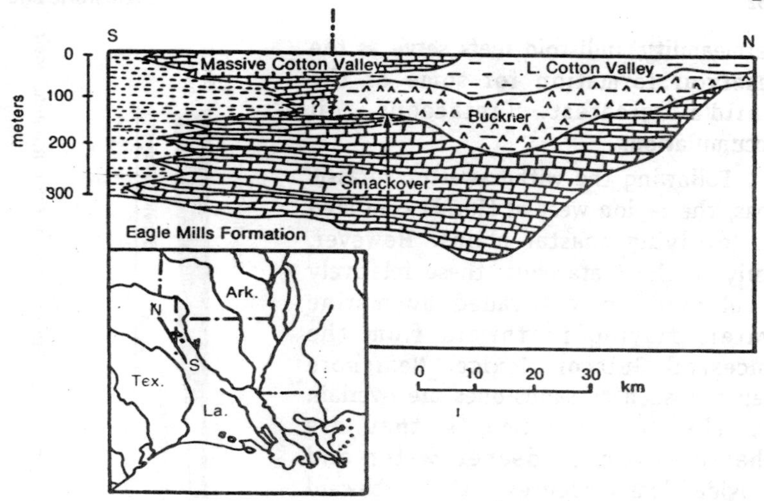

Fig. 6.4. Cross-section of Early Upper Jurassic strata in the subsurface of the northern Gulf Coast.

full share of flooding. Early in Cretaceous time, the Atlantic Coastal Plain, which had been experiencing erosion since the beginning of the Mesozoic, began to subside. At about the same time, the Appalachian belt that lay to the west was elevated. On the subsiding coastal plains, alternate layers of marine and deltaic terrestrial deposits accumulated and were gradually built into a great wedge of sediments that thickened seaward. Today, the thinner, eastern border of the wedge is narrowly exposed in New Jersey, Maryland, Virginia, and the Carolinas. But the greatest volume of Cretaceous sediments was deposited farther east, along the present continental shelf.

To the south, the Florida region was a shallow submarine bank during the Cretaceous. The *oldest* strata consist of limestones, but later in the period clastics were bought into the area by streams flowing from the southern Appalachians. Gradually, these source areas were worn down, and carbonate deposition prevailed.

The Cretaceous is noteworthy as a period of the Mesozoic in which carbonate reefs were particularly extensive. Among the invertebrates that contributed their skeletal substance to these reefs, a group of pelecypods called *rudists* were immensely important. In many Cretaceous reefs, the shells of these creatures form the basic framework of the reef. Because of their high porosity and permeability, rudistoid reefs serve as the reservoir formation for some of the world's greatest Cretaceous oil accumulations.

Following the withdrawal of Jurassic seas, the region west of Florida consisted of low-lying coastal plains. However, early in the Cretaceous, these relatively level lands were invaded by marine waters moving northward from the ancestral Gulf of Mexico. Nearshore deposits such as sandstones are overlain by the finer sediments that are characteristic of deeper water and provide clear evidence of the northward migration of the shoreline. The advance of the Cretaceous sea was not uniform. An extensive regression occurred near the end of the first half of the period; however, flooding resumed in Late Cretaceous time, when a wide seaway occupied a tract from the Gulf of Mexico to the Arctic Ocean. Among the Upper Cretaceous formations deposited in this inland sea, chalk was particularly prevalent. Chalk is a white, fine-grained, soft variety of limestone that is composed largely of the microscopic calcareous platelets (called coccoliths) of golden brown algae. It is common among beds of Cretaceous age in many other parts of the world. Indeed, the Cretaceous takes its name from *creta*, the Latin word for chalk.

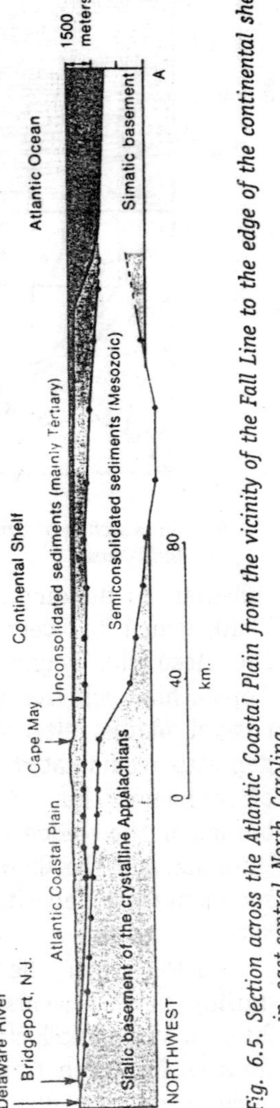

Fig. 6.5. Section across the Atlantic Coastal Plain from the vicinity of the Fall Line to the edge of the continental shelf in east-central North Carolina.

To the West

As the eastern border of North America was experiencing the tensional effects of separation from Europe and Africa, compressional forces prevailed in the west. While the newly formed Atlantic Ocean was widening, North America moved westward, overriding the Pacific Plate. Thus deformation of western North America does have a relationship to events in the east. It has been shown, for example, that the pace of tectonic activity in the North American Cordillera was most intense during the time when sea floor spreading was most rapid in the Atlantic.

During the Mesozoic, a steeply dipping subduction zone developed along the western margin of North America. The advancing Pacific Plate carried not only oceanic basalts and sea floor sediments to the subduction zone but microcontinents as well. The microcontinents have been incorporated into the Cordillera as allochthonous terrains. These terrains are distinctively coherent in age, lithology, and fossil content and are thought to be fragments of a once larger continent that existed somewhere in the Pacific Ocean Basin. Thus, the western border of North America, grew, not only by accretion of geosynclinal materials but also by the incorporation of large chunks of continental crust formed elsewhere and conveyed to our shores by sea floor spreading. At times, some of these chunks were too much for the subduction zone to swallow. They choked the subduction zone and causes shifts in its orientation.

The Cordilleran Geosyncline as it existed during the Mesozoic can be divided into a western eugeosynclinal belt containing thick volcanic and siliceous deposits and a wide eastern miogeosynclinal tract adjacent to the more stable interior of the continent. Nearly 800 meters of clastic sediments and volcanics exposed in southwestern Nevada and southeastern California attest to the instability of the eugeosyncline. Geologists speculate that the belt may well have resembled the Indonesian Island Arc of today.

Triassic

The initial orogenic event of the Cordilleran region is difficult to date precisely, although evidence for deformation at or near the Permian-Triassic boundary can be found from Alaska to Nevada. In the United States, the event has been named the *Sonoma Orogeny*. Its effects can be studied in west-central Nevada. Like the Antler Orogeny, which preceded it, the Sonoma Orogeny was caused by the collision of an island arc system with the southwestern margin of North America. It was quite a smashup, as oceanic and island arc rocks were thrust scores of kilometers eastward against the continental margin. The island arc itself became permanently welded to the western edge of the continent, thereby increasing its girth by 200 to 300 km.

The Triassic rocks of the far western part of the Cordilleran Geosyncline include great thicknesses of volcanics and graywackes presumably derived from the island arc. There is a question, however, of whether or not these rocks were deposited where they are now found or whether they are part of a displaced terrain, called the *Sonoma Terrain*, that actually originated in some unknown part of the Pacific.

Early Triassic sediments of the miogeosyncline consist of sandstones and limestones of mostly shallow marine origin. The thickest section of these marine strata occurs in southeastern Idaho,

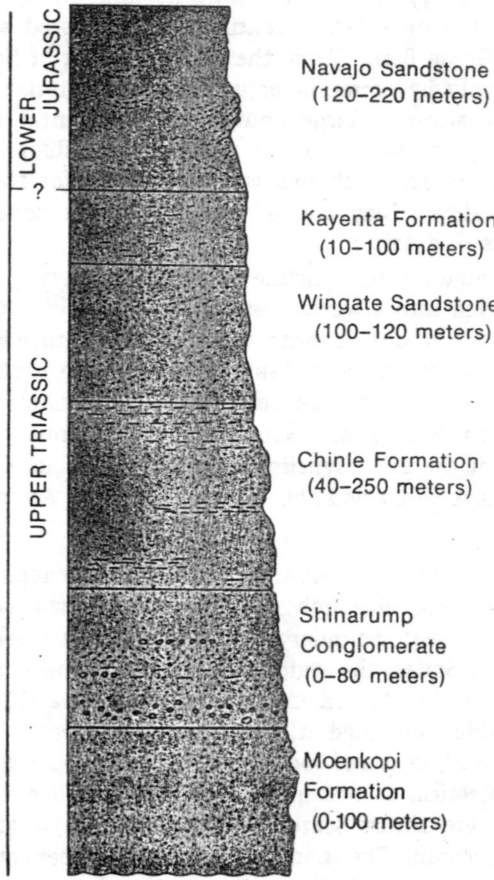

Fig. 6.6. Generalized geologic section of Upper Triassic and Lower Jurassic sedimentary rocks of central Utah.

Where nearly 1000 meters of Lower Triassic sediments accumulated. Eastward from the miogeosyncline, these marine beds interfinger with continental red beds. Although Triassic seas remained in Canada during the Middle Triassic, in the United States they regressed westward, leaving vast areas of the former sea floor subject to erosion. Upper Triassic formations rest directly above the unconformity resulting from this erosion. The Upper Triassic series consists mostly of continental deposits transported by rivers flowing westward across an immense alluvial plain. There were also upland source areas in western and southern Nevada, Arizona, and California. For example, the lowermost strata consist of the sandy *Moenkopi Formation*, followed by the pebbly *Shinarump Conglomerate*, derived from uplifted neighbouring areas in Arizona, western Colorado, and Idaho. Above the Shinarump lie the vividly coloured shales, silts, and sandstones of the *Chinle* and *Kayenta* formations. The sediments of these two formations were deposited in stream valleys and lakes. They alternate with sand dune accumulations of the *Navajo* and *Wingate* formations. The formations are beautifully exposed in the walls of Zion Canyon in southern Utah. They display sweeping cross-bedding, such as is formed in sandy deposits transported by winds. (However, such so-called "festoon cross-bedding" is not always caused by winds, for submarine dunes may be very similar in form.)

The Painted Desert of Arizona is developed mostly in Chinle rocks. The formation is known throughout the world for the petrified logs of conifers it contains. Each year, thousands of tourists examine these logs, now turned to colourful agate, in the Petrified Forest National Monument. Apparently, during times of flood, the trees were left on sand bars or trapped in log jams and covered by sediment. Percolating solutions of underground water subsequently replaced the wood with silica.

Jurassic to Early Tertiary Tectonics

Because orogenic events in the Cordillera of the western United States overlapped one another in time of occurrence and location, it is difficult to apply names to them and list them in chronologic sequence. Perhaps at the risk of oversimplification, however, it is possible to examine the general cause of the major episodes of orogeny and describe the style of deformation. The fundamental cause of Cordilleran orogenic activity during the Mesozoic was the continuing eastward underthrusting of oceanic lithosphere beneath the continental crust of the North American Plate. That underthrusting varied in rate, in inclination, and, to a small degree, in direction. It resulted in eastward-shifting phases of deformation, which initially affected the eugeosyncline on the far west, then the miogeosyncline on the east, and finally the margin of the craton itself.

The deformational and magmatic activity associated with the eugeosynclinal tract is sometimes termed the *Nevadan Orogeny*. Beginning in the Triassic, and increasing in tempo during the Jurassic and Cretaceous, sea floor sediments, graywackes, slates, cherts, and volcanics that had been swept into the subduction zone were severely folded, faulted, and metamorphosed. The intensely crumpled and altered rock sequences that were affected by compression between the converging plates are appropriately termed mélange, which means a jumble. The Franciscan Fold Belt of California provides a good example of a mélange. In addition to the deformation of the sedimentary rock sequences, enormous volumes of granodioritic rocks were generated above the subduction zone and intruded overlying rocks repeatedly during the Jurassic and the Middle and Late Cretaceous. The Sierra Nevada, Idaho, and Coast Range Batholiths are impressive reminders of the vast scale of this Nevadan magmatic activity.

Somewhat before the Sierra Nevada Batholith was emplaced in the Cretaceous, another series of deformations had begun east of the eugeosyncline. This second phase of the tectonic development of the Cordillera primarily affected miogeosynclinal rocks over a time span from Middle Jurassic to earliest Cenozoic. This new orogeny has been named the *Sevier*, and its style of deformation was distinctive. In response to compression transmitted eastward from the subduction zone, miogeosynclinal strata were sheared from underlying Precambrian rocks and broken along parallel planes of weakness to form multiple, imbricated, low-angle thrust faults. The French word *decollement* (unsticking) has been used to describe this kind of structure, in which older rocks are thrust upon younger in multiple, nearly parallel slabs of crust. It has been estimated that the compressional structures produced during the Sevier Orogeny resulted in over 100 km of crustal shortening in the Nevada-Utah region. In addition to the major thrust faults, several large folds are known, as well as intricately folded strata that were deformed within the major thrust units.

Fig. 6.7. Mesozoic batholiths in west-central North America.

Although the term Sevier is sometimes reserved for deformational episodes in the Nevada Utah region, a similar style of deformation occurred to the north in Montana, British Columbia, and Alberta. Each of the major ranges in this region is a fault block composed of Paleozoic miogeosynclinal strata that have been thrust toward the east along westwardly dipping fault surfaces. The most famous of these faults is the Lewis Thrust, along which Proterozoic rocks of the Belt Group were carried 65 km toward the east.

Magmatic activity along the far western edge of the North American Plate had diminished near the end of the Cretaceous, and by Early Tertiary much of the major thrusting of the miogeosynclinal region had subsided. Once again, deformational events shifted eastward to the cratonic region where now are located the Rocky Mountains of New Mexico, Colorado, and Wyoming. These more eastwardly disturbances are sometimes called the Laramide Orogeny. Although thrust faults do occur in the region of Laramide activity, the more characteristic kind of structures are broadly arched domes, basins, and anticlines. Strata composing the domes and anticlines are draped over central masses of Precambrian igneous and metamorphic rocks. In several instances, erosion has stripped away the cover of strata, exposing the central crystalline mass. Resistant layers of inclined beds that surround the central cores stand as mountains today. Many of these uparched areas appear to have been produced by movements along underlying faults in the basement rocks. It has been suggested that these controlling structures were in turn developed by drag when the eastward-moving subducted oceanic lithosphere scraped along the sole of the cratonic margin of North America.

Most of the structures of the present Rocky Mountains are the result of the Laramide phases of orogeny. The landscape we see today, however, is the result of repeated episodes of Cenozoic, erosion and uplift. Erosion, acting on the geologic structures already present, sculptured the final scenic design.

Jurassic Sedimentation

During the Jurassic, when the mélange was being formed in the eugeosyncline, the miogeosyncline was the site of a less dramatic kind of sedimentation. Early Jurassic deposits consist of clean sandstones, such as the *Navajo Sandstone*. Large-scale cross-bedding is well developed in the Navajo. However, thin beds of fossiliferous limestones and evaporites occur locally and indicate that at least parts of the formation were deposited in water. Indeed, recent interpretations suggest that much of the cross-bedding in the Navajo resulted from water currents that formed sand dunes on the floors of shallow marine areas. The Navajo and associated sand bodies were probably deposited in a nearshore environment, and it is likely that some of the deposits were part of a coastal dune environment. They do not seem to have been laid down on the floors of a vast interior desert, as has been often postulated.

Judging from studies of cross-bedding orientations, the source area for these clean sandstones was probably the Montana-Alberta craton. Somewhat older but similar quartz sandstones were recycled and spread southward into Wyoming and Utah, and then westward across Nevada.

Marine conditions became more widespread in the Middle Jurassic, when the entire west-central part of the continent was flooded by a wide seaway that extended well into central Utah. This great embayment has been dubbed the *Sundance Sea*. Its deposits, largely derived from the Mesocordilleran Highlands, are spread widely across the Northern Great Plains, where they make up the Sundance Formation. The *Mesocordilleran Highlands* were upland tracts trending north-south roughly along the border between the miogeosyncline and eugeosyncline. The highlands first appeared late in the Triassic and by Late Jurassic had grown into a persistent chain of islands and uplands that stretched from Alaska to Central America. On the flanks of the highlands, the fluctuating nature of the Sundance Sea is indicated by interfingering of its marine strata with land-laid deposits. Apparently, the sea was never able to connect with contemporary embayments that spread northward from the ancestral Gulf of Mexico. Gradually, near the end of the Jurassic, the seas withdrew, leaving a vast, swampy plain across which meandering rivers built wide floodplains of mud, gravel, and silt. These deposits compose the *Morrison Formation*, which extends across millions of square kilometers of the American West. Enclosed within these floodplain deposits are the bones of more than 70 species of dinosaurs, including some of the largest land animals ever to have existed.

Cretaceous Sedimentation

During the Cretaceous, the Pacific border region of North America was a land of lofty mountains formed during the orogenesis that had begun in the Jurassic. Erosion of these ranges brought sediments downward into adjoining rapidly subsiding basins, many of which were open to the Pacific Ocean. In some places, over 15,000 meters of volcanics and clastics accumulated. Cycles of folding and intrusion occurred repeatedly, with rather exceptional deformations taking place during Middle and then Late Cretaceous time.

Eastward of the eugeosyncline, the Cretaceous began with the advance of marine waters both northward from the ancestral Gulf of Mexico and southward from arctic Canada. These Early Cretaceous invasions did not meet, so that an area of dry land existed in Utah and Colorado. Their advance was reversed by a general regression that resulted in the unconformity used in separating the Cretaceous into an early and late division. The flooding that followed in the Late Cretaceous was the greatest of the entire Mesozoic Era. This time, the embayment from the north joined with the southern seaway and effectively separated North America into two large islands.

Sedimentation in Cretaceous epicontinental seas was largely controlled by local conditions. Along the Gulf Coast, limestones and marls (clayey limestones) accumulated. North of Texas, however, the great bulk of sediments consisted of clastics supplied by streams flowing off the western highlands. Typically, these deposits are sandstones that gradually thin and grade eastward into shales. Sandstones of the *Dakota Group* are particularly well known. The dark brown sandstone layers of the Dakota are exposed in many places along the eastern front of the Rocky Mountains, where the inclined beds form prominent ridges called hogbacks. In parts of the Great Plains, the Dakota sandstones are an important source of underground water.

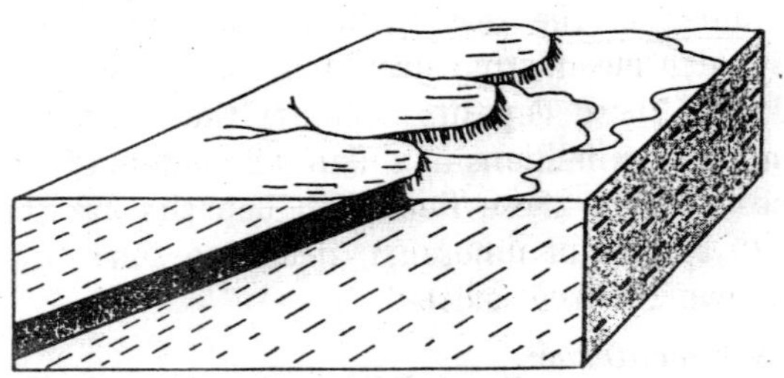

Fig. 6.8. A hogback ridge.

The carbonate formations of the miogeosyncline include soft, clayey limestones and chalky shales of the *Niobrara Formation*. The Niobrara has yielded the remains of a variety of marine creatures, including enormous numbers of oysters, a giant Cretaceous diving bird (*Hesperonis*), marine reptiles, and the large flying reptile *Pteranodon*. Toward the end of the Cretaceous, the seas that supported these creatures began a slow withdrawal. This regression of the Cretaceous epicontinental sea from central North America was contemporaneous with the Laramide deformation that produced the Rocky Mountains. Coal-bearing deltaic and other continental sediments initiated a new phase of sedimentation on the old sea floor.

To the North

Because of the problems inherent in attempting geologic work in frigid polar climates, data on the geology of the northern geosyncline have been limited. However, within the last decade, the exploitation of petroleum reservoirs in Alaska has contributed new information about some areas along the northern margins of the continent.

For the most part, Mesozoic sediments accumulated in two basins along the northern geosynclinal belt: the Brooks Basin of northern Alaska and Sverdup Basin farther east. The Mesozoic record for the Brooks Basin begins with Triassic clastics and limestones, some of which are fossiliferous. Jurassic and Cretaceous beds consist of volcanic rocks, sandstones, and shales.

Mesozoic formations of the Sverdrup Basin include both marine and nonmarine sandstones and shales. Some of the formations contain petroleum. Sandstones and siltstones predominate within the Triassic system. Similar clastics, interspersed with coal beds, were deposited during the Jurassic. Volcanic rocks, continental clastics, red beds, and additional coal beds characterize the Cretaceous rock record.

Eurasia And The Tethys Geosyncline

Extending eastward from Gibraltar and across southern Europe and Asia to the Pacific is the great Alpine-Himalayan mountain belt. Most of the rocks composing the ranges within the belt were laid down in an important geosynclinal trough known as the *Tethys*. Long after their deposition, these rocks were deformed into mountain ranges as northward-moving segments of Gondwanaland collided with Eurasia. Here and there, however, the waters remained and persist today as the Mediterranean, Caspian, and Black Seas.

The depositional history of the Tethys Geosyncline can be traced well back into Early Paleozoic time. There is even some evidence of crustal mobility in parts of the belt as long ago as Late Precambrian. However, just as most of the depositional and tectonic activity for the Appalachian Geosyncline transpired within the Paleozoic, for the Tethys, the Mesozoic and Cenozoic are the most eventful eras.

During the Triassic Period, the Tethys was the site of limestone deposition. In various locations, marine invertebrates flourished.

Pelecypods, crinoids, reef corals, and algae are well represented in the fossil record. Of particular importance are the abundant Triassic ammonoid cephalopods found in the Alpine region, because these mollusks are the principal fossils used in Triassic correlation. North of the Tethys, a highland area known as the Vindelician Arch developed and separated the geosynclinal depositional environment from the quite different sedimentation of north-central Europe. In that northern region, the Triassic record begins with reddish, nonmarine clastic sediments that resemble there deposited under arid conditions during the preceding Permian Period. These largely fluvial deposits are overlain by shoreline sands, marls, evaporites, and limestones deposited during a temporary marine invasion of central Europe. The sea did not remain long. Nor was it able to reach as far north as Great Britain, where a nonmarine sequence known as the New Red Sandstone was being deposited. As the sea withdrew, continental sedimentation much like that of the earliest Triassic resumed.

Marine conditions were far more widespread during the Jurassic than they had been during the Triassic. Shallow seas spilled out of both the Tethys and the Atlantic and the Atlantic and spread across Europe, leaving a rich sedimentary record in the basins that lay between the old Hercynian uplands. Although the predominant Jurassic sediment is limestone, there is often a gradation to fine clastics adjacent to the highlands. Eventually, marine conditions extended from the Tethys across Russia and into the Arctic Ocean. The invasion was short-lived, however, and the shallow seas gradually drained from the continent by the end of the Jurassic. Marine conditions persisted in the Tethys, and a complicated pattern of facies changes developed in response to blockfaulting along the floor of the geosyncline.

The Cretaceous is the most eventful geologic period of European Mesozoic history. During this time, northward-moving Africa began to close the gap that was the Tethys Sea. The Alpine Geosyncline was subjected to an episode of powerful compressional folding. As evidence of this event, a zone of *ophiolites* can be traced along the southern margin of Europe. Ophiolites are greenish igneous rocks believed to have once been part of the crust of the ocean that existed between Gondwanaland and Laurasia. The ophiolites are found in association with radiolarian cherts and lithified deep sea oozes, which were very likely also part of that sea floor. The movement of the oceanic crust beneath Eurasia as the African Plate approached created the ophiolite zone as well as the deformation and volcanism that characterized the southern margin of Eurasia during the Cretaceous.

The Late Cretaceous is noteworthy not only for its mountain-building events but also as a time of extensive marine transgressions. More of Europe was inundated during the Late Cretaceous than at any other time since the Cambrian. A great embayment from the Tethys worked its way northward until it ultimately joined with a similar southward encroachment from the region of the North Atlantic. One result of the linking of the two seaways was a rapid interchange of marine organisms from the two formerly separated regions. In the Tethys-Alpine region, the blockfaulting that had begun during Late Jurassic continued until about Middle Cretaceous. At that time, folding became the predominant structural style. The axes of the anticlines tended to parallel the east-west trend of the Tethys. Most of these great welts remained submerged, but the tops of some of the folds rose above the level of the waves, forming elongate islands along the north side of the narrowing Tethys seaway. Erosion of the anticlinal islands produced clastic debris that was periodically swept downward by turbidity currents into the adjoining basins, to be deposited as a thick sequence of interbedded sandstones and shales. These sediments and associated volcanic rocks foretell the coming of tumultuous events during the Cenozoic.

An event that determined the geographic configuration of the northern coastlines of Spain and France occurred in Europe between Late Jurassic and Late Cretaceous time. As indicated by paleomagnetic data, there was no Bay of Biscay separating eastern France from Spain prior to the Late Jurassic. The Bay of Biscay opened as the Iberian Penninsula (Spain and Portugal) rotated about 35° from its former line of contact with France. The upheaval of the Pyrenees was another consequence of this rotation. The dating of the rotational event is confirmed by faults of Late Jurassic age on the northern border of the Bay of Biscay and by the fact that the oldest sediments in the bay are Late Cretaceous in age.

East of the Alpine region, the Tethys extended as a great loop past Burma and southeastward into Indonesia. The sediments of the Triassic and Jurassic of the eastern Tethys consisted of shales, fossiliferous limestones, and dolostones. During the Cretaceous there was volcanic activity throughout what is now the Himalayan region. Some areas seem to have experienced marked subsidence, and clastic sediments resembling those moved by submarine landslides or turbidity currents merged with volcanic tuffs, breccias, and radiolarian cherts.

Gondwanaland Continents

Africa

By the beginning of the Triassic, separations had already developed between the eastern coast of America and northwest Africa. However, there was as yet no rift between Africa and South America, and therefore there was no South Atlantic Ocean. Most of the portion of Gondwanaland that we now call Africa was relatively stable throughout the Mesozoic, with only relatively minor transgressions along the northern and eastern borders. During the Cretaceous, the Tethyan waters spread across what is today the northern tier of countries from Algeria to Egypt. Fine muds and carbonates were the most prevalent deposits.

Although the Mesozoic rock record in the more interior parts of Africa is often poor, some regions do yield particularly interesting geologic information. One such region is the Karoo Basin

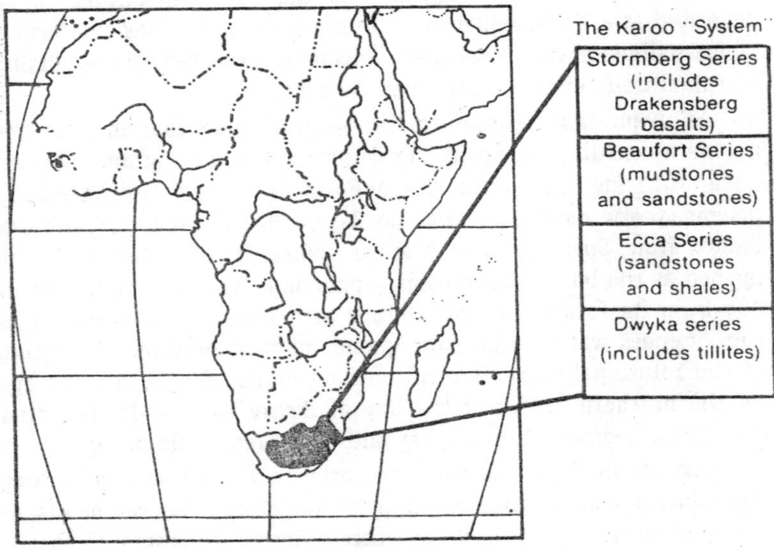

Fig. 6.9. Location of the Karoo Basin of South Africa and the four principal units of the Karoo "System."

at the southern end of Africa. The Karoo was a continental basin that was formed late in the Carboniferous and received swamp, lake, and river deposits until late in the Triassic Period. Rocks of the Karoo are known to paleontologists around the world for their wealth of fossilized mammal-like reptiles. The fauna included large and small herbivores, diverse carnivores, and a range of insectivores and omnivores. As the Triassic drew to a close, the reddish silts and sands of the Karoo sequence were covered by flow after flow of lowviscosity lava, which gushed onto the surface from fissures and, less commonly, from volcanoes in the southeast. The old Karoo landscape was buried beneath more than 1000 meters of basalt. Geologists speculate that these great outpourings of lava were associated with the pulling away of the Gondwanaland segments that once adjoined South Africa. Such fragmentation would very likely have caused severe fracturing of the foundations of the continents and provided multiple avenues for the upwelling of the molten rock. The extrusions continued well into the Jurassic Period. Contemporaneous lava floods and volcanism also occurred on the separating land masses of South America, Australia, and, somewhat later, Antarctica.

Africa appears to have been more at peace with itself throughout the remainder of the Jurassic and the Cretaceous. Marginal seas extended along the eastern edges of the continent. Periodic advance and retreat of these seas brought an alternation of marine and nonmarine beds.

Australia and New Zealand

During the Triassic, Australia was still attached to Antarctica but was pulling northward. In this earliest Mesozoic period, both Australia and Antarctica were predominantly emergent. Terrestrial sandstones and shales resembling those of the Permian were deposited. Except for brief marginal embayments, deposition of lake and stream sediments continued to dominate, until the middle of Early Cretaceous. At that time, a marine transgression brought sandstones and chalk beds into the interior.

During the Mesozoic, New Zealand was geologically far more restless than Australia. It lay astride the Pacific mobile belt and thus experienced deposition of enormous thickness of Mesozoic geosynclinal sediments and volcanics. Near the end of the Jurassic, a series of orogenic events occurred as a result of the collision and subduction of the Pacific Plate beneath the island continent.

In the course of the orogeny, mountains were erected, and deep, unstable trenches developed. The activity continued into the Cenozoic and, intermittently, even to the present day.

India

During the Mesozoic, India moved steadily northward on its remarkable voyage to Laurasia. The passage was for the most part quiet, but it was to have a tumultuous ending. The land mass experienced only relatively minor marginal incursions of the sea. Across the interior, erosion of upland areas spread terrestrial clastic sediments into the lowlands and plains. Some of these continental formations have yielded a wealth of dinosaur bones and superb fossils of plants. By Cretaceous time, India was nearing the Tethyan Geosyncline. Here and there, the floor of the Tethyan trough was buckled into elongate ridges that were early indications of the coming tectonic storm. While these ridges were developing, the northwestern half of India was flooded with immense quantities of low-viscosity basaltic lava. These now solidified lavas are known as the *Deccan Traps*; they are also termed "flood basalts." Radiometric dates obtained from the basalts indicate that the outpourings continued from Cretaceous time well into the Early Cenozoic. It is likely that these basalt floods record the passage of India across a fixed "hot spot" in the mantle.

South America

Much like Africa, South America stood well above sea level at the beginning of the Mesozoic. Along the western margin of the continent, a lengthy geosyncline was already in existence during the Triassic. Not unexpectedly, it was differentiated into a eugeosyncline on the Pacific side and a miogeosyncline along the edge that bordered the craton. Today the boundary between these two tracts would lie along the central ranges of the Andes. In some areas a geanticline similar to the Mesocordilleran Highlands of North America lay between the eugeosynclinal and miogeosynclinal tracts. Graywackes, conglomerates, siliceous sediments, and lavas accumulated in the eugeosynclinal belt, whereas carbonates and shales predominated in the neighbouring miogeosyncline. In addition to these marine tracts, broad basins in the interior accumulated continental deposits. Particularly remarkable are the eolian and fluvial sands and silts that spread across the southeastern region of the continent during the Triassic.

The beds contain a rich fauna of Triassic vertebrates. Lava flows, chronologically equivalent to those of the Karoo rocks in Africa, are also prevalent in the upper part of this stratigraphic sequence.

By Jurassic time, the initial narrow split between South America and Africa had widened into a configuration resembling the present-day Red Sea. New sea floor formed along the spreading center of the developing South Atlantic. The western side of South America was at the leading edge of the westward-moving tectonic plate and was being underthrust by an opposing oceanic plate, much as was the case in the far west of the United States.

Near the end of the Jurassic, the Andean belt experienced deformation and volcanism. This activity appears to have begun in the south and then sporadically shifted with time toward the north. The climax of orogenic deformation occurred in Late Cretaceous and Early Tertiary and was accompanied by regional uplift that led to widespread withdrawal of the seas. Here and there, evaporites were deposited in isolated basins that for a time retained marine waters.

The frequency and intensity of deformational and volcanic activity along the Andean Geosyncline increased during the Cretaceous. The subduction zone (created by the movement of the Pacific Plate against South America) received both Pacific floor sediments and older, crustal materials from the continents. These materials were melted at great depths. The melts in turn worked their way back toward the surface as great intrusions and outpourings of andesitic basalt. Deformation and igneous activity continued well into the Early Cenozoic and formed many of the structures now seen in the towering peaks of the Andes.

Antarctica

Antarctica was predominantly emergent throughout the Mesozoic. Eastern areas of the now frigid land mass were sites of continental deposition. Beds of volcanic ash and lava flows occur frquently betwnee these lake and stream deposits.

The nature of sedimentation was markedly different in western Antarctica. Outcrops along the Antarctica Peninsula consist of volcanic and clastic formations that are very similar to those of equivalent age in the Andes. Very likely, the strata of the Antarctica Peninsula represent part of a continuous mobile belt that extended up through the Scotia Arc and along the western margin of South America.

Economic Resources

Uranium Ores

Rocks of the Mesozoic System, like those of earlier eras, contain a wealth of important mineral resources. Notable among such resources are the nuclear fuels. In the United States, uranium ores are derived chiefly from continental Triassic and Jurassic rocks of New Mexico, Colorado, Utah, Wyoming, and Texas. The chief ore mineral, known as *carnotite*, was deposited in the form of uranium salts within the pore spaces of fluvial sandstones. Apparently, the abundance of organic material in the sediment enhanced the precipitation of the uranium salts. In some instances, petrified logs have provided amazing concentrations of not only uranium but also vanadium and radium. These are striking exceptions, however, and the radioactive ores are mostly of very low concentration.

There is at present concern that the richer sources of fissionable uranium 235 in this country are likely to be exhausted by about the same time as our petroleum reserves are depleted. Among the alternate sources of nuclear energy now being studied are uranium 238 and thorium 232, which can be converted to fissionable isotopes by a process termed *breeding*. This will permit the use of the more abundant low-grade uranium ores. Of course, breeder reactors are not yet perfected and may not prove sufficiently safe or as enduring as other possible energy sources.

Until either the breeder reactor or devices to utilize solar energy are perfected, the nations of the world will continue to rely heavily on fossil fuels. Mesozoic rocks are hosts for these critical resources also. The Jurassic, for example, is an important coal-producing system. The larger foreign mines are located in Siberia, China, Australia, Tasmania, and Spitzbergen. In North America, thick seams of Jurassic coal occur in British Columbia and Alberta. Cretaceous coal underlies more than 300,000 sq km of the Rocky Mountain region. Much of this coal is now being vigorously mined, particularly because of its relatively low content of environmentally offensive sulfur.

Fossil Fuels

Mesozoic rocks in certain favourable areas around the world supply large quantities of oil and gas to energy-hungry industrial nations. The oil provinces of the Middle East and North Africa probably contain more oil than the combined reserves of all other countries. Middle East petroleum comes primarily from thick sections of Jurassic and Cretaceous sediments that had accumulated in the Tethys seaway. Other areas of petroleum production from Mesozoic rocks occur in the Rocky Mountains, Alaska, Arctic Canada, the Gulf Coastal states, western Venezuela, southeast Asia, beneath the North Sea, and beneath the eastern offshore area of Australia. Not withstanding the size of some recent discoveries in rather forbidding parts of the world, it appears likely that the present rates of oil and gas consumption will cause exhaustion of the world's resources in less than a century. Therefore, oil and gas must be replaced by other energy sources in the near future, especially if we are to conserve these valuable materials for the chemical industry.

Ores

Metalliferous deposits were formed widely throughout the active orogenic belts of the Mesozoic world. A variety of metals now found in the Rocky Mountains, the Pacific states, and British Columbia were emplaced during batholithic intrusions that accompanied Jurassic and Cretaceous mountain building. Among these are deposits were the gold-bearing quartz veins known as the "Mother Lode." The California gold rush of 1849 was a consequence of the discovery of gold-bearing gravels eroded from the Mother Lode. The copper, silver, and zinc veins of Butte, Montana, and Coeur d' Alene mining districts were emplaced as a result of Cretaceous igneous activity. One belt of porphyritic copper-bearing rock extends from Denver to the Four Corners area. The zone has become known as the "porphyry copper belt." The intrusions that produced the belt range in age from Cretaceous to Early Cenozoic. Triassic rocks yield copper in Germany and Russia. Not all the ore deposits resulted from igneous activity. In England and Alsace-Lorraine, there are important Jurassic iron ores that are sedimentary in origin.

Nonmetallic deposits of the Mesozoic include sulfur and salt. Both are produced from the salt domes mentioned earlier. Diamonds are also obtained from Mesozoic igneous rocks. Siberia's diamond-bearing intrusions are believed to have penetrated the upper crust during the Triassic and Jurassic. Similar "diamond pipes" in Africa are probably of Cretaceous age.

Plate Tectonics and Ore Deposits

In recent years, geologists have begun to see numerous correlations between patterns of mineral distribution and locations of present and former tectonic plate boundaries. Plate boundaries are likely sites for movements of hot aqueous fluids that bear important metals in solutions. With changes in temperature or pressure, oı on contact with reactive rocks, such hydrothermal solutions will precipitate their dissolved wealth. The process may operate at both convergent and divergent plate boundaries. Both Cordilleran and Andean mobile belts are examples of convergent boundaries. An example of hydrothermal mineralization at a divergent boundary is provided by the Red Sea, which has "pools" of very hot and exceptionally salty water along its bottom. The pools are rich in iron, manganese, zinc, and copper. The brines apparently have percolated upward through the young oceanic crust, dissolving the metals enroute. Metals brought to the surface in this way in the past may have been conveyed in thin layers across immense tracts of the ocean floor by sea floor spreading. Sediments containing metallic ions extracted from the sea water itself may have enriched the accumulation. Ultimately, sea floor spreading would move collision with another plate, providing an opportunity for their inclusion as ore bodies within the deformed belt.

Of course, the formation of ores in orogenic belts is a far more complicated process than this conceptual model suggests. Every ore body requires very special local conditions and events for its emplacement. Many ore deposits have no relationship to plate boundaries. However, in the case of the metallic deposits of North America and South America, a relationship is possible, for both of these regions form overriding convergent boundaries with the plates of the Pacific.

7

Mesozoic Biosphere

The *biosphere* is the world of life. We can view the biosphere as consisting of all living things and those parts of the continents, oceans, and atmosphere with which living things interact. In this chapter we direct our attention to the biosphere of the Mesozoic Era. As is true today, the distribution, evolution, and abundance of life during the Mesozoic were strongly influenced by climate. Climate, in turn, was affected by the changing locations of continents, major marine transgressions and regressions, and the formation of mountain ranges.

Mesozoic Climates

Cool climates seem to have characterized many continental areas during the final days of the Paleozoic Era. The vastness of the Pangaea II supercontinent, the upheaval of mountains, general uplifts, and withdrawal on inland seas in many regions were contributing causes of the generally cooler conditions. Gradually, however, the climate warmed, and glaciers in Africa, Australia, Argentina, and India began to melt away as these continents began to drift away from the south pole. In general, the 160 million years of the Mesozoic seemed to have been blessed with warm and rather equable climates.

During the Triassic the continents were still tightly clustered. The paleoequator extended from central Mexico across the northern bulge of Africa. The Triassic was a time of general emergence of the continents. Mountains, thrust upward at the end of the Paleozoic, inhibited the flow of moist air into the more centrally located regions, causing widespread aridity. Evaporites, dune

Fig. 7.1. Triassic paleogeographic map showing position of equator and land masses, as well as the distribution of evaporites and coal.

sandstones, and red beds accumulated at both high and low latitudes and attest to relatively dry and warm conditions.

Reconstructions, based in part on paleomagnetic studies, suggest that during the Jurassic the continents were at the approximate latitudinal positions they occupy today. Marine waters extended northward into a great trough formed by the opening of the Atlantic Ocean, and in many places shallow inland seas spilled out of the deep basins onto the continents. An arm of the Pacific Ocean extended westward as the great Tethys Sea. It is not unreasonable to assume that warm westward flowing equatorial currents may have penetrated far into the Tethys. These same equatorial currents, deflected by the east coast of Pangaea II, were shunted to the north along coastal Asia and to the south along

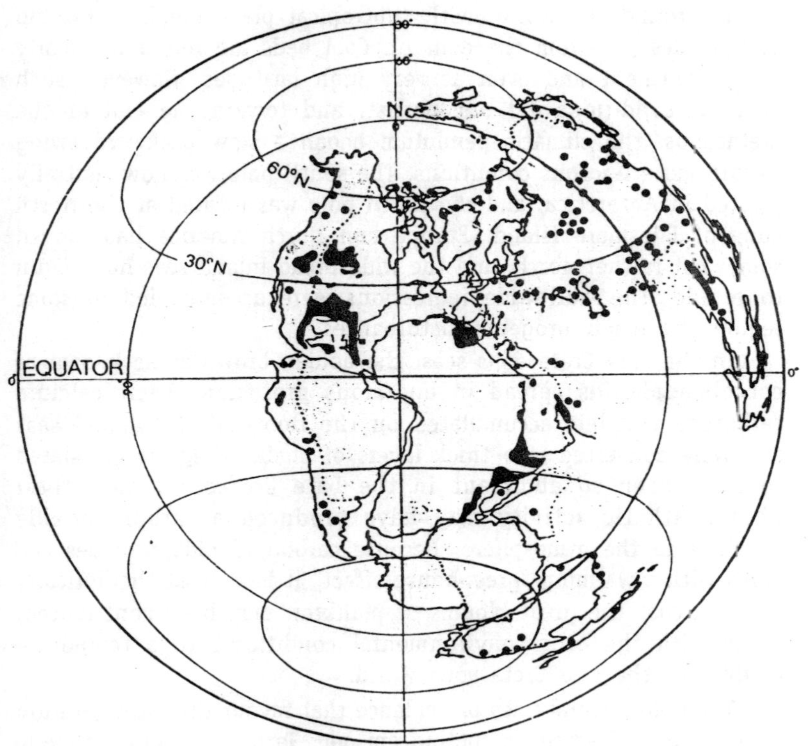

Fig. 7.2. Jurassic paleogeographic map showing position of equator and land masses, and distribution of evaporites and coal.

northeastern Africa and India. To complete the cycle, the cooled currents may have returned to the equator along the west side of the Americas. The presumed ocean and wind currents, and the rather extensive coverage of seas, both upon and adjacent to continents, perpetuated the mild climates that seem to characterize the Jurassic. Glacial deposits of this age are simply not known, and coal beds occur in Antarctica, India, China, and Canada. Paleobotanical evidence suggests that tropical conditions prevailed over regions that are now in temperature zones. Even dinosaurs lived in temperate areas somewhat north of the present arctic circle.

During most of the Cretaceous, climatic conditions were a continuation of the generally warm and stable conditions that had prevailed during the Jurassic. A remarkably homogeneous flora spread around the world, with subtropical plant families thriving in latitudes 70° from the equator. Coal beds are found on nearly every continent and even at very high latitudes. However, such pleasant conditions did not persist, and toward the end of the Cretaceous, the climatic pendulum began a slow backward swing toward more rigorous conditions. The south pole was now centrally located in Antarctica, and the north pole was located at the north edge of Ellesmere Island. Europe and North America had moved somewhat farther north, and the widespread inland seas had begun to recede. The worldwide regressions were accompanied in some regions by major orogenic disturbances.

In the Late Cretaceous seas, the golden brown algae known as coccolithophorids spread in enormous profusion. Their calcium carbonate platelets accumulated on the floors of the inland seas and were converted into thick layers of chalk. They are estimated to have been so abundant in the Late Cretaceous that their photosynthetic activity may have produced a carbon dioxide shortage in the atmosphere. Because carbon dioxide provides the earth with a warming "greenhouse effect" it is at least theoretically possible that the great blooms of plankton may have contributed, along with the other environmental conditions, to a temporary chilling of the late Cretaceous world.

There are several lines of evidence that favour the interpretation of a terminal Cretaceous cooling episode. Terrestrial plants provide some clues. The tropics-loving cycads underwent a sharp reduction, and ferns declined in both North America and Eurasia. Hardier plants, such as conifers and angiosperms, extended their realms. Evidence comes also from paleotemperature studies based on the oxygen isotope method described earlier. The oxygen isotope ratios obtained from open ocean planktonic calcareous organisms indicate a decline in ocean temperatures beginning about 80 million years ago. If there was indeed a dip in world wide annual mean temperatures, it might have had a deleterious effect on animal life. For this reason, some paleontologists speculate about its relationship to the extinctions that occurred at the end of the Mesozoic.

The Mesozoic Flora

The existence of animal life on earth is ultimately dependent on plant life. This generalization is valid today, and it was equally valid during the Mesozoic. Then, as now, plants made up the broad base of the food pyramid. Their nutritious starches, oils, and sugars made possible the evolution and continuing existence of animals. Plants are a fundamental part of the most important self-sustaining ecologic system on earth. The operation of the system is dependent upon oxygen and carbon dioxide. Animal respiration provides the carbon dioxide needed for plant photosynthesis, whereas plants supply—as a by-product of photosynthesis—the oxygen needed by animals. In the geologic past, variations in plant productivity may have caused corresponding changes in the amount of carbon dioxide and oxygen in the atmosphere. Such variations may have favoured the evolution of some organisms over others and may even have been responsible for the demise of particular groups.

Marine Plants

Because plants that live in the oceans are suspended in water, they do not require the vascular and support systems that characterize land plants. The majority of marine plants are therefore unicellular, although they may grow in impressive colonies and aggregates. The major groups of marine plants are part of that vast realm of floating organisms called plankton; because they are plants, they are further designated *phytoplankton*. As indicated on

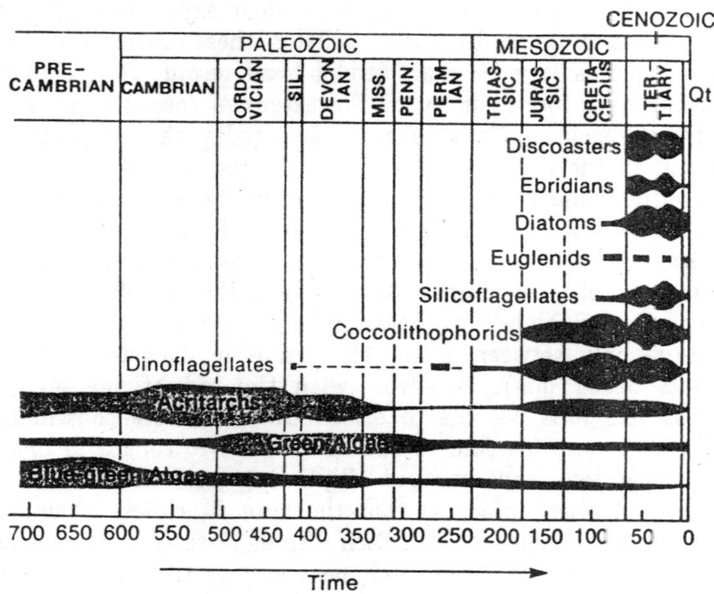

Fig. 7.3. Geologic distribution and abundances of phytoplankton.

the chart, phytoplankton that did not secrete mineralized coverings predominate in the pre-Mesozoic eras. These include both blue-green and green algae, as well as a group of cellulosecovered unicellular algae of uncertain affinity that are loosely called acritarchs. Of more immediate importance to us are the coccolithophorids, dinoflagellates, silicoflagellates, and diatoms, which were the more abundant phytoplankton groups of the Mesozoic. The expansion of the coccolithophorids and dinoflagellates really began in the Early Jurassic.

Dinoflagellates are frequently encountered as fossils and are important aids in Mesozoic and Cenozoic stratigraphy. From the Jurassic on, they were among the primary producers in the marine food chain. Dinoflagellates are unicellular organisms having a cellulose cell wall like that in pollen. For propulsion, the cell wall is equipped with two flagella: One is longitudinal and whiplike, and the other is transverse and ribbon-like. During their life cycle, dinoflagellates develop a motile plankton form and a cyst phase that is formed within the motile organism. The cyst form is extremely resistant to decay, and, indeed, only dinoflagellate cysts are known as fossils.

The coccolithophorids also began their expansion during the Early Jurassic. Unlike the dinoflagellates, these calcium carbonate–secreting organisms have a splendid fossil record. Their abundant remains have formed many of the extensive coccolith limestones of the Mesozoic and Early Cenozoic. Even today, they are frequently present in the deep sea sediment known as calcareous ooze. The coccolithophorid organism is one of several varieties of unicellular golden brown algae. These algae deposit calcium carbonate internally on an organic matrix and construct tiny, shieldlike structures called coccoliths. Once formed, the coccoliths move to the surface of the cell and encase it in calcareous armor. Coccoliths measure only between 1 and 16 microns in diameter (1 micron equals 0.001 mm). However, with the aid of the electron microscope, it is possible to discern their intricate construction. With such magnification, coccoliths are seen to consist of one, or sometimes two, superimposed elliptical plates. These plates are concave on one surface so that they can fit closely around the outside of the spherical cell. Each disc or plate is itself composed of still smaller crystalloids, and these may be triangular, rhombic, or variously shaped by the organism. The crystalloids are uniformly arranged, usually in a circular, radial, or spiral plant that is often astonishing in its beauty and precision. Because they are frequently fossilized, have experienced frequent evolutionary changes through time, and are readily dispersed by oceanic currents, coccoliths have become extremely useful in stratigraphic correlation.

The earliest known silicoflagellates and diatoms appeared in the Middle Cretaceous. Along with other phytoplankton, they experienced a decline at the end of the Cretaceous, and then all groups expanded again into the early epochs of the Cenozoic. The silicoflagellates and diatoms are, along with the coccolithophorids, members of the Phylum Chrysophyta.

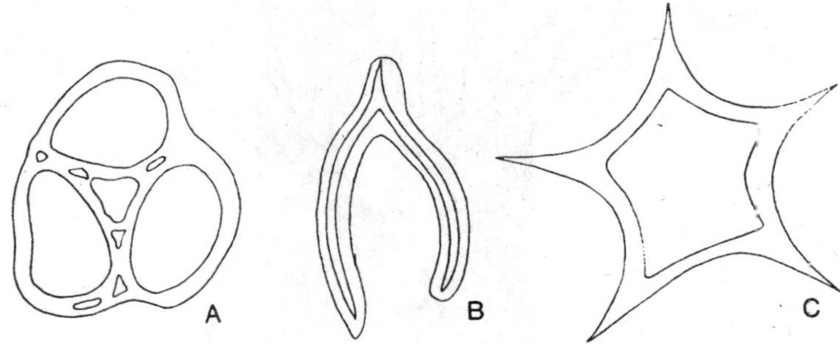

Fig. 7.4. Three Mesozoic silicoflagellates. A—Corbisema (Cretaceous-Eocene). B—Lyramula (Cretaceous). C—Vallacerta (Cretaceous).

As suggested by their somewhat ponderous name, the silicoflagellates are flagella-bearing algae that secrete a siliceous internal "skeleton." Radiating spines characterize most genera, whereas others are roughly stellate. They range in size from 10 to 150 microns.

Like the silicoflagellates, diatoms also secrete siliceous covering. The covering is called a *frustule*. It is usually composed of an upper and lower part that fit together like a lid on a box. The test may be circular, cylindric, triangular, or a variety of other shapes and is usually quite beautiful. Today, marine diatoms are most prevalent in the cooler regions of the oceans. In the past, a proliferation of diatoms was often associated with volcanic activity. Apparently, the silica supplied to sea water as fine volcanic ash stimulated diatom productivity.

Terrestrial Plants

For the vertebrate paleontologist, the "Age of Dinosaurs" may seem a splendid way to designate the Jurassic Period. However, a poleobotanist might well argue that the "Age of Cycads" would be

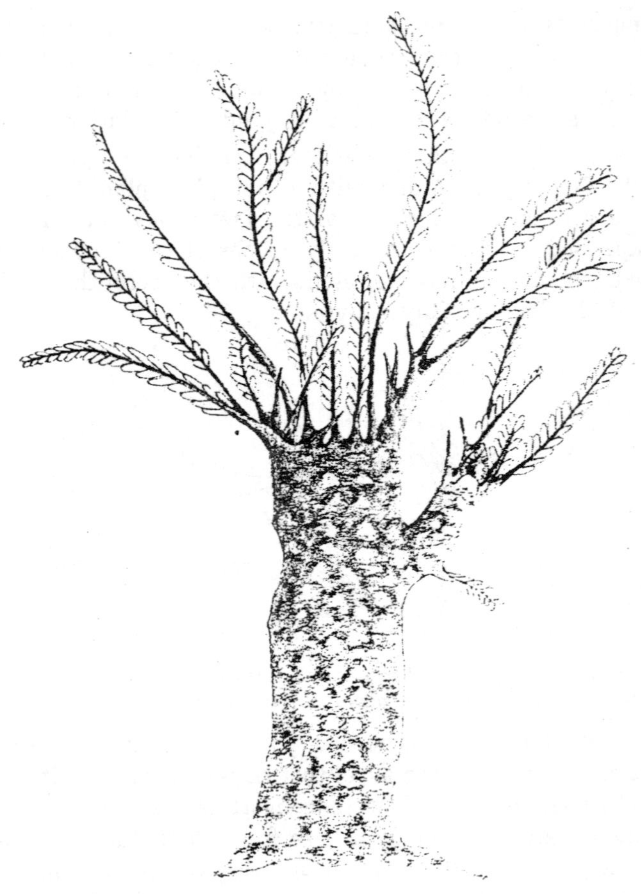

Fig. 7.5. Restoration of a cycadophyte (Williamsonia) from Jurassic strata of India.

equally appropriate. The cycads, more formally known as the Class Cycadophyta, are seed plants in which true flowers have not been developed. Jurassic cycads included tall trees with rough, columnar branches marked by the leaf bases of earlier growths and by crowns of leathery pinnate leaves. Actually, the Cycadophyta include two related groups, the cycadeoids ("fossil cycads") and Cycadales ("true cycads"). Although the adult plants in both of these groups were superficially similar, they differed in reproductive structures and in other anatomic details. Cycads experienced a marked decline in the Late Cretaceous, and only a few have survived to the present time. One such survivor is the sago "palm."

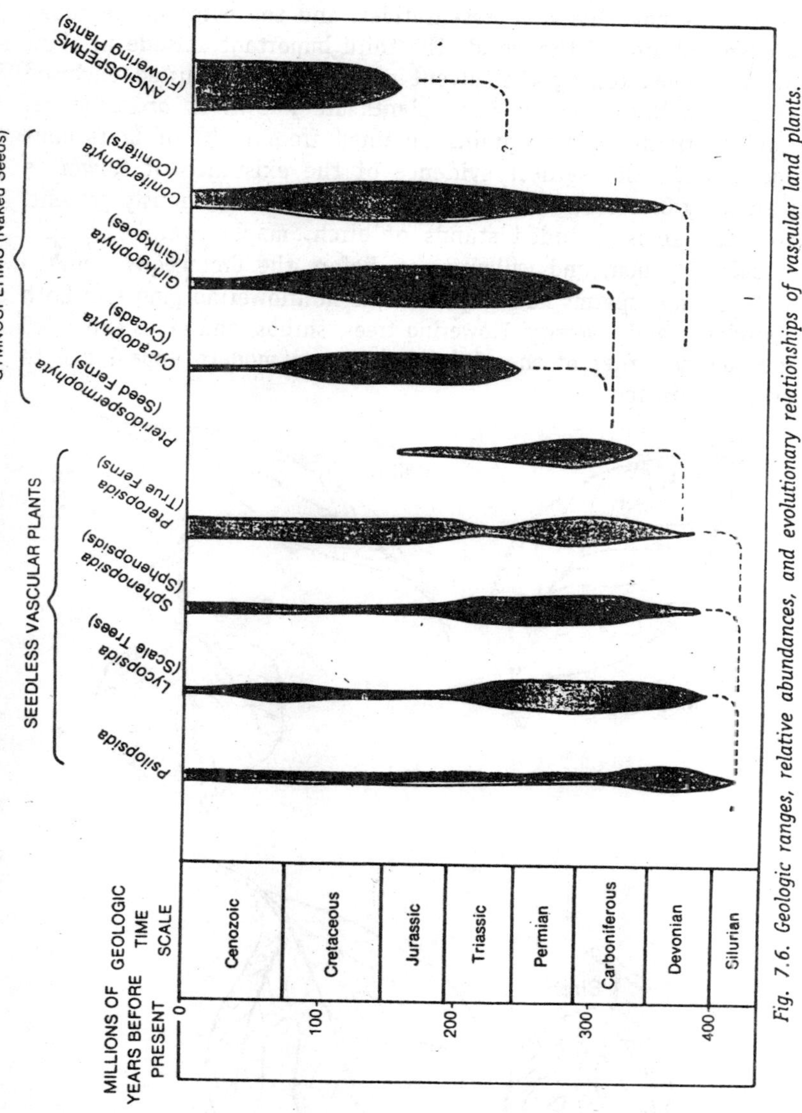

Fig. 7.6. Geologic ranges, relative abundances, and evolutionary relationships of vascular land plants.

It has already been stated that there are three important episodes in the evolution of land plants. The first stage led to the development of the spore-bearing, leafy, treelike plants. The second witnessed the evolution of nonflowering, pollinating seed plants such as cycads, ginkgoes, seed ferns, and conifers. All but the seed ferns have living representatives, and the conifers are today widespread around the world. The third important episode in plant history is marked by the advent of species that carried enclosed seeds and bore flowers. Such plants are known as *angiosperms*. Angiosperm-like pollen grains obtained from rocks of Cretaceous age provide the earliest evidence of the existence of flowering plants. By Cretaceous time, angiosperms were conspicuously present. Forested areas included stands of birch, maple, walnut, beech, sassafras, poplar, and willow trees. Before the Cretaceous came to a close, angiosperms had surpassed the nonflowering plants in both abundance and diversity. Flowering trees, shrubs, and vines expanded into every corner of the globe and gave a modern appearance to every landscape.

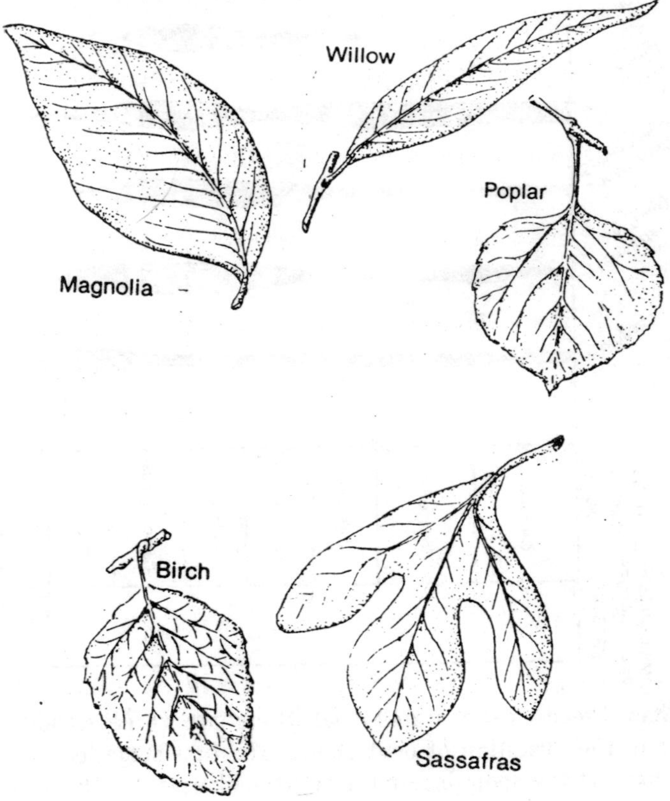

Fig. 7.7. Leaves of angiosperms found fossilized in Cretaceous Rocks.

Either directly or indirectly, the evolution of Mesozoic insects, birds, reptiles, and mammals was strongly influenced by this remarkable floral revolution. Angiosperms produced to insure survival of the plant embryo but that also insured the survival of animals that used this bounty as food.

The relationship between insects and flowering plants is well known. The angiosperms, by encouraging insect visits, are able to utilize insects as delivery agents for pollen. This provides a far more efficient means of pollen dispersal than the more random wind pollination, which requires each plant to disperse tens of millions of pollen grains rather than a few dozen. The insect pollinators are encouraged to visit particular flowers in order to obtain the nectar and pollen needed for food. Having had success with a flower of a certain form, colour, and scent, they move off to find others of the same kind in order to repeat the favourable experience. In this way, pollen is transported on the hair, legs, and bodies of insects from plant to plant of the same species. The selective competition for efficient pollinators has induced a constantly changing range in variations among both plants and insects. In the angiosperms, the need for each plant to be recognizably different resulted in a spectacular floral variety that has persisted from the Cretaceous down to the present.

The Invertebrates

Continental Invertebrates

Although paleontologists have assembled an enormous body of information about the marine invertebrates of the Mesozoic, relatively little is known about continental groups. The pulmonate, or air-breathing, snails are rarely found. Freshwater clams and snails are more commonly fossilized. Freshwater crustaceous, especially such groups as *branchiopods* and nonmarine species of ostracods, are frequently found in Mesozoic lake bed deposits. It is reasonable to assume that many varieties of worms existed, but they left few traces. The spiders, millipedes, scorpions, and centipedes that had been abundant in Carboniferous forests undoubtedly were present also in the Mesozoic, although their remains are elusive. Among the insects, flies, mosquitoes, caddis flies, earwigs, wasps, bees, and ants are known from rocks as old as Jurassic. Many of the best fossils are collected from the Solnhofen Limestone, an unusual Jurassic formation in Bavaria. Even the Solnhofen collection is not truly representative, however, because most of the remains are of larger species. This observation has caused paleontologists to assume that smaller insects were eaten by fish or were decomposed.

The fossil record for Cretaceous insects is also sadly inadequate. Insects preserved in amber of Cretaceous age are known from Canada and Alaska. Specimens include bees, wasps, ants, beetles, flies, and mosquitoes. Butterflies, moths, termites, and fleas occur in rocks of the Early Cenozoic but have not yet been found in rocks as old as Cretaceous.

Marine Invertebrates

At the end of the Paleozoic, many families of marine invertebrates either declined or suffered extinction. The Mesozoic resurgence of marine organisms was somewhat tardy, as indicated by the limited nature of Early Triassic faunas. However, after a few groups had become well established, marine life expanded dramatically. The pelecypods (Bivalvia) became increasingly prolific from Middle Triassic on and eventually surpassed the brachiopods in colonization of the sea floor. Among the most successful pelecypods were the oysters, which were represented by such genera as *Gryphaea* and *Exogyra*. Some members of the oyster group became giants of their kind. Other pelecypods grew conical shells that strikingly resembled the horn corals of the Paleozoic. In these forms, called *rudists*, the left valve formed a small lid that closed the open end of the cone. Rudists, which are known from the Jurassic and Cretaceous, formed impressive reef structures.

In those parts of the Mesozoic marine environment where the water warm, relatively clear, and shallow, corals proliferated. During the Late Jurassic, for example, the Tethyan Geosyncline was the site of major coralline evolution and reef building. The corals of the Mesozoic are termed *scleractinids*. Unlike their Paleozoic predecessors, they did not use calcite to build their skeletal structures but rather the crystallographically different (but compositionally identical) aragonite. As is the case today, scleractinids had rather restrictive environmental requirements. Most could exist only in clear water of normal salinity no deeper than about 50 meters, with temperatures no lower than about 20°C. One reason corals require shallow water is that they have a symbiotic relationship with an alga that lives within in coral polyp and is dependent on sunlight. Corals seem to have lived much farther north in the Mesozoic than they do today, suggesting a more northerly location for the paleoequator during that era.

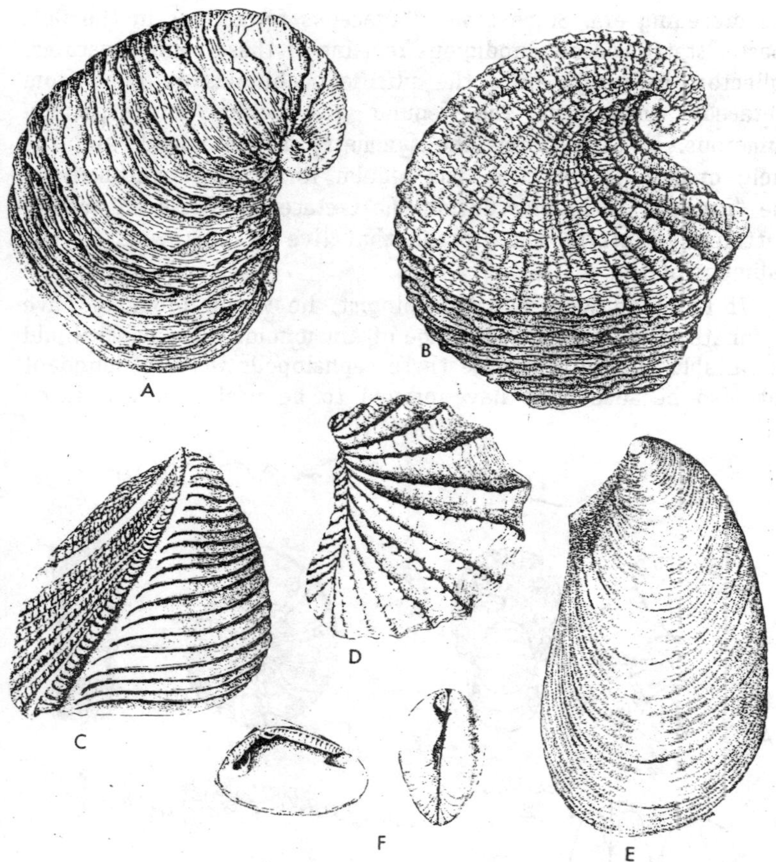

Fig. 7.8. Mesozoic pelecypods. A—Exogyra ponderosa (Cretaceous). B—Exogyra cancellata (Cretaceous). C—Trigonia thoracica (Cretaceous). D—Trigonia costata (Jurassic). E—Inoceramus labiatus (Cretaceous). F—Nucella percrassa.

The great reefs of stony corals offered food and shelter to a host of other kinds of oceanic life. Persisting groups of brachiopods clung to the reef structures, as did rudists, algae, bryozoans, and other sedentary creatures. Gastropods grazed ceaselessly along the reef structures, whereas crabs and shrimp scuttled about seeking food in the recesses and cavernous hollows of the reefs. In the quieter lagoonal areas behind the reefs and on the floors of the epicontinental seas, starfish, sea urchins, and crinoids often thrived.

Those cousins of the crinoids, the echinoids, became far more diverse and abundant in the Mesozoic than they had ever been in the preceding era. Some Lower Cretaceous formations in the Gulf Coastal states contain prodigious remains of these spiny creatures. Collectors in Europe prize the silicified echinoids obtained from Cretaceous chalk beds. The "regular" sea urchins were especially numerous. In these forms the symmetry is pentameral, and the shell, or *test,* is spherical. The regular forms were overtaken by the "irregular" echinoids during the Cretaceous. These are mostly flattened, bilateral sea urchins that live as burrowers in the sediment of the sea floor.

If Poseidon were a paleontologist, he would probably have designated the Mesozoic. "The Age of Ammonoids." The name would be suitable not only because these cephalopods were so abundant but also because they have proved to be useful in worldwide

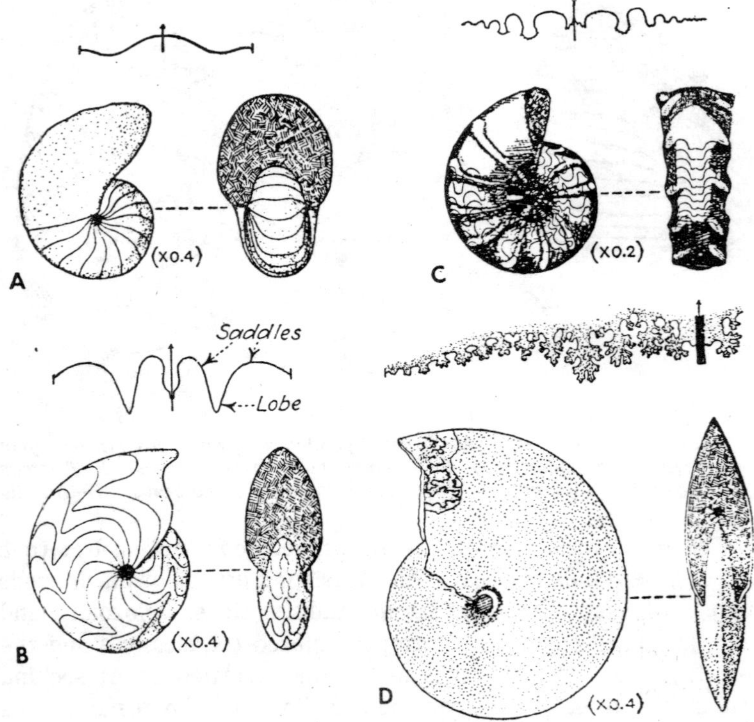

Fig. 7.9. Sutures of cephalopod conchs. A—A nautiloid cephalopod (arrow at midventral line pointing toward conch opening). B—Ammonoid cephalopod with goniatitic sutures. C—Ammonoid with ceratitic sutures. D—Ammonoid cephalopod with ammonitic sutures.

correlation. Zones developed on the basis of ammonoid guide fossils permits correlation of Mesozoic time-rock units with a level of precision surpassing that of radiometric techniques. You may recall that two orders of cephalopods arose during the Paleozoic Era. These were the Nautiloidea, having relatively smooth sutures, and the Ammonoidea, having wrinkled sutures. *Sutures* of cephalopods are lines formed on the inside of the conch where the edge of each chamber's partition, or *septum*, meets the inner wall. Wrinkled sutures are simply a reflection of septa that, like the edges of a pie crust, are fluted. Based on the complexity of the suture patterns, the Ammonoidea can be subdivided into goniatites, which lived from Silurian through Permian time, ceratites, which were abundant in Permian and Triassic marine areas, and ammonites. Ammonites, although represented in all three Mesozoic periods, were most prolific during the Jurassic and Cretaceous.

Knowledge of the exact suture pattern of ammonoid cephalopods is necessary for their identification and hence their use in correlation. The function of the septal fluting that produced the suture patterns provides an interesting subject for speculation. Most paleontologists favour the theory that, like the corrugated steel panels used in buildings, fluted septa provide greater strength. Comparison studies based on the living cephalopod *Nautilus* have revealed that the gas-filled chambers exert only a slight toward pressure, whereas the water pressure on the outside of the conch wall is considerable. The septal fluting may have helped the animal to withstand the differences in pressure. Proponents of this concept call attention to the fact that ammonoid conchs (unlike nautiloid conchs) tend to thin toward the adult chambers. Apparently, in order to compensate for that thinning, and presumably weaker, conch wall, the septa in many species became more closely spaced and more intricately fluted.

The greater variety of Mesozoic ammonoids is a reflection of their success in adapting to a variety of marine environments. They seem to have expanded not only within the shallow epicontinental seas but also in the open oceans. During the Cretaceous, many ammonites became aberrant in shape, so that the normally planispiral forms were joined by species with open spirals, straightened conchs, and even some that coiled in a helicoid fashion, like that of a snail. However, near the end of the Cretaceous, the entire diverse assemblage began to decline and, rather mysteriously, became extinct by the end of the era. Only their close relatives, the nautiloids, managed to survive.

Another group of cephalopods that became particularly common during the Mesozoic were the squidlike belemnites. The belemnite conch was inside the animal and resembled a cigar in shape. The pointed end was at the rear of the animal, and the forward part was chambered. Most belemnite conchs were less than a half meter in size, but some attained lengths of over 2 meters. A few remarkable specimens from Germany are preserved as thin films of carbon and clearly show the 10 tentacles and body form. X-ray study of these fossils has revealed details of internal anatomy as well. Much like the modern-day squid, the belemnites were probably able to make rapid reverse dashes by jetting water out of a funnel located at the anterior end.

The belemnites were highly successful during the Jurassic and Cretaceous. Triassic belemnites may very well have been the ancestors to the squids, which were also numerous during the Jurassic and Cretaceous. Octopods, because they lack a conch, are inadequately recorded in the Mesozoic. However, their presence is affirmed by an imprint of an octopus found in strata of Late Cretaceous age from Lebanon.

Marine gastropods were also abundant during the Mesozoic. Many are found in sediments that represent old beach deposits. Then, as now, capshaped limpets grazed slowly across wave-washed boulders while a variety of snails with tall, helicoid conchs crawled about on the surfaces of shallow reefs. For the most part, the gastropod fauna had a decidedly modern appearance and included many colourful and often beautiful forms that have present-day relatives.

Modern types of marine crustaceans, such as crayfish, lobsters, crabs, shrimp, and ostracods, were abundant by Jurassic time. Of these crustaceans, the ostracods have become particularly useful to stratigraphers. At some localities, barnacles grew prolifically on rocks and reef structures.

Among the single-celled protozoan animals that crawled or floated about the Mesozoic seas were the radiolarians and the foraminifers. Radiolarians and the foraminifers. Radiolarians make their open, delicate, spinose skeletons from opaline silica. In some regions today, radiolarian and diatom skeletal remains accumulate on the sea floor to form extensive deposits of siliceous ooze and in the past have contributed to the formation of chert beds.

The tests of foraminifers are generally more durable than those of radiolarians. The -"forams," as they are often called, left an imposing Mesozoic fossil record and one that is of the greatest importance in stratigraphic correlation. They are especially important in the exploration for petroleum. Because of their small size and strong tests, large numbers of foraminifera can be obtained unbroken from the small pieces of rock recovered while drilling for oil. They are then used in tracing stratigraphic units from well to well. Forams are also sensitive indicators of water temperature and salinity and have provided data useful in reconstructing ancient environmental conditions.

Foraminifers were only meagerly represented in the Triassic but began to proliferate in the Jurassic and Cretaceous. Their expansion continued well into the Cenozoic. Nearly all were bottomdwelling species until Cretaceous time, when plankton groups began their colonization of the upper levels of the ocean in prodigious numbers. Among the planktonic foraminifers, such genera as *Rotalipora* and *Globotruncana* contributed their empty tests to the calcareous sea floor sedimentary beds, some of which were later lithified into Cretaceous chalks and marls.

The Vertebrates

The Triassic Transition

The general unrest, broad uplifts, and upheavals that occurred during the Carboniferous and Permian periods caused regressions of epicontinental seas, resulted in a variety of continental environments, and generally provided the environmental stimulus needed to maintain the spread and diversification of land vertebrates. Although marine faunas change rather abruptly in passing from the Paleozoic to the Mesozoic, there is considerably more continuity in land faunas. The labyrinthodont amphibians, for example, continued into the Triassic before becoming extinct. The cotylosaurs, or "stem reptiles," also were able to cross the era boundary. Other groups that continued from the Permian were the therapsids. The most progressive of these mammal-like reptiles, the lctidosauria, succeeded their Permian precursors and lived successfully on into the Late Triassic as contemporaries of primitive mammals.

Many new reptile types appeared in the Triassic. Among these were the ancestors to the first turtles. Triassic turtles were basically similar to their living descendants except that they retained teeth in their jaws. The Triassic was also the geologic period during which various lineages of marine reptiles began to appear. The rhynchocephalians, represented today by the tuatara of New Zealand, were abundant. Most interesting of all, however, were reptiles known as archosaurs. The *Archosauria* are a large and important group of reptiles that include the living crocodiles and the extinct flying reptiles, dinosaurs, and thecodonts. The thecodonts have a distinguished place in vertebrate history, for they were the ancestors of the dinosaurs.

The Thecodonts

Early Thecodonts, as exemplified by *Hesperosuchus* and *Euparkeria*, where small, agile, lightly constructed reptiles with long tails and short forelimbs. They had already developed the unique habit of walking relatively erectly on their hind legs. This bipedal mode was an important innovation. Bipedalism permitted thecodonts to move about more speedily than their sprawling ancestors. Because their forelimbs were not used for support, they could be employed for catching prey; even more important, they could be modified for flight. Thus, the thecodonts were the ancestors not only of dinosaurs but of flying reptiles and birds as well.

Fig. 7.10. Hesperosuchus from the Triassic of the south-western United States.

Fig. 7.11. Restoration of the Lower Triassic thecodont, Euparkeria.

Not all thecodonts were nimble, bipedal sprinters; some reverted to a four-footed stance and evolved into either armored land carnivores or large crocodile-like aquatic reptiles called *phytosaurs*. Occasionally in the history of life, initially unlike organisms from separate lineages gradually become more and more similar in form. The once distinctly different groups, in fact, change over many generations so that they are better adapted to a particular environment. The evolutionary process responsible for the trend toward similarity in form is called *convergence*. Phytosaurs and crocodilians are good examples of evolutionary convergence. Indeed, the most visible distinction between the two groups is the position of the nostrils, which are at the end of the snout in crocodiles which are at the end of the snout in crocodiles but were just in front of the eyes in phytosaurs.

The Dinosaurs

Of all the reptiles that now live on this planet or have lived on it in the past, few are more fascinating than dinosaurs. Dinosaurs are the most awesome and familiar of prehistoric beasts. These headlines of the Age of Reptiles compose not one order, but two, each having evolved separately from the thecodonts. The two orders are the Saurischia ("lizard-hipped") and the Ornithischia ("bird-hipped"). As suggested by these names, the arrangement of bones in the hip region provides the criterion for the twofold classification. The reptile pelvis is composed of three bones on each side. The uppermost bone is the ilium, which is firmly clamped to the spinal column. The bone extending downward and slightly backward is the ischium. Forward of the *ischium* is the *pubis*. In the saurischians, the arrangement of the three pelvic bones is

Fig. 7.12. Rutiodon, a Triassic phytosaur.

triradiate, as it was in their thecodont ancestors. However, in the ornithiscians the pubis is swung downward and backward so that it is parallel to the ischium, as in birds.

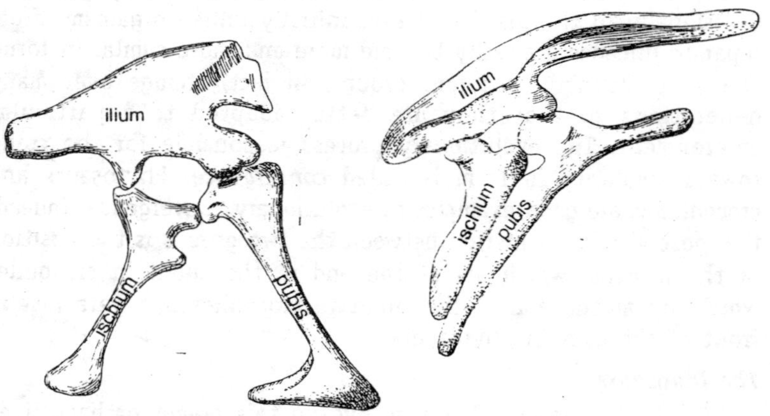

Fig. 7.13. Basis for the division of "dinosaur" into two groups. On the left is the arrangement of pelvic bones in the Saurischia and on the right, the arrangement of ornithischia.

The earliest dinosaurs were nearly all saurischians. Most were relatively light, nimble, carnivorous bipeds known from fossils discovered in Triassic beds of both South and North America as well as China. *Coelophysis*, for example, was a small, hollow-boned early saurischian found in the Chinle Formation of New Mexico. These birdlike reptiles, called coelurosaurs, continued into the Jurassic and Cretaceous. *Ornithomimus* must have looked very much like an ostrich, with its long neck, toothless jaws, and small head.

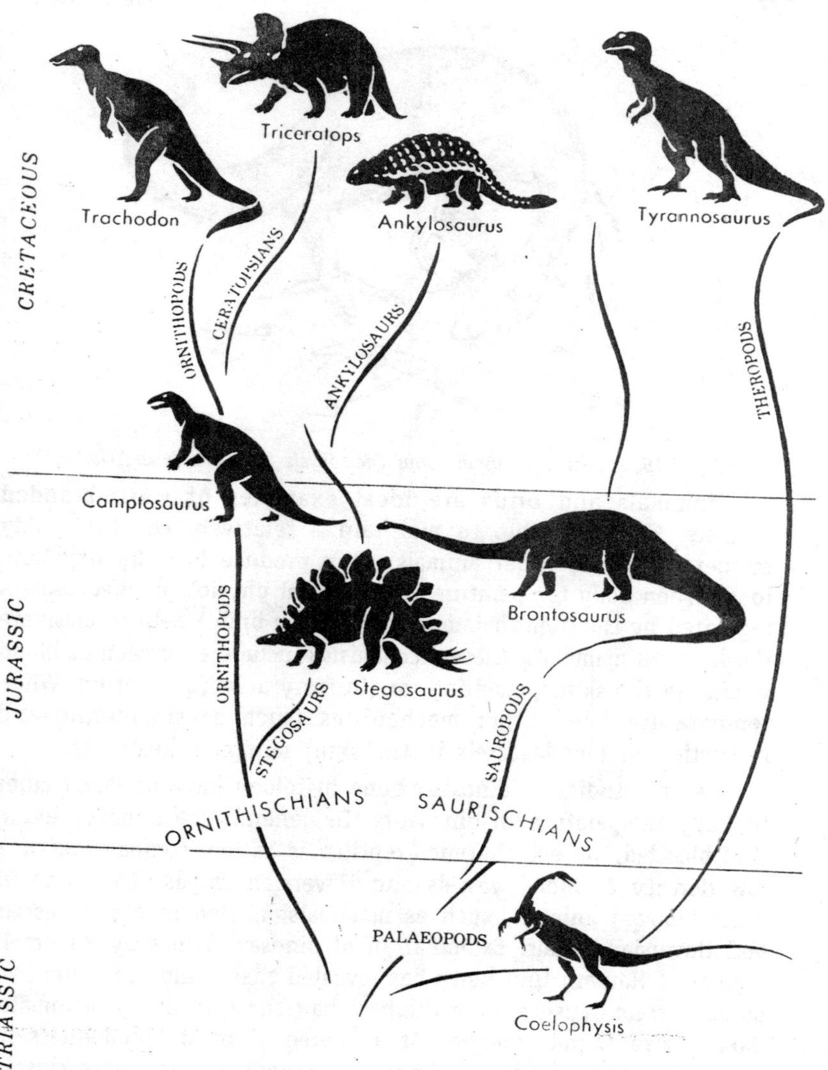

Fig. 7.14. Evolution of the dinosaurs.

Because it lacked teeth, paleontologists speculate that it lived on the eggs laid by its contemporaries; however, it could just as easily have eaten smaller vertebrates. It is possible that *Ornithomimus* and its kin were capable of a certain amount of body temperature regulation—"warm-bloodedness."

Fig. 7.15. The carnivorous dinosaur Coelophysis from the Upper Triassic.

Mammals and birds are ideal examples of warm-blooded animals. They are able to maintain a relatively constant body temperature. Like other animals, they produce heat by oxidizing food. When body temperature rises, special physiologic mechanisms regulated by the hypothalamus (part of the brain) help to dissipate the heat. In mammals, these mechanisms include expansion of blood vessels in the skin, perspiring, or (in furry animals) panting. When temperature falls, other mechanisms (such as shuddering and restriction of blood vessels in the skin) minimize heat loss.

Recent studies of dinosaur bone histology have provided clues to body temperature in dinosaurs. In general, the bone of living cold-blooded, or ectothermic, reptiles is rather compact, with a low density of blood vessels and Haversian canals. The bone of warm-blooded animals, such as mammals, is rich in blood vessels and Haversian canals. Examination of dinosaur bones by Robert T. Bakker of Harvard University has revealed that many dinosaurs had blood vessels densities even higher than those in living mammals. Bakker also found evidence of a degree of warm-bloodedness in the predator-prey ratios of dinosaurs, Dinosaur ratios more closely resembled the situation for warm-blooded than for cold-blooded populations. He noted further that the presence of dinosaur remains at locations that were cool during the Mesozoic can be more readily explained if the creatures were, at least in part, endothermic. These new ideas are now being actively examined. Their validity is based on a limited amount of fossil evidence, and thus there is a danger of overinterpretation. With the present information, it does seem that a case can be made for true endothermy, at least among the coelurosaurs.

The larger carnivorous saurischians (including coelurosaurs) are called *theropods*. Of more dramatic dimensions than the coelurosaurs were the large carnosaurians like *Allosaurus* of the Jurassic and the Cretaceous dinosaurs *Deinonychus* and *Tyrannosaurus*. The last-named beasts attained lengths of over 13 meters and weighed in excess of 4 metric tons. Carnosaur hind limbs were robust and muscular. Great curved claws for tearing flesh protruded from each of three toes, whereas a nearly functionless fourth toe bore a smaller claw. The forelimbs were, in our view, ridiculously small. As befits an animal that must kill with its jaws and teeth, the head of the carnosaur was large. Doubly serrated teeth, up to 6 inches long, lined the powerful jaws. These were truly spectacular predators.

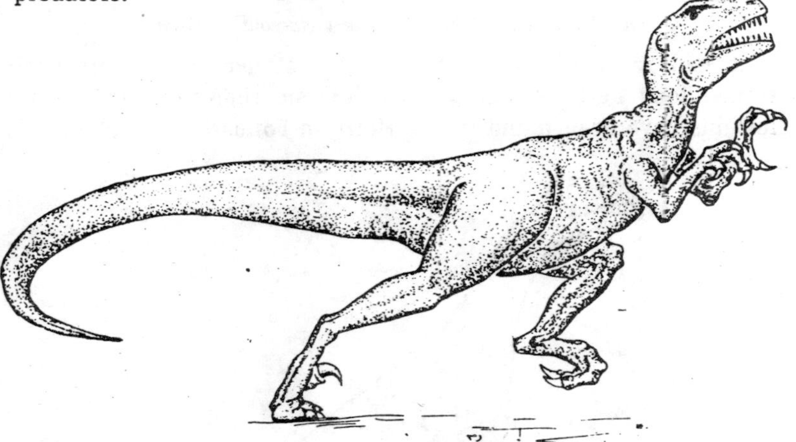

Fig. 7.16. The Cretaceous dinosaur Deinonychus. This reptile was about 8 ft. long. It possessed a large skull, and the margins of the jaws were set with serrated, saber-like teeth.

The saurischian group included herbivorous sauropods as well as flesh-eating theropods. The ancestry of the sauropods can be traced to the Late Triassic "protosauropod" Known as *Plateosaurus*. From this smaller, partially bipedal form, the more typical giant sauropods appeared at the beginning of the Jurassic and survived right up to the end of the Cretaceous. They are the animals that people first think of when the hear the world "dinosaur". The best-known sauropods were enormous long-necked, long-tailed beasts

Fig. 7.17. Tyrannosaurus, a late Cretaceous carnosaur.

that had returned to the four-legged stance to support their tremendous bulk. A well-known Jurassic representative whose remains have been found in the Morrison Formation of Colorado is

Fig. 7.18. Plateosaurus, the Late Triassic ancestor of the giant sauropods.

Fig. 7.19. The enormous Jurassic sauropod Brontosaurus.

Brontosaurus, the "thunder beast." This favorite of school children (as well as producers of Hollywood movies) measured almost 20 meters in length and weighed about 30 metric tons. *Diplodocus*, a contemporary, was less bulky and had greater length.

For many years paleontologists have speculated that, even with their massive, pillar-like legs these largest of land animals could not have supported their own weight continuously. It was therefore surmised that they dwelt in the buoyant waters of lakes and streams. However, this long-held theory of sauropods habits has recently been challenged. According to the new view, brontosaurs roamed through the forests much like modern elephants. For sauropods, there were probably advantages in being large. Great size affords protection and slows changes in body temperature. The ratio of surface area to mass for an animal decreases as size increases. Consequently, the large animal has a proportionately smaller surface for heat loss, and, just as a large pot of water loses its heat more slowly than a small pot, so does the large animal lose its heat more slowly than a small animal.

The other major dinosaur line, the Ornithischia, evolved near the end of the Triassic and thrived throughout the remaining Mesozoic. Ornithischians were plant eaters. The teeth in the forward part of the jaws were replaced by a beak suitable for cropping vegetation. The group included both quadrupedal and bipedal varieties, with the bipedal condition considered more primitive. Even the most advanced quadruped ornithischians had such short fore limbs that their descent from bipedal forms seems certain.

The bipedal group of ornithischians is known as *ornithopods*. Their evolutionary history began in the Triassic with relatively

Fig. 7.20. Camptosaurus, a small to medium-sized (6 to 8 feet long) herbivorous dinosaur. Although there were no teeth in the beaklike forward part of the jaws, the remainder of the jaws was equipped with a tight mosaic of row upon row of teeth suitable for grinding plant food.

small species that lived primarily on dry land. A representative large Jurassic ornithopods is *Camptosaurus*. This was a bipedal dinosaur of medium size with a heavy tail, short fore limbs, and long hind legs. The articulation of the jaw was arranged to bring the teeth together at the same time, an arrangement frequently seen in herbivores down through the ages. Leaves and stems were cropped by the forward, beaklike part of the jaws and passed backward to the cheek teeth for chopping and chewing. From camptosaurid-like ancestors, the larger Cretaceous ornithopods developed. Among these was *Iguanodon*, one of the first dinosaurs to be scientifically described. *Iguanodon*, sometimes called the "thumb-up" dinosaur because of the horny spike that substituted for a thumb, is thought to have been a gregarious animal that moved about in "herds." Evidence for this comes from a Belgian coal mine where 29 of these individuals were found together as a result of having fallen into an ancient fissure.

By Cretaceous time, most members of the ornithopod group had moved to lake environments—perhaps to avoid encounters with the ferocious carnosaurs. The most successful of these amphibious reptiles were ornithopods known as trachodonts or hadrosaurs, or more commonly "duck-bill dinosaurs". Mummified remains of these creatures clearly show the appearance of the skin. The aquatic ornithopods possessed webbed feet and a flattened tail for use in

Fig. 7.21. Iguanodon, a herbivorous ornithischian dinosaur from the Lower Cretaceous of Europe.

swimming. From the skeletons that have been unearthed, paleontologists have reconstructed the droll-looking face, in which the forward part of the jaws was often flattened and widened to resemble the bill of a duck. Behind the toothless forward part of the jaw were closely appressed lines of teeth that seem well suited for chewing coarse vegetation.

An interesting peculiarity of some hadrosaurs was the development of bony skull crests containing tubular extensions of the nasal passages. The function of these bizarre structures is of special interest to paleontologists. It has been suggested recently that the cranial crests were used to catch the attention of sexual partners of the same species and could be further employed as vocal resonators for promoting an individual's success in obtaining a breeding partner.

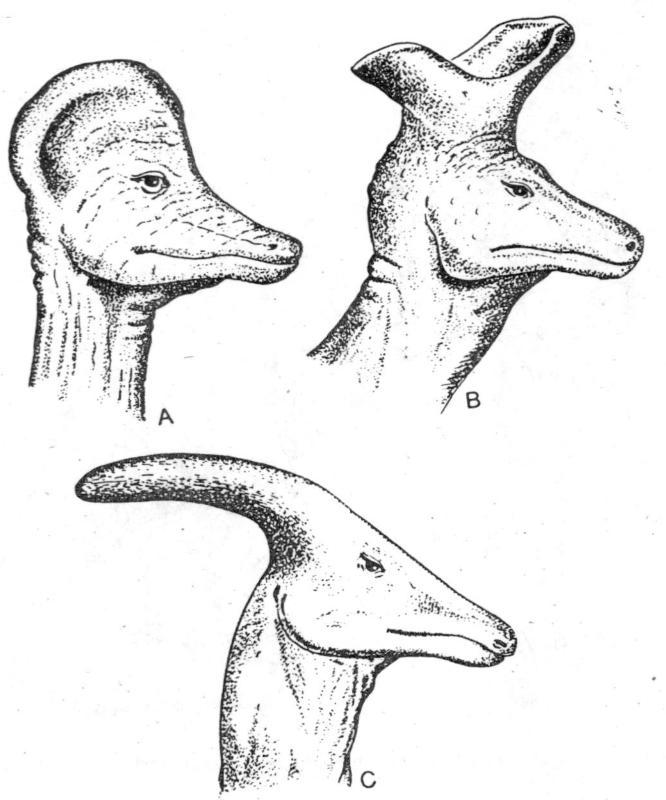

Fig. 7.22. Bizarre skull crests developed in hadrosaurs. A—Corythosaurus. B—Lambeosaurus. C—Parasaurolophus.

Not all ornithopods had skull crests, and in some the crests lacked nasal tubes. One crestless ornithopod that certainly qualifies as a legitimate "bone head" was *Pachycephalosaurus*. The skull in pachycephalosaurians consisted mostly of solid bone with only a small space to accommodate the unimpressive brain.

The best known of the quadrupedal ornithischians are the *stegosaurs*. Stegosaurs had two pairs of heavy spikes mounted on the tail. These were used for defense. However, their more identifying feature was the double row of alternating pentagonal plates that stood upright along the back. Scientists have debated the purpose of these plates for many years. One theory that is currently being considered is that the plates functioned in the regulation of body temperature by serving as body-heat dissipaters. Indeed, the

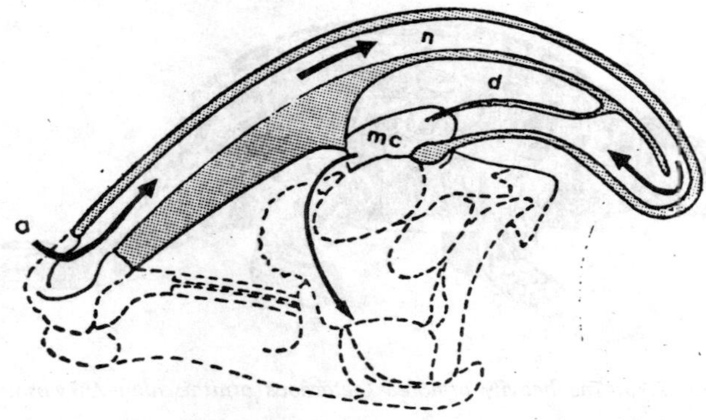

Fig. 7.23. Internal structure of the skull crest of Parasaurolophus cyrtocristatus. The superficial bone of the left side has been removed to expose the left nasal passage (n). Air enters nostrils at a, move up and around partition in crest, and from there moves down and back to internal openings in the palate.

arrangement, size, and shape of the plates, and their probable rich supply of blood vessels, would favor the thermoregulatory suggestion, although the plates may have served as camouflage or protection as well.

Fig. 7.24. Stegosaurus.

During the Cretaceous, stegosaurs were succeeded by the heavily armored *ankylosaurs*. These bulky, squat ornithiscians were completely covered by closely fitted bony plates that covered the

Fig. 7.25. The heavily armored Cretaceous ornithischian Ankylosaurus.

entire length of their 6-meter backsides. A great ball of bone at the end of the tail in *Ankylosaurus* could be used as a bludgeon against an unwary foe.

The fourth group of quadrupedal ornithiscians are the *ceratopsians*. These beasts take their name from horns that grew on the face of all but the earliest forms. Typically, ceratopsians possessed a median horn just above the nostrils, and in some species and additional pair projected from the "fore-head." The head was quite large in proportion to the body and displayed a shieldlike bony frill at the back of the skull roof that served as protection for the neck region and as a place of attachment for powerful jaw and neck muscles. All ceratopsians possessed a parrot-like beak. Judging from the scars that mark the shield bones of ceratopsians, they were often attacked by the great carnosaurs. No doubt they frequently emerged the victor. During the Late Cretaceous, ceratopsians moved eastward from Asia across the land connection to North America and inhabited the region that lay to the west of the epicontinental sea.

The Pterosauria

Again and again in the history of life, the descendants of a small group of animals that were initially adapted to a narrow range of ecologic conditions have dispersed into a great variety of environments. As they responded through evolutionary processes to the formerly unoccupied living space, the animals changed in ways that made them better suited to their new surroundings. This process, known as *adaptive radiation*, is nicely demonstrated by the Mesozoic reptiles. The adaptive radiation originated with the stem reptiles of the Late Carboniferous and ultimately produced the enormous variety of large and small, herbivorous and carnivorous, dry land and aquatic animals of the Mesozoic. The radiation did not end with terrestrial vertebrates, however, for during the Mesozoic, reptiles invaded the marine environment and even managed to overcome the pervasive tug of gravity and achieve soaring flight. Those reptiles that took to the air were the *pterosaurs*. They were the first flying vertebrates, and, although they may appear a bit graceless, their existence from Early Jurassic until late in the Cretaceous attests to their adaptive success.

The typical pterosaur had a rather large head and eyes and long jaws, which in most forms were lined with thin, slanted teeth. The bones of the fourth finger were lengthened to help support the wing, whereas the next three fingers were of ordinary length and terminated in claws. The wing was a sail made of skin stretched between the elongate digit, the sides of the body, and the rear limbs. There were two general groups of pterosaurs. The more primitive were the rhamphorhynchoids, which had long tails, and the advanced were tail-less pterodactyloids. The latter group is exemplified by *Pteranodon*, species of which had an astonishing wingspan of over 7 meters. The body of *Pteranodon* was about the size of that of a goose. The skeleton was lightly constructed, as is fitting for an aerial vertebrate. The animals probably glided along much like oversized sea birds, snapping up various sea creatures in their toothless jaws.

Fig. 7.26. The Jurassic pterosaur Rhamphorhynchus.

Relative to body size, pterosaurs had somewhat larger brains than some of their land-dwelling relatives. Perhaps this was a result of the higher level of nervous control and coordination needed for flight. They may also have had a degree of endothermy or warm-bloodedness. Strong evidence for this theory has been obtained by the Russian paleontologist A. Sharov. Sharov discovered a superbly preserved fossil of a Jurassic flying reptile, which he named *Sordus pilosus*. The name means "hairy devil" and is an appropriate reference to the growth of hair or hairlike feathers that covered the body and limbs. Remarkably, this insulating cover is actually preserved in the fine Jurassic lake shales in which the creature was entombed.

A Return to the Sea

The marine habitat is one in which the archosaurs were not notably successful. Only one archosaurian group—the sea crocodiles—were able to invade the oceanic environment. Other reptilian groups, such as the ichthyosaurs, plesiosaurs, mosasaurs, and sea turtles, however, were very successful in their return to the productive seas of the Mesozoic. Many made their diet of the myriads of sharks and bony fishes that had preceded them in populating the seas. The Cretaceous witnessed the modernization of the fish group, culminating in the appearance of the most modern of bony fishes, the *teleosts*.

Not unexpectedly, the invasion of the marine environment required many modifications in form and function. Paddle-shaped limbs and streamlined bodies evolved to allow efficient movement through the water. Because these reptiles were unable to abandon air breathing and reconvert to gills, their lungs were modified for greater efficiency. In those sea-going reptiles that were unable to lay their eggs ashore, reproductive adaptations provided for birth of the young while at sea.

Marine reptiles that depended upon paddleshaped limbs for locomotion were already present during the Triassic Period. One group, the *nothosaurs*, were just beginning to take on adaptations that would be perfected in their likely descendants, the plesiosaurs. The nothosaurs were joined in the Triassic by a group of mollusk-eating, flippered reptiles known as *placodonts*. These bulky animals had distinctive pavement type teeth in the jaws and palate, which they used for crushing the shells of the marine invertebrates upon which they fed.

By far the best known of the paddle swimmers were the *plesiosaurs*. Their earliest remains are found in Jurassic strata.

Fig. 7.27. Bulky mollusc-eating Triassic marine reptile Placodus.

Sometimes nicknamed "swan lizards," they had short, broad bodies and large, many-boned flippers. In some species the neck was extraordinarily long and was terminated by a smallish head. Slender, curved teeth, suitable for ensnaring fish, lined the jaws. *Elasmosaurus*, a well-known Cretaceous plesiosaur, attained an overall length in excess of 12 meters.

In addition to the long-necked types, there were many plesiosaurs characterized by short necks and large heads. It is likely that the shortnecked plesiosaurs were aggressive divers. *Kronosaurus*, a giant, short-necked form from the Lower Cretaceous of Australia, had a skull size that probably exceeds that of any known reptile. It measured 3 meters in length.

The most fishlike in form and habit of all the marine reptiles were the Triassic and Jurassic *ichthyosaurs*. In many ways, they were the reptilian counterparts of the toothed whales of our present-day oceans. Ichthyosaurs had fishlike tails in which vertebrae

Fig. 7.28. Restoration of an ichthyosaur, a marine reptile similar to a modern porpoise in form and habits.

extended downward into the lower lobe, boneless dorsal fins to help prevent sideslip and roll, and paddle limbs for steering and braking. The head was a suitably pointed entering wedge for cutting rapidly through the water. A ring of bony plates surrounded the large eyes and may have helped protect them against water pressure. Clearly, these were active predators with good vision and the ability to move swiftly through water.

The *mosasaurs*, a group of giant monster ilizards, were a highly successful group of Cretaceous sea dwellers. A typical mosasaur looked somewhat like a large moray eel to which four flippers had been added. The creatures propelled themselves through the water by the sculling action of their long, vertically flattened tails and the rhythmic undulations of their long bodies. The lower jaw had an extra hinge at midlength, which greatly increased its flexibility and gape. Mosasaurs were primarily fish eaters, but some frequently dined on large mollusks. Conchs of cephalopods have been found with puncture marks that precisely match the dental of their mosasaurian foes.

Perhaps less spectacular than the mosasaurs, but far more persevering, were the sea turtles. In this group, also, we find a trend toward increase in size. The Cretaceous turtle *Archelon*, for example, reached a length of nearly 4 meters. As an adaptation to their aquatic habitat, the carapace of the marine turtle was greatly reduced and the limbs were modified into broad paddles.

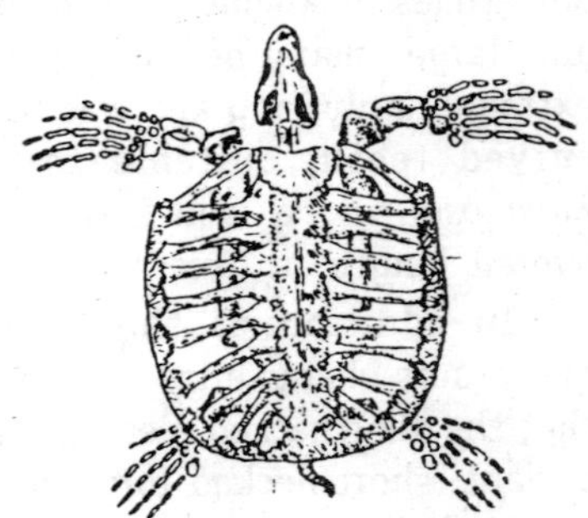

Fig. 7.29. Skeleton of the giant Cretaceous marine turtle Archelon.

As briefly indicated earlier, the marine crocodiles were the only members of the archosaurian group that entered the sea. They became relatively common during the Jurassic, but only a few remained by Early Cretaceous. It is quite likely that they did not fare well in competition with the mosasaurs.

The Birds

From the time of Darwin, naturalists have been aware of the structural similarities between birds and reptiles. These similarities have prompted the statement that birds are only glorified reptiles that have gained wings and feathers and lost their teeth. However, such statements depreciate the marvelous attainments of birds, not the least of which are their superior powers of flight and high level of endothermy. Both of these attributes are related to the transformation into feathers of what once were reptilian scales.

There is little doubt that birds descended from Triassic thecodonts, which were already birdlike in their hollow bone structure and bipedal stance. Some thecodonts may even have had thermal insulation. Middle Triassic lake beds in Turkestan have yielded the remains of a small thecodont that was covered with long, overlapping, keeled scales that might very well have served to trap an insulating layer of air next to the body. The scales seem to be an ideal transitional form for the feathery insulation of birds.

The earliest bird thus far discovered, *Archaeopteryx*, is a perfect link between the bipedal thecodonts and modern birds. *Archaeopteryx* was a creature about the size of a crow. With the exception of its distinctly fossilized feathers, its features were still largely reptilian. The jaws bore thecodont teeth, and the creature had a long, lizardlike tail that bore feathers. Unlike the wings of modern birds, in which the bones of the digits coalesce for greater strength, the primitive wings of *Archaeopteryx* retained claw-bearing free fingers for climbing and grasping. The lightweight sternum lacked a keel, indicating that the heavy muscles needed for sustained flight were lacking. Clearly, *Archaeopteryx* was not a vigorous aviator but rather a forest dweller that glided from one tree branch to another.

Small, delicate, hollow-boned animals are not readily preserved, and the fossil record for birds—especially Mesozoic birds—is not good. The Cretaceous provides the next partial glimpse of bird evolution. Toothed birds otherwise resembling terms and gulls are occasionally found in the Cretaceous deposits of the inland chalky seas. Some Cretaceous birds, such as *Hesperornis*, became excellent

Fig. 7.30. The ancestral bird, Archaeopteryx, from the Jurassic of Bavaria.

swimmers. They retained feathers and other birdlike characteristics but lost their wings and relied upon their webbed feet for swimming. Marine sediments provided the rapid burial needed to preserve these aquatic birds. On land, preservation of Mesozoic birds was rare indeed, and their fossil record is quite inadequate.

The Mammalian Vanguard

While the Mesozoic reptiles were having their heyday, small, furry animals were scurrying about in the undergrowth and unwittingly awaiting their day of supremacy. These shrewlike creatures were the primitive mammals. On the basis of rare and often minuscule remains, they are known from all three systems of the Mesozoic. Among the earliest of the mammals were the *morganucodonts*, jaws, teeth, and skull fragments of which have been recovered in Late Triassic rocks of south Wales. It is evident from these rather scrappy remains that morganucodonts still retained many vestiges of reptilian structures. However, the articulation of the jaw to the skull was mammalian, and the lower jaw was functionally a single bone, as in mammals.

There were at least six additional groups of Mesozoic mammals. Each is recognized primarily on the basis of tooth morphology.

Fig. 7.31. Restoration of Morganucodon, an early mammal from the Late Triassic of Wales.

Among those known from the Jurassic and Cretaceous are the *docodonts, triconodonts, symmetrodonts, multituberculates* and *pantotherians*. The docodonts had multicusped molar teeth, which suggests that they may have been the stock from which present-day monotremes evolved Very likely, they fed upon insects, as did many of these primitive mammals. The triconodonts can be recognized by cheek teeth in which three cusps are aligned in a row. Some were as large as a cat and may very well have preyed upon smaller vertebrates. Symmetrodonts had molars constructed on a more or less triangular plan. As suggested by their name, multituberculates had teeth with many cusps. They may have been the first entirely herbivorous mammals. Their chisel-like incisors and the gap between the incisors and cheek teeth give them a decidedly rodent-like appearance.

Fig. 7.32. The rodent-like multituberculate Taeniolabis.

From an evolutionary point of view, the pantotheres are the most important of the mammalian vanguard, for in the Late

Cretaceous they gave rise to the marsupials and placentals. Marsupials, for example, have an asymmetrically triangular pattern of cusps that was evidently inherited from the pantotheres. In general, pantotheres were small, ratlike animals with long, slender, toothy jaws. It is likely that they preyed not only upon insects but also upon small lizards and mammals.

For the mammals, the Mesozoic was a time of evolutionary experimentation. For 100 million years, they effectively and unobtrusively lived among the great reptiles while simultaneously improving their nervous, circulatory, and reproductive anatomy. Equipped with an exceptionally reliable system for control of body temperature, they were better able to survive in cooler climates than many of their cold-blooded contemporaries. As the reptile population declined near the end of the era, mammals quickly expanded into the many habitats vacated by the saurians.

A Second Time of Crisis

Just as the end of the Paleozoic was a time of crisis for animal life, so also was the conclusion of the Mesozoic. Primarily on land but also at sea, gradual extinction overtook many seemingly secure groups of vertebrates and invertebrates. In the seas, the ichthyosaurs, plesiosaurs, and mosasaurs perished. The ammonoid cephalopods and their close relatives, the belemnites, as well as the rudistid pelecypods, disappeared. Entire families of echinoids, bryozoans, planktonic foraminifers, and calcareous phytoplankton became extinct. On land the most noticeable losses were among the great clans of reptiles. Gone forever were the magnificent dinosaurs and soaring pterosaurs. Turtles, snakes, lizards, crocodiles, and the New Zealand reptile *Sphenodon* (the "tuatara lizard") are the only reptiles that survived the great biologic crash. Altogether, in the Late Cretaceous, extinctions eliminated about a fourth of all known families of animals.

The question of what caused the decimation in animal life at the end of the Mesozoic continues to intrigue paleontologists. Scores of theories, some scientific and many preposterous, have been offered. Those that have the most credibility attempt to explain simultaneous extinctions of both marine and terrestrial animals and seek a single or related sequence of events as a cause. In general, the theories tend to fall into two broad categories. The first of these relies on some sort of abnormally high amounts of cosmic radiation, which might either destroy organisms or damage their reproductive capabilities. Some proponents of this concept feel that reversals of the earth's magnetic field may have temporarily eliminated protection from cosmic radiation. Unfortunately, the theory has some aspects that are difficult to defend. It does not plain why marine organisms, which could descend a few meters and escape radiation damage, were decimated at levels far in excess of those among unprotected land plants. Also, several known periods of polar reversals do not correlate to episodes of mass extinction. In addition, the Mesozoic extinctions spanned a time interval far greater than that encompassed by documented magnetic reversals.

Another possibility for radiation damage to the Cretaceous biosphere might have been the arrival of a blast of lethal rays from a nearby supernova. Supernovas are colossal stellar explosions that radiate about as much energy as 10 billion suns and would affect the surfaces of planets located over 100 light years away (1 light year being about 8,898,000,000,000, km). Astronomers estimate that, on the average, a supernova may occur within 50 light years of the earth every 70 million years. If such an event did take place, there seems little doubt that plant and animal communities would be severely affected. For the present, however, we must await firm evidence that such a stellar explosion did occur nearby about 65 million years ago.

The other category of extinction hypotheses calls upon phenomena that are intrinsic to the earth itself. One concept within this second group is of particular interest in that it explains both marine and terrestrial extinctions in a manner that is ecologically reasonable and that correlates with geologic events of the Late Mesozoic. As indicated in the previous chapter, the Late Cretaceous was a time of extensive marine transgressions. Several lines of evidence indicate that these marine invasions were the result of displacement of ocean water by rising midoceanic ridges. The uplift of the ridges might have been a consequence of acceleration in the rate of sea floor spreading. Whatever the cause, the enormous expanses of inland seas on most of the continents helped to moderate and stabilize world climate. Times were good for living things, which experienced remarkable increases in variety and abundance. And then this Mesozoic "Garden of Eden" experienced hard times. The inland seas regressed. Perhaps the regressions were caused by a slowing of sea floor spreading rates; whatever the cause, there is ample geologic evidence that seas did indeed regress. The result was an episode of harsher climatic conditions, prolonged drought, general cooling, and increased seasonality. The highly diverse forms of animal life, however, had become adapted to the previous, more stable environment. Many lineages could not adjust to the changes and were exterminated. Their dermis affected associated organisms in varying degrees and resulted in a wave of extinctions among ecologically dependent species higher in the food pyramid.

Much of the latter theory is speculation. However, there is now a large body of evidence that indicates a strong correlation between episodes of extinction and major withdrawals of epicontinental seas. The association of those withdrawals with plate tectonic theory provides a new twist to an idea long favoured by geologists.

8

Cenozoic Era (The Paleogene World)

The close of the Cretaceous Period marked a major transition in earth history. Never again did coccolithophores abound to the degree that they formed massive chalk deposits; scarcely any belemnoids survived, and ammonoids, rudists, and marine reptiles were gone from the seas. What remained were marine taxa that persist as familiar inhabitants of modern oceans, including bottom-dwelling mollusks, teleost fishes, and other groups that will be described in this chapter. On the land, the flowering plants of the Paleogene resembled those of latest Cretaceous time in many ways, but animal life differed dramatically from that of the previous period. Taking the place of the dinosaurs were the mammals, which were universally small and inconspicuous at the start of the Paleogene interval but in many ways resembled modern mammals by the period's end.

The most profound geographic change during Paleogene time was a refrigeration of the earth's polar regions, which resulted in a chilling of the deep sea and, ultimately, in continental glaciation, a phenomenon that will be discussed in the next chapter. Of great importance of North America were Paleogene mountain-building events in the west that foreshadowed subsequent Neogene uplifts of ranges like the Sierra Nevada and Rocky Mountains. In southern Europe, the Paleogene elevation of the Alps was of Comparable significance.

The sediments that record these and other Cenozoic events are for the most part unconsolidated, or soft, although most Neogene carbonates and some siliciclastics are lithified. Cenozoic deoposits are widespread in Europe and the distincution between their fossil biotas and those of Mesozoic deposits was recognized early in the history of geologic investigation. Subdivision of the Cenozoic Era is less clear-cut; today, many geologists divide the era into two periods: *Paleogene Period*, which includes the *Paleocene*, Eocene, and *Oligocene* epochs, and the *Neogene period*, which includes the *Miocene*, *Pliocene*, *Pleistocene*, and *Recent* (or *Holocene*) epochs. This Paleogene-Neogene classification has become increasingly popular during the past two decades and has thus been adopted in this book. Traditionally, however, the Cenozoic Era has been divided into two periods of quite different lengths : the *Tertiary*, which enocompasses the interval from the Paleocene through the Pliocene, and the *Quaternary*, which includes only the Pleistocene and Recent epochs (an interval of less than 2 million years). The advantage of the Paleogene-Neogene division is that it yields two periods of more similar duration.

Fortunately, the question of how to divide the Cenozoic into periods is not of great consequence, because geologists tend to focus less on the periods of this era than on its epochs. The reason is that the Cenozoic epochs, though shorter than Paleozoic or Mesozoic periods, are so recent that their sedimentary records are well displayed and can be analyzed in great details—especially now that the history of deep-sea sediments and their fossil biotas have come to light through the use of coring techniques.

The first formally recognized Paleogene epoch was the Eocene, which was established in 1833 by Charles Lyell on the basis of deposits found in the Paris and London basins. Lyell named and described the Eocene Series in his great book *Principles of Geology*, a work that also served to popularize the uniformitarian view of geology. As it was initially designated, the Eocene Series constituted a thick stratigraphic interval that corresponded to what is now recognized as the entire Paleogene System. In other words, the Eocene Series originally included the sediments now assigned to the Paleocene and Oligocene series. It was not until 1854 that Heinrich Ernst von Beyrich distinguished the Oligocene Series from the Eocene in Germany and Belgium on the basis of fossils, and then, in 1874, W. P. Schimper established the Paleocene Series on the basis of distinctive fossil assemblages of terrestrial plants in the Paris Basin.

Worldwide Events

We will depart from the format of the preceding five chapters by discussing life of the Paleogene Period under the heading of " Worldwide Events." The logic underlying this approach is that Paleogene life is so familiar to us that it requires no special introduction; its most interesting features are the major expansions and contractions of certain taxonomic groups on a worldwide scale. Nor will we devote a special section to Paleogene paleogeographic, since the Paleogene lasted no more than 41 million years—aspan in which paleogeographic changes occurred on such a small scale that they are best considered under the heading "Regional Events."

Evolution of Marine Life

The present marine ecosystem is for the most part populated by groups of animals, plants, and single-celled organisms that survived the extinction at the end of the Mesozoic Era to expand during the Cenozoic. Many benthic foraminifers, sea urchins, cheilostome bryozoans, crabs, snails, bivalves, and teleost fishes survived in sufficiently large numbers to assume prominent ecologic positions in Paleogene seas. Perhaps the biggest beneficiaries of the terminal Cretaceous extinction, however, were the reef-building corals, which had relinquished their dominant reef-building role to the rudists in mid-Cretaceous time but reclaimed it following the rudists' extinction. Unrortunately, few Paleocene coral reefs have been found, partially because of the relative rarity of exposed Paleocene strata representing tropical conditions. The rarity of Paleocene coral reefs also seems to indicate that corals did not recover quickly from the terminal Cretaceous mass extinction. During the warm Eocene Epoch that followed the Paleocene, however, corals were again widespread.

The calcareous phytoplankton, including the coccolithophores, suffered severe losses at the end of the Cretaceous Period but rediversified somewhat during the Paleogene. These forms as well as the diatoms and dinoflagellates, which were not as adversely affected, have accounted for most of the ocean's productivity throughout the Cenozoic Era, just as they did in Cretaceous time.

Although many elements of Paleogene marine life closely resembled those of Late Cretaceous age, some forms were dramatically new. Perhaps the most distinctive marine organisms of this period were the *whales,* which during the Eocene Epoch evolved from carnivorous land mammals and quickly achieved success as large marine predators. Joining the whales as replacements for the reptilian "sea monsters" that were the top carnivores of the Mesozoic Era were enormous sharks. Unlike the whales, however, sharks descended from similar creatures that lived during Cretaceous time.

It would appear that the marine ecosystem expanded during Paleogene time to including new niches along the fringes of the oceans. *Sand dollars*, for example, which are the only sea urchins able to live along sandy beaches evolved at this time from biscuit-shaped ancestors, and new kinds of bivalve mollusks also invaded exposed sandy coasts. Both of these new groups have successfully inhabited shifting sands by virtue of their ability to reburrow quickly when washed from the sand by waves. Other newcomers to the ocean margins were penguins, a group of swimming birds of Eocene origin and the possibly the *pinnipeds*, a group that includes walruses, seals and sea lions. It is widely believed that the pinnipeds evolved before the beginning of the Neogene Period, although this group left no known Paleogene fossil record.

Evolution of Terrestrial Plants

For land plants, the story is quite different. The transition to the Paleogene was apparently not marked by any drastic change in the character of terrestrial floras; instead, the great radiation of flowering plants simply continued. In the process, modern families of flowering plants were ones that are alive today, and although many modern plant genera has not get evolved, forests plants evolved. By the beginning of Oligocene time, some 37 million years ago, about half of all genera of flowering had taken on a distinctly modern appearance. The oldest known rose species, for example, is of Late Eocene age.

One major evolutionary event that did take place during the Paleogene interval was the origin of the *grasses*. Although these usually low-growing flowering plants were present before the end of the Paleocene, they did not reach their full ecologic potential until Late Oligocene and Miocene time. Early grasses were apparently confined to wooded or swampy areas. Like the modern sedges that form marshlands along continental coastlines, the mode of growth of early grasses did not allow their leaves to grow continuoulsy and thus to recover from heavy grazing by animals of the sort that inhabit open country in large numbers. It was only by virtue of an adaptive breakthrough—the origin of continuous growth, the process that forces us to cut our lawns every week or two—that grasses were ultimately able to invade open country with great success. Once they were able to survive the effects of heavy grazing by animals, grasses quickly spread over vast expanses of the earth to form grasslands.

Another interesting evolutionary point consider is that, with the enormous numbers of individual grass plants populating marshes and grasslands, effective reproduction would be difficult if these plants required pollination by insects. Thus, it comes as no surprise that grasses are wind-pollinated plants. It should be noted, however, that grasses can also carpet large areas by means of budding, a nonsexual reproductive process in which individual plants put out underground runners that sprout new plants, thus rapidly and densely colonizing the soil.

Early Paleogene Terrestrial and Freshwater Animals

The mammals, having inherited the world from the dinosaurs, underwent a remarkably rapid adaptive radiation during the early part of the Cenozoic era. It was probably through both competition and predation that the dinosaurs prevented the mammals from undergoing any great evolutionary expansion during Mesozoic time.

At the start of the Paleocene Epoch, when they first had the world to themselves, most mammals were small creatures that resembled modern rodents; at this time, it would appear that no mammal was substantially larger than a good-sized dog. Furthermore, most mammal species tended to remain generalized in both feeding and locomotory adaptations; those that dwelt on the ground generally retained a primitive limb structure in which the heels of the hind feet and the "palms" of the front feet touched the ground during movement. Perhaps 12 million years later, however, by the end of Early Eocene time, mammals had diversified to the point that most of their modern orders were in existence. Bats already fluttered through the night air, for example, and, as we have seen, large whales swam the oceans.

Also included among the Paleocene mammals were groups that survived from Cretaceous time, such as marsupials, multituberculates, and placental mammals called insectivores. Other Paleocene mammals have been assigned by some experts to the *Primates*, the order to which modern humans belong. Although these small animals were quite different from monkeys, apes, or humans in many respects, by Early Eocene time they did climb with grasping hindlimbs and

Fig. 8.1. Reconstruction of the early primate Cantius, a small genus of Early Eocene time. This arboreal animal had large toes on its hind feet and nails much like our own, and it apparently jumped from limb to limb.

forelimbs that foreshadowed our own hands and feet. By mid-paleocene time true mammalian carnivores—members of the living order Carnivora—had emerged. This is the order to which nearly all living carnivorous placental mammals belong. By the end of Paleocene time, the earliest members of the horse family had evolved as well; these animals were no larger than small dogs, but by the end of the epoch larger herbivorous mammals, some the size of cows, had also appeared.

The Eocene Epoch was marked by a continued increase in the variety of mammals. The number of mammalian families doubled to nearly 100, approximating that of the world today. In addition, more modern varieties of hoofed herbivores began to appear. Most animals of this kind are known as *ungulates* and these are divided into *odd-toed ungulates*, (living horses, tapirs, and rhinos) and *even-toed ungulates*, or cloven-hoofed animals (cattle, antelopes, sheep, goats, pigs, bisons, camels, and their relatives). The odd-toed ungulates expanded before the even-toed group did, but primitive even-toed ungulates were also present early in the Eocene of the modern even-toed types camels and relatives of the present-day chevrotain (the Oriental "mouse deer") evolved before the end of the epoch. The earliest members of the *elephant* order also appeared during Eocene time; The oldest known genus, *Moeritherium*, was a bulky animal about 3 meter (~10 feet) long, that possessed rudimentary tusks and a short snout rather than a fully developed trunk. The *rodents*, which had originated in the Paleocene, continued to diversify as well, but their success may have been attained at the expense of the archaic multiuberculates, which were

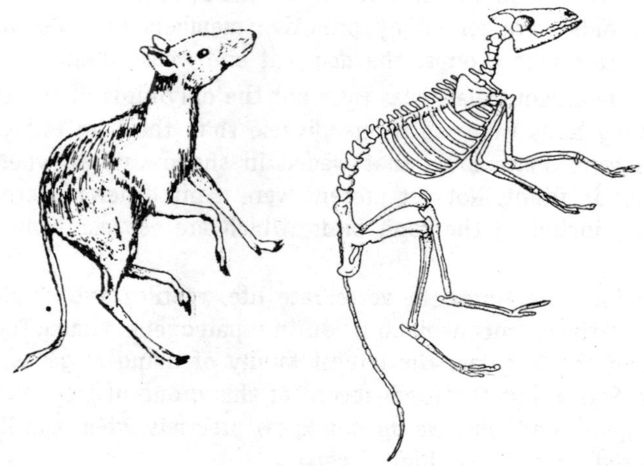

Fig. 8.2. Diacodexis, an early even-toed ungulate, or cloven-hoofed herbivore. Hyracotherium was an early odd-toed ungulate. The limb structure of Diacodexis shows that it was an unusually adept runner and leaper for Early Eocene time.

Fig. 8.3. Moeritherium, an early member of the elephant group. This elongate animal stretched to a length of about 3 meters (~10 feet). During the Eocene Epoch, it probably wallowed in shallow waters and grubbed for roots or other low-growing vegetation.

also specialized for gnawing seeds and nuts. As the rodents expanded during the Eocene, the multituberculates declined, finally becoming extinct early in the Oligocene Epoch.

Among the animals that preyed on the herbivorous mammals described above were groups that had their evolutionary origin in the Paleocene Epoch. These include the superficially doglike *mesonychids* and the *diatrymas,* which were huge flightless birds with powerful clawed feet and enormous slicing beaks. The diatrymas disappeared toward the end of the Eocene Epoch. At that time the mesonychids were joined by primitive members of three familiar modern carnivore groups: the *dog, cat* and *weasel* families.

The monstrous diatrymas were not the only birds of the Eocene, but flying birds were much less diverse than they are today. Most species were shore birds that waded in shallow water when they were not in flight. Not yet present were many other modern kinds of birds, including the song birds, which are especially numerous today.

As for other forms of vertebrate life, reptiles and amphibians were relatively inconspicuous during paleogene time. The first record of the Ranidae, the largest family of living frogs, is in the Eocene Series, but the fossil record of this group of fragile animals is not good, and thus we do not know precisely when the Ranidae originated or attained high diversity.

There is no question that the insects took on a modern appearance with the origin of several modern families in Paleogene

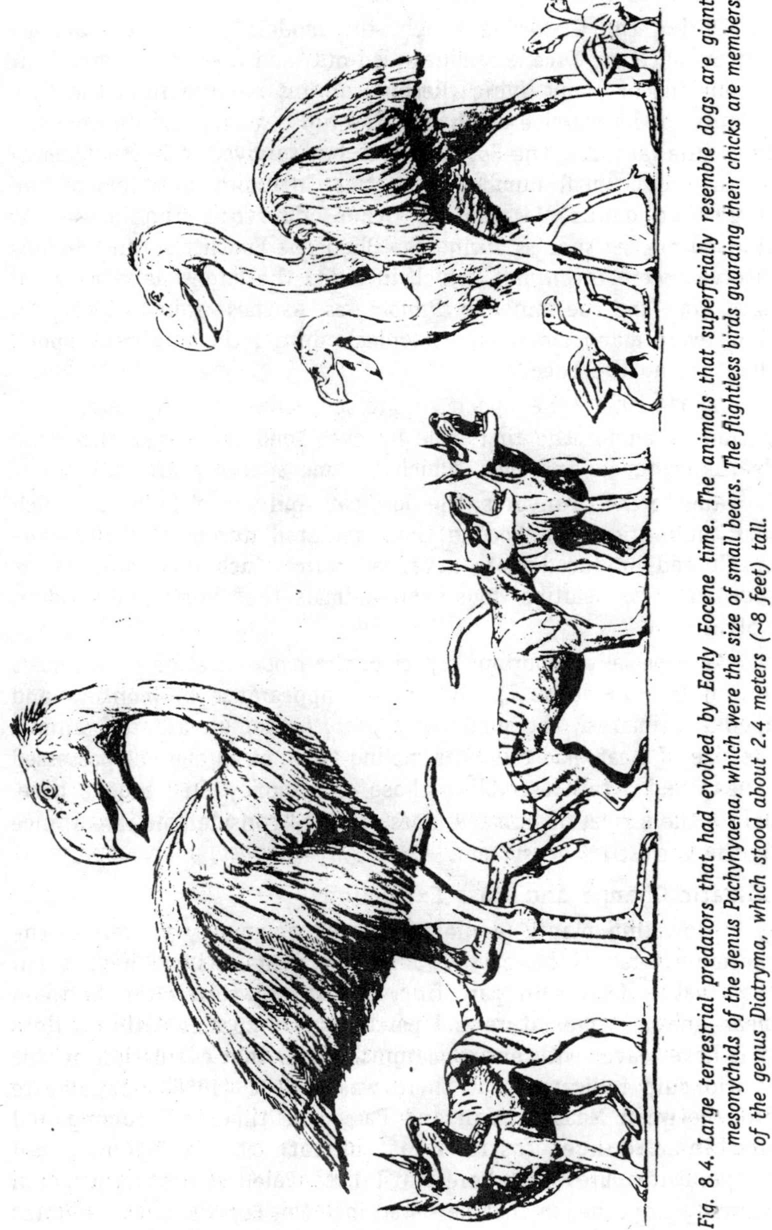

Fig. 8.4. Large terrestrial predators that had evolved by Early Eocene time. The animals that superficially resemble dogs are giant mesonychids of the genus Pachyhyaena, which were the size of small bears. The flightless birds guarding their chicks are members of the genus Diatryma, which stood about 2.4 meters (~8 feet) tall.

time. Several Oligocene forms, which have been preserved with remarkable precision in amber, closely resemble living species.

Mammals of the Oligocene Epoch

During the Oligocene Epoch, the modernization of mammals continued. Many Eocene families died out, and there were important expansions of many living (Recent) groups ranging from the tiny rodents to the massive elephants. The horse family had disappeared from Eurasia during the Eocene Epoch but survived in North America by way of a small number of species, including members of the three-toed genus *Mesohippus*. Other odd- toed ungulates that enjoyed greater success during the Oligocene Epoch than in previous intervals were the *rhinos*, which included the largest land mammal of all time and the rhinolike *titanotheres*. As these animals illustrate, there were many more large mammals during the oligocene Epoch than during the Eocene.

As the Oligocene Epoch progressed, odd-toed ungulates were outnumbered for the first time by even-toed ungulates, including deerlike animals and pigs, which became specially diverse.

Among the carnivores, the dog, cat, and weasel families, which had their origins in Eocene time, rediated during the Oligocene Epoch and produced more advanced forms, including large saber-toothed cats, bearlike dogs, and animals that resembled modern wolves.

An especially important aspect of the modernization of mammals during the Oligocene Epoch was the appearance of *monkeys* and apelike primates. The genus *Aegyptopithecus*, an arboreal animal the size of a cat, had teeth resembling those of an ape but possessed a head and a tail resembling those of a money. In Neogene time, before the arrival of humans, apes attained considerable importance in the terrestrial ecosystem.

Climatic Change and Mass Extinction

Flowering plants (angiosperms) are commonly viewed as the thermometers of the past 100 million years. As we have seen, their value derives in part from the strong correlation between mean annual temperature and percentage of species within a flora that have leaves with entire margins. Recall that evaluation of this relationship indicates that there was a 10°C (18°F) temperature drop between Maastrichtian and Paleocene time in Wyoming and Montana. Younger fossil floras indicate that relatively cool temperatures prevailed here until Late Paleocene time, when a warming trend began that persisted into the Eocene. Data for floras of the Mississippi Embayment show a similar pattern.

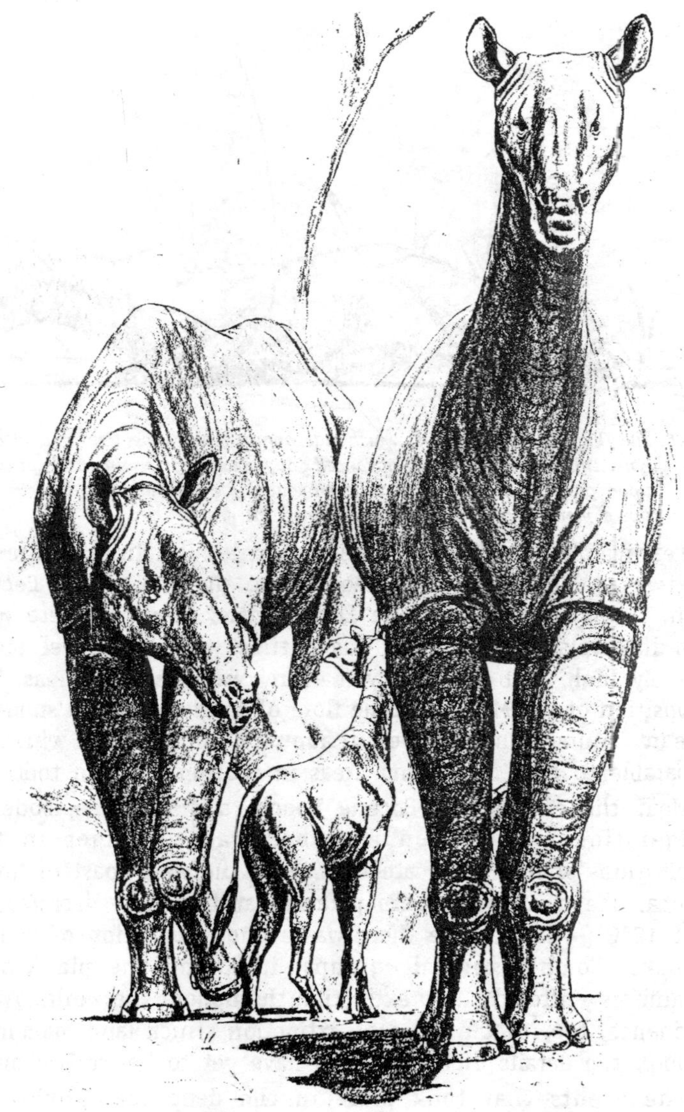

Fig. 8.5. Indrichotherium (formerly Baluchitherium), which, as far as we now know, was the largest mammal ever to walk the earth. This Oligocene giant, which belonged to the rhinoceros family, stood about 5.5. meters (~18 feet) at the shoulder. This is the height of the top of the head of a good-sized modern giraffe.

Fig. 8.6. The Oligocene cat Dinictis, which was approximately the size of a modern lynx. This animal possessed advanced adaptations for running and springing upon prey. Its canine teeth were elongated for stabbing, but Dinictis was not a member of the true saber-toothed cat group.

Leaf-margin data for several North American localities also reveal that two cool episodes took place before the end of the Eocene Epoch. During Middle and Late Eocene time, however, there were warm intervals as well. This was a time when sea level stood relatively high, and so there were many warm, shallow seas. The composition of the Middle Eocene flora of southern Alaska suggests a mean annual temperature of about 22°C (~72°F), which is comparable to that of lowland areas in southern Mexico today.

Near the end of the Eocene Epoch, a dramatic episode of refrigeration occurred on a global scale. Changes in the compositions of floras indicate that along the west coast of North America, the mean annual temperature at this time declined by about 12°C (~ 22°F). Here there was a large extinction of marine mollusks. Some taxonomic groups, including the planktonic foraminifers, suffered heavy extinction throughout the world. There is evidence, too, that a pulse of extinction struck land mammals, although the details of this episode have yet to be worked out.

The events that took place in the deep sea provide an explanation for the episode of cooling and extinction that occurred in Late Eocene time. Study of deep-sea cores reveals not only a major extinction of abyssal benthic foraminifers and ostracodes but also an increase of the ratio of oxygen 18 to oxygen 16 in the skeletons of the foraminifers in equatorial regions and near Antarctica. This increase apparently reflected the first growth of

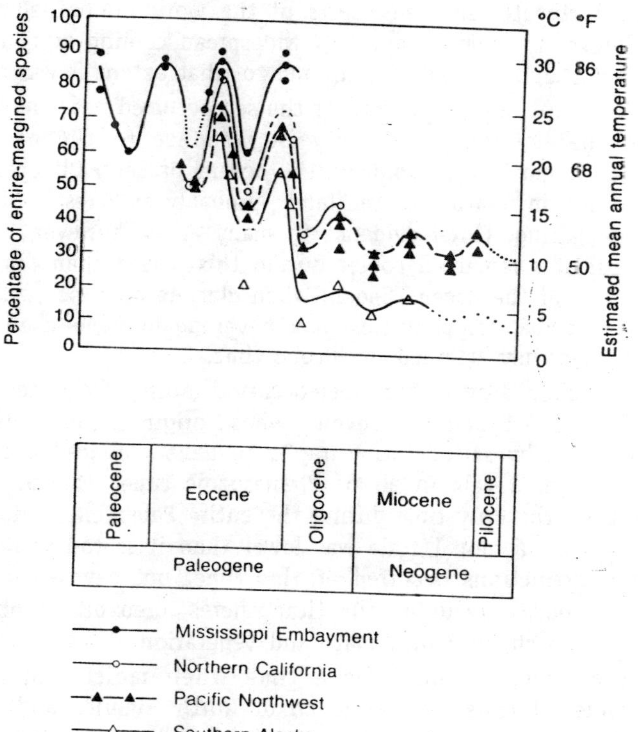

Fig. 8.7. Curves showing estimates of Cenozoic temperature changes in four areas of North America. The curves are based on percentages of entire-margined leaves in fossil floras. The fact that the curves follow parallel paths where they overlap in time suggests that the trends hold for large areas of the earth's surface.

glacial ice on and adjacent to Antarctica, with the lighter oxygen isotope (^{16}O) accumulating preferentially in this ice. The growth of glacial ice on Antarctica was a prelude to the greater expansion of high-latitude ice sheets that took place near the end of the Cenozoic Era during what is informally known as the recent Ice Age.

It was during the Late Eocene cooling episode that the psychrosphere—the cool bottom layer of the ocean—came into being as cold, dense polar water began to sink to the deep-sea . The short interval of isotopic change that characterizes deep-sea cores suggests that the psychrosphere formed in less than 100,000 years; the temperatures of bottom waters over large areas of the deep sea dropped by perhaps 4 to 5°C (~7 to 8°F).

It is assumed that the upwelling of psychrospheric waters affected climates in many parts of the world. Nonetheless, it is not known to what extent this widespread cooling was an effect of the Antarctic glacial buildup and to what extent it was a cause.

A modest global lowering of the sea occurred near the Eocene-Oligocene boundary, probably in response to the growth of Antarchtic glaciers. In contrast the oceans of Early Oligocene time stood high in relation to the land—probably as a result of partial glacial malting. Floral evidence in many areas, however, indicates that climates remained cooler during this period than during the early part of the Eocene Epoch, when glaciers had not yet formed. In fact, global temperatures may never again have risen to the levels that they attained in Eocene time.

A sudden drop in sea level occurred during the latter part of the Oligocene Epoch. This event, whose origin is not known, has been shown by seismic stratigraphy to have produced one of the lowest ocean levels in all of Phanerozoic time. This appears to have been the only time during the entire Paleogene Period when sea level on a global scale was lower than it is today. No heavy marine extinctions occurred at this time, but several groups of land mammals, including the titanotheres, died out, probably in response to changes in climate and vegetation.

The Eocene Epoch was a time when moist tropical and subtropical forests cloaked much of North America and Eurasia, but the cooling events of late Eocene and Oligocene time altered this condition forever. Paleobotanical studies show that during the Oligocene Epoch, in response to cooler and drier conditions, savannahs—grassy plains with scattered trees and shrubs—spread across large areas of major continents. At the same time, moist subtropical and tropical forests came to be confined largely to low latitudes, where they remain to this day in the from of jungles and rain forests. Because many of the evolutionary changes associated with this climatic transition took place in Miocene time.

Regional Events

In examining important regional events of the Paleogene Period, we will travel to the ends of the earth—first to the South Pole, where Antarctica became separated from Australia and developed its icy cover, and then toward the North Pole, where land areas of North America and Eurasia were more closely connected than they are today. Next we will look at the history of the Cordilleran region of North America, where the Laramide orogeny yielded many structures that remain conspicuous in the Rocky Mountains today. Before leaving North America, we will examing the nature of deposition along the Gulf Coast, where large volumes of petroleum are trapped in Paleogene sediments. Our final stop will be Europe, where many fossiliferous deposits accumulated and the modern structure of the Alps began to form.

Antarctica and the Origin of the Psychrosphere

The only major event of continental rifting that took place exclusively during the Cenozoic Era was the separation of Australia from Antarctica, which occurred during the Eocene Epoch. With the inception of this rifting early in Eocene time, Australia began to move northward, and the resulting alteration of wind and water circulation near the South Pole had consequences that extended over much of the planet.

Even before its separation from Australia, Antarictica had been centered over the South Pole, but it remained warm because its shores were bathed in relatively warm waters from lower latitudes. Oxygen isotope analyses of foraminifers from deep-sea cores suggest, however, that sea temperatures adjacent to Antarctica dropped during Eocene time as Australia drifted northward. A major cause of this cooling was the formation of the circum-Antarctic current that, up to the present time, has flowed clockwise around Antarctica partly because of the prevailing winds that result from the earth's rotation in the opposite direction. During Oligocene time, despite the fact that narrow continental connections remained with Australia and South America, the circum-Antarctic current was established near the sea surface, and the apparent consequence was that warm waters from low latitudes failed to reach Antarctica. As a result, the continent and the waters adjacent to it, being largely isolated in a polar position, cooled down.

In response to this cooling—and perhaps also in response to worldwide cooling resulting from some other, unknown factor—the first sea ice began to form around Antarctica near the end of the Eocene, and waters of near-freezing temperature sank to the deep sea and spread northward, forming the psychrosphere. The direct result was an extinction of bottom-dwelling foraminifers and other larger inhabitants of the deep-sea floor. At the same time—and perhaps as a result of the Arctic cooling—the earth experienced more widespread climatic deterioration. Floras reveal that temperatures plummeted at this time even in the terrestrial habitats of the Northern Hemisphere.

Changes in oxygen isotopes within sea water, as recorded in fossil foraminifers, reveal that a slow buildup of ice occurred on Antarctica during Oligocene time. Once the Antarctic "Cooling system" was in place, the mean annual temperatures at high latitudes never again reached their Eocene levels; in fact, the climatic deterioration at the end of the Eocene fostered further deterioration later in Cenozoic time. The recent Ice Age of the Northern Hemisphere, would never have occurred if global temperatures had not already been lowered.

Australia, on its journey northward during the Paleogene Period, experienced only minor marine encroachment along its margins—except during Late Eocene time, when global transgression reached its Paleogene peak. In the next chapter, we will review the history of Australia during Neogene time, when the migration of that continent caused it to experience major climatic modifications.

The Top of the World: Changing Positions of Land and Sea

Major geologic and geographic changes also took place during the Paleogene Period within 30° latitude of the North Pole. Many of these changes can be read from the ages of various segments of the deep-sea floor. The Arctic deep-sea basin was in existence during the Cretaceous Period, but until nearly the end of Cretaceous time, it remained separated from the Atlantic, except by way of shallow seas, because North America Greenland, and Eurasia were still united as a single landmass. Then, as we saw in the previous chapter, the Mid-Atlantic Rift proceeded northward along two forks, splitting Greenland from North America on the west and Eurasia on the east. Finally, in mid-Paleogene time, the pattern of plate movement simplified as the western arm of rifting ceased to be active, ending the relative movement of Greenland away from North America. Since that time, the rifting of the northern Atlantic has been limited to the area between Greenland and Scandinavia. It has been suggested that plate movements in this area, like those in the south polar region contributed to the origin of the psychrosphere. The suggestion is that early in Paleogene time the Arctic Ocean was isolated from larger oceans to the south and thus retained its frigid waters. Then, late in Eocene time, rifting between Greenland and Scandinavia proceeded far enough to allow a channel to open up between the Arctic basin and the Atlantic, causing dense, frigid Arctic waters to spill into the larger ocean as part of the psychrosphere.

Meanwhile, throughout the Cenozoic Era, continental crust has continued to separate the Arctic and Pacific basins; Alaska and Siberia are connected by a stretch of continental crust despite the fact that this connected segmet now lies submerged beneath the Bering Sea. During much of the Cenozoic Era, this segment, now lies submerged beneath the Bering sea. During much of the Cenozoic Era, this segment, which is known as the *Bering land bridge*, Stood above sea level, serving as a land corridor between North America and Eurasia. Throughout Paleogene time, this corridor remained open, allowing mammals and land plants to migrate, between Asia and North America. As we will see, however, during the Neogene Period, more frequent inundations of the sea prevented the exchange of species between the old and new worlds.

Tectonics of Western North America

Mountain-building activity in the Cordilleran region of North America Continued into the Paleogene Period, but with a number of changes. Figure 8.8, which summarizes the orogenic history of the eastern Cordilleran region, shows that the Sevier episode occupied almost the entire Cretaceous Period. In latest Cretaceous time, however, a new style of tectonic activity was initiated, and it persisted through the Paleocene and well into Eocene time. The episode characterized by this new style is known as the Laramide orogeny.

The Laramide orogeny was not unusual in the north or in the south. In the north, extending from the United States into Canada, there remained an active belt of igneous activity and inland from it, an active fold-and-thrust belt. Thrust sheets of enormous proportions are spectacularly exposed in the Canadian Rockies. A similar pattern of tectonism persisted both in the southern United States and in Mexico.

The unusual features of the Laramide orogeny were in the central part of the western united States, where a broad area of tectonic quiescence extended from central Utah through Nevada to California. East of this inactive region there was a strange pattern of tectonism in which large blocks of crystalline basement rock were uplifted in a belt extending from Montna to Mexico. The largest of these blocks were centered in colorado, where the

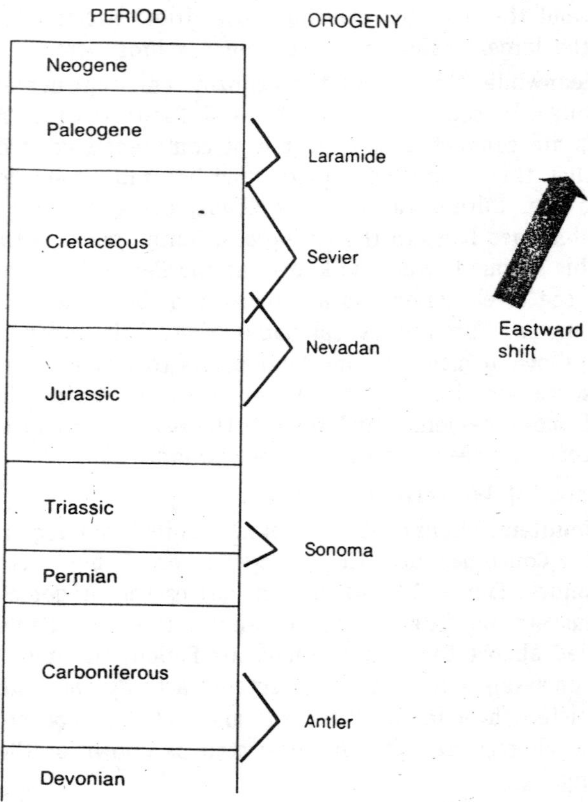

Fig. 8.8. Summary of major orogenic events in the eastern Cordilleran region. Between Jurassic and Paleogene time, orogenic activity migrated eastward.

Ancestral Rocky Mountains had formed as basement uplifts more than 200 million years earlier, late in the Paleozoic Era.

What created the unusual tectonic pattern that characteriszed the central part of the Cordilleran region? Note that the Paleogene basement uplifts were for the most part positioned well to the east of Sevier orogenic activity and recall, in addition, that an eastward migration of orogenic activity culminating in the Sevier orogeny took place during the Mesozoic Era. The widely favoured explanation for the earlier eastward shift applies to the Laramide shift as well: A central segment of the subducted plate that passed beneath North America began to penetrate the mantle at a still lower angle, extending a great distance eastward before becoming deep enough to melt and to create igneous activity within the overlying crust.

Father west, along the coast, the Great Valley of California continued to receive marine sediment. While northern California and the Sierra Nevada region remained as highlands. A separate basin in Washington and oregon received deep-water sediments and layers of pillow lava. Here the Olympic Range began to form along a sharp inward bend of the subduction zone.

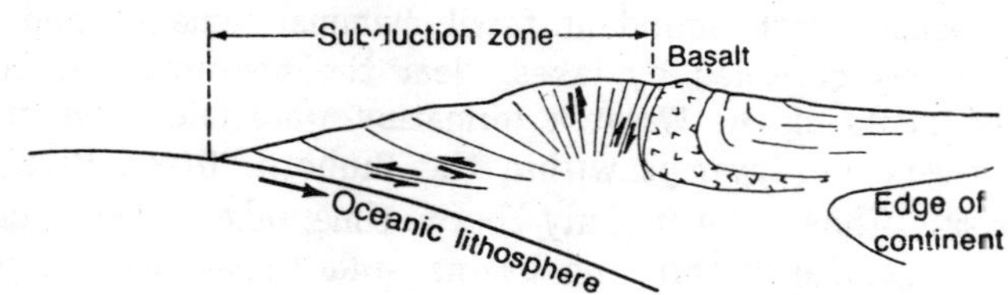

Fig. 8.9. Formation of the Olympic Range in an embayment of the Pacific border of the state of Washington. Subduction piled sediments against basaltic rocks.

Having reviewed the general pattern of Paleogene orogenic activity in the Cordilleran region, let us now focus on the eastern belt of uplifts, which stretches from Montana to New Mexico. Because this is the region in which the central and southern Rocky Mountains developed during Neogene time, Paleogene events that preceded the uplift of this segment warrant special attention. Deformation here began in latest cretaceous time with the origin of north-south trending ranges and basins; in Utah and Wyoming, these structures lay along the eastern margin of the northern fold-and-thrust belt. The basement uplift farthest to the east formed the Black Hills of South Dakota. In Colorado, many ranges were formed by the elevation of large basement uplifts along thrust faults. It has been suggested that these uplifts were produced by a slight clockwise rotation of the Colorado Plateau, which behaved as a rigid crustal block, absorbing some of the convergence along the subduction zone to the west. It is not known why there was little volcanism in this central region.

Today, of course, many peaks and ridges of the central and southern Rocky Mountain uplifts stand at very high elevations; the front Range uplift, for example, rises far above the high plains of eastern Colorado. It is inappropriate, however, to evaluate the effects of the Laramide orogeny simply by viewing the elevations of the Rocky Mountains today, since, as we will see, the high elevations of the modern Rockies reflect post-Laramide uplift. During the Laramide orogeny, erosion nearly kept pace with uplift in most areas of the United States, resulting in a regional topography that was less rugged than that which characterizes the area today. Basins in front of elevated areas were receiving large volumes of rapidly eroding material.

By the end of Early Eocene time, the regional north-south pattern had weakened, and individual basins were experiencing independent histories. Most of these basins received alluvial and swamp deposits with abundant fossil mammal remains, and some were at times occupied by lakes. Near the beginning of Eocene time, sediments of the Wasatch formation were laid down in and around rivers and swamps within the Bighorn, Green River, and Uinta basins. Then, later in Early Eocene time, lakes came to occupy most of the areas within the basins, and these lakes survived, sometimes in restricted areal extent, throughout much of the Eocene Epoch. The famous Green River deposits accumulated in and around the margins of these lakes. The transition from Wasatch sediments to those of the Green River is well displayed today in many poorly vegetated areas of Wyoming and Utah. Plant remains in these sediments reveal that Eocene climates in this region were quite different from those of today, having been warm (perhaps subtropical) at times.

Wilkins Peak Member of the Green River Formation includes alluvialfan, braided-stream, salt-flat, and lake deposits. These lake deposits, which are extremely well laminated, have commonly been termed "oil shales" because algal material within them has broken down to yield vast quantities of petroleum. Unfortunately, this petroleum is disseminated throughout the rock and has thus proved difficult to extract. Nonetheless, the Green River deposits form the largest body of ancient lake sediments known, and they may eventually serve as a valuable source of fuel. The fine undistrubed lamination of the Green River lake deposits accounts for the remarkable preservation of a host of animal and plant fossils, including delicate creatures such as frogs and insects.

Another interesting group of rocks found in the region of Paleogene basins and uplifts are the volcanics that form the Absaroka Range in western Wyoming and Montana. A large portion of Yellowstone National Park lies within the Absarokas; here, the still-active geysers and hot springs serve as evidence that igneous activity has not ceased completely. Recall that Yellowstone seems to represent a hot spot in which igneous activity is localized. During Eocene time, this area stood at the eastern margin of the volcanic belt of the Pacific northwest. Volcanism at this time was episodic, and each episode yielded a variety of rocks, including volcanic igneous units as well as sedimentary units composed of volcanic debris. All of these volcanic episodes, however, were catastrophic, destroying entire forests and we must assume, the animal life within them. Fossilized leaves, needles, cones and seeds reveal the presence of lowlands with subtropical vegetation, like that simultaneously occupying the Green River Basin, and of slightly cooler uplands. Today, the remnants of the Eocene forests can be

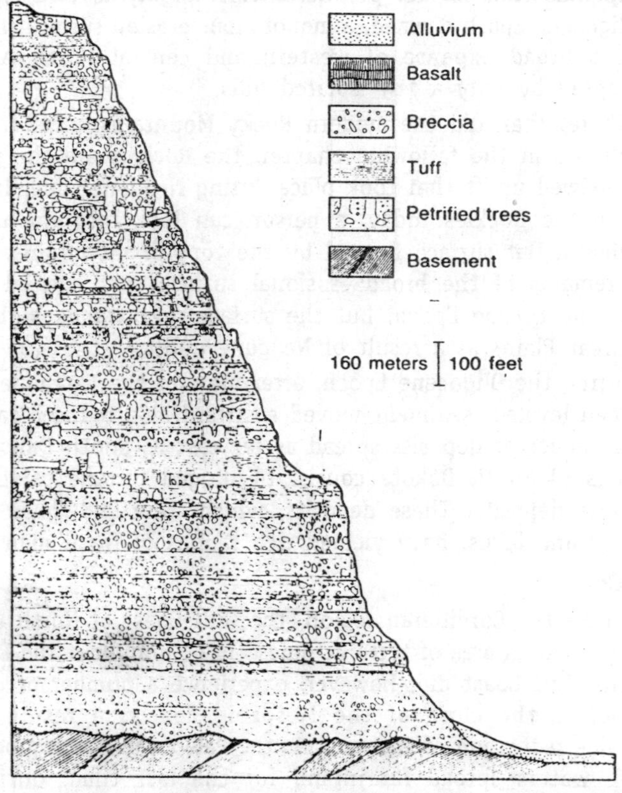

Fig. 8.10. Succession of 27 petrified forests now exposed along the Lamar Valley in Yellowstone National Park. Each of these Eocene forests was destroyed by a volcanic eruption.

seen at high elevations in Yellowstone National Park, where trees are preserved upright as stumps buried by lavas, mud flows, and flood-deposited volcanic debris. More than 20 successive forests, all of which were killed in this way, can be identified along the Lamar Valley in Wyoming.

As the Eocene Epoch drew to a close, the level of volcanism declined sharply in every part of the northwestern United States except Oregon and Washington. In a few areas to the east, including the Absarokas, volcanic activity persisted but was weak and sporadic. In addition, most of the depositional basins that lay between Montana and New Mexico were filled with sediment by the end of Eocene time. The Laramide orogeny was completed, and the uplands that it had produced were largely leveled; thus, as the Oligocene Epoch dawned, a monotonous erosion surface stretched across a broad expanse of western and central North America, interrupted by only a few isolated hills.

Where, then did the modern Rocky Mountains come from? As we will see in the following chapter, the Rockies are the product of a renewed uplift that took place during Neogene time. In many areas in the Rockies today, a person can look into the distance and view a flat surface formed by the tops of mountains. This is what remains of the broad erosional surface that existed at the end of the Eocene Epoch, but the surface now stands high above that Great Plains as a result of Neogene uplift.

During the Oligocene Epoch, after most of the Laramide uplifts had been leveled, sediment moved eastward at a reduced rate, but a thin veneer of deposits spread as far east as South Dakota. The Badlands of South Dakota consist of rugged terrane carved from Oligocene deposits. These deposits, which were laid down largely in rivers and lakes, have yielded rich faunas of fossil mammals.

Gulf Coast

Unlike the Cordilleran region, the Gulf Coast of North America has remained an area of tectonic quiescence throughout the Cenozoic Era. The Gulf Coast did, however, experience a number of changes just before the close of the Mesozoic Eon as a result of the worldwide regression of the seas. As the Paleocene Epoch progressed, marine waters spread far inland for the last time, during the Phanerozoic Eon. In fact, North Atlantic waters spread all the way to the Dakotas in the form of the Cannonball Sea, a body of water named for a stratigraphic unit known as the Cannonball Formation. Subsequently, the seas retreated to the Mississippi Embayment, where a thick sequence of Eocene marine sediments accumulated and finally, during the Oligocene Epoch, the seas withdrew to the approximate position of the present shoreline. During Oligocene time, the waters rose again, but they never spread as far inland as they had been during most of Eocene time. The total thickness of Paleogene sediments near the present coastline of the Gulf of Mexico exceeds 5 kilometers (~3miles), largely because of the enormous quantities of sediment carried to the region by the Mississippi river system. The Cenozoic clastic wedge in this region, though largely buried, has been studied in great detail during the highly successful search for pertroleum there.

Europe

The separation of northern Europe from Greenland during the Paleogene Period had consequences that extended far to the east. Northern lrenland and western Scotland experienced intensive igneous activity, which appears to have been related in some way to the presence of the new rift that separated Europe and Greenland. Lava flows more than 1.5 kilometers thick (~1 mile thick) accumulated in these parts of the British Isles. The most scenic of these flows form Giant's Causeway in northern lreland.

Another phenomenon that was probably related to rifting was a sagging of the North Sea basin, in the center of which thousands of meters of sediment accumulated during Paleogene time. The North Sea also lapped upon the neighbouring continents, contributing extensive marine deposits to southeastern Britain and to the European mainland between northern France and Denmark.

Paleogene transgressions and regression produced an alternation of marine and nonmarine units in the London and Paris basins. Although it is a marginal unit, the London Clay of late Early Eocene age yields a rich terrestrial flora that contains about 350 species, many of which are related to species now found in the jungles of the Malay Peninsula. Also present are the fossil remains of marine crocodiles. Unlikely as its may seem considering it frigid nature today, the North Sea warmed up in Eocene time to become more or less tropical. Southeastern England appears to have had an average annual temperature of about 25°C (~78°F) compared to the modern figure of 10°C (~50°F). There were two causes: global warming during the Paleocene-Eocene transition and the establishment of a connection, in the area of the present English Channel, with warm waters to the south.

It was largely through his studies of the Paleogene mammal faunas in nonmarine beds of the Paris Basin that Baron Georges von Cuvier established the science of vertebrate paleontology at the beginning of the nineteenth century. Nonetheless, the most remarkable terrestrial fossil faunas of the European Eocene—and perhaps of the entire Phanerozoic record—are found in the lignites, or brown coals, of the Geisel Valley Germany, which lies close to the North Sea. These units are the product of decaying vegetation in Middle Eocene swamps, where even fragile flowers are preserved, fossil leaves often retain their green chlorophyll, beetles display their original iridescent colours, and frogs are so well mummified that their skin is preserved with the original cells intact and still bearing nucle! Here the dense organic layers of lignite have excluded oxygen and decay-causing oxygen-dependent bacteria, offering us a unique glimse of biological detail for biota more than 40 million years old.

Farther east, during much of Eocene and Oligocene time, and great seaway, the *Turgai Strait*, extended along the eastern margin of the Ural Mountains from the Tethys to the Arctic Sea. The presence or absence of land connections across the Turgai Strait determined whether mammals could migrate between Asia and Europe during the Paleogene Period, just as the presence or absence of the Bering land bridge governed interchange between North America and Asia and the rifting of Greenland governed migration between Europe and land areas to the west. During Late Paleocene and Early Eocene time, mammal faunas of Europe were remarkabley similar to those of North America. Many genera including the earliest horse, *Hyracotherium*, inhabited both continents. Apparently, at this time the rifting had not yet fully separated Europe, Greenland, and North America, and passage was still possible. Europe and Asia, on the other hand, were parts of a single craton but nonetheless had distinctive mammalian faunas because of the presence of the Turgai Strait. During the Oligocene Epoch, the similarities between the faunas of Asia, Europe, and North America fluctuated with the coming and going of corridors for migration.

A major transgression that occurred in Late Eocene and Oligocene time ushered in an invasion of North Sea waters that extended across all of Europe to the Turgai Strait. This invasion, which flooded northern France more extensively than had earlier Paleogene marine incursions, was responsible for an especially thick accumulation of marine deposits in the *Rhine Graben*, a fault-block basin that actually represents one arm of a triple junction of rifts. The history of the Rhine graben began in late Mesozoic time, when it appears that a mantle plume (a hot spot) became stationed beneath the crust, forming a dome. The typical three-pronged rift developed early in the paleogene Epoch, and the Rhine Graben, which was the longest of these prongs, subsided during the cenozoic Era to receive thousands of kilometers of alluvial-fan and river deposits as well as marine sediments that resulted from the gerat Oligocene transgression.

The Oligocene transgression in Europe was ended by a regression that permanently drained the Turgai Strait before the end of Oligocene time. This regression, which left the European continent standing above sea level, was the global event that caused the world's oceans to drop to one of the lowest levels of the entire Phanerozoic.

Mediterranean

In the Mediterranean region, many events of Paleogene time were associated with the westward movement of Africa in relation to Eurasia. Early in Paleogene time, the Adriatic plate began to approach the European continent. The subduction beneath the Adriatic plate produced extensive igneous activity and folding-and-thrusting, and the Alps began to assume their present form. A foredeep developed north of the Alpine suture; here, flysch accumulated until mid-Oligocene time, when molasse deposition began. The Alpine orogenic events were only one part of widespread mountain building that extended beyond the Caucasus to the Himalayas.

Meanwhile, along the southwestern border of the Mediterranean, the Atlas Mountains began to form as a result of the movement of the African plate in relation to Eurasia. Early in the Paleogene Period, marine deposition had prevailed in this part of North Africa, but it ceased during the latter part of the Eocene Epoch, when folding and thrusting formed the structures known as the High Atlas. In contrast, Libya and Egypt, which lay to the east, experienced an interval of tectonic quiescence at this time. Here, sedimentary units accumulated throughout Paleogene time. During the Eocene Epoch, thick bodies of limestone were deposited. Many of these consist primarily of the remains of large, disc-shaped foraminifers called nummulitids. The Sphinx and the great pyramids of Egypt were made of blocks of nummulitic limestone, which is relatively resistant to chemical weathering in the arid climate of the Sahara. This marine deposition was short-lived, however. As the Eocene progressed, the pattern of sedimentation here changed, owing to the northward transport of vast quantities of siliciclastic sediment from the interior of Africa. This pattern of sedimentation continues to the present day through the growth of the Nile delta. The thick Fayum deposits of Egypt record progradation of siliciclastics around the mouth of the ancestral Nile. The older Fayum sediments, which are of Late Eocene age, yield fossil marine faunas that include sharks and early whales. By latest Eocene time, however, progradation had progressed to the point at which marine faunas had given way to nonmarine biotas. These comprised huge tree trunks and a rich variety of mammals that constitute one of the world's greatest mammalian faunas of Oligocene age.

As brief as it was, the marine deposition that prevailed in the Mediterranean region during much of Eocene time was uncharacteristic of Africa. This continent, which now stands unusually high above sea level, also experienced only very minor marginal flooding by marine waters throughout the Paleogene Period.

9

Cenozoic Era (The Neogene World)

Because it includes the present, or Recent Epoch, the Neogene Period holds special interest for us. In examining the Neogene, we will learn how the modern world took shape—that is, how the ecosystem acquired its present configuration and how important topographic features assumed the forms we are familiar with today.

The boundary between the Paleogene and Neogene Systems has no great historical significance inasmuch as no mass extinction marks the close of one and the start of the other. During the Neogene itself, however, major changes in life and in the physical features of the earth have occurred even though the period has spanned only about 24 million years. The most far-reaching biotic changes were the spread of grasses and weedy plants and the modernization of vertebrate life. In addition, snakes, songbirds, frogs, rats, and mice expanded dramatically, and humans evolved from apes. In the physical world, the Rocky Mountains and the less rugged Appalachians took shape during Neogene time, as did the imposing Himalayas. The Mediterranean Sea also dried up and rapidly formed again during the neogene. The most widespread physical change on earth, however, was climatic. Continental glaciation began in the latter part of Neogene Hemisphere. Although this interval is commonly thought of as corresponding to the Pleistocene Epoch, it actually began late in the Pliocene Epoch and presumably will continue long into the future. The spreading of glaciers from polar regions is episodic, and what is formally referred to as the Recent Epoch is almost certainly only the latest of numerous interval between pulses of severe glaciation. Today, ice caps in the far north remain poised to spread southward, as they have done repeatedly during the past 2 million years or so.

All four Neogene epochs—the Miocene, Pliocene, Pleistocene, and Recent—were named by Charles Lyell, who introduced them in the 1833 volume of his *Principles of Geology*—although the Pleistocene was known as the "Newer Pliocene" until 1837. Lyell distinguished the epochs of the Neogene Period on the basis of his observations of marine strata and fossils in France and Italy, noting that Pleistocene strata were characterized by molluscan faunas in which about 90 percent of the species are still alive in modern oceans. Lyell found that Pliocene strata contained fewer surviving species and that miocene strata contained fewer still. It was not until later in the nineteenth century, however, that glacial deposits were recognized on land and found to correlate with the marine Pleistocene record.

Worldwide Events

In this chapter, as in the preceding one, we will review major biological events in the context of global environmental changes. The reason once again is that the animals and plants that currently inhabit the earth are, in general representative of more ancient Cenozoic life, which thus requires no special introduction. Instead, we will examine Neogene forms of life in the context of the world ecosystem and will learn how this ecosystem changed as a result of the climatic fluctuations that were so profound during the Neogene Period. Also, as in the previous chapter, we will not devote a special section to paleogeography, which has undergone little change during the brief Neogene interval.

Life in Aquatic Environments

Charles Darwin observed long ago that invertebrate lifetends to evolve less rapidly than vertebrate life; thus, the short Neogene Period has produced only modest evolutionary change in the marine invertebrate fauna.

Less is known about freshwater Neogene life, which has consisted largely of soft-bodied species not readily preserved in the fossil record. We will, however, note the expansion of one major group of freshwater phytoplankton that has produced preservable hard parts: the diatoms.

Adaptive radiations in the sea

Not surprisingly, the most dramatic evolutionary development in Neogene oceans has been the expansion of a group of vertebrate animals, the whales. During the Miocene Epoch, a large number of whale species came into existence; among them were the earliest representatives of the group that includes modern sperm whales, which are carnivores with large teeth, and also the group that includes modern baleen whales, which feed by straining zooplankton from sea water. Dolphins, which are specialized whales, also made their first appearance early in Miocene time.

At the other end of the size spectrum for pelagic life, the globigerinacean foraminifers, which barely survived the terminal

Fig. 9.1. Reconstruction of the marine fauna represented by fossils of the Middle Miocene Calvert Formation of Maryland. Representing the whale family were early baleen whales (Pelocetus), which strained minute zooplankton from the sea (upper left); long-snouted dolphins (Eurhinodelphis–lower left); short-snouted dolphins (Kentriodon–upper center); and carnivorous sperm whales (Orycterocetus–right).

Fig. 9.2. Sharks include the hammerhead Sphyrna, shown here under attack by a sperm whale, sand sharks of the genus Odontaspis (bottom center), the six-gilled shark Hexanchus (left center), and the giant white shark Carcharodon, one of which is killing a whale. Among the inhabitants of the sea floor were large bivalve molluscs of the scallop family (lower right).

Eocene mass extinction, expanded again early in the Miocene Epoch. Remarkably, the new array of genera resembled the globigerinacean genera that had perished in the Eocene mass extinction. Neogene globigerinaceans serve as valuable index fossils for oceanic sediments.

On the sea floor, evolutionary changes from Paleozoic time were relatively minor. One important development of Miocene time, however, was the appearance of the first *algal ridges* on coral reefs. For millions of years, coralline algae had contributed to reef growth but lacked the capacity to form a substantial ridge facing powerful waves. Protected by algal ridges, coral reefs have since Miocene time thrived along coasts pounded by heavy surf.

Expansion of freshwater diatoms

Unlike the marine diatoms, which began their evolutionary expansion in the Mesozoic Era, the dominant group of freshwater diatoms, the *Pennales*, did not evolve until early in Cenozoic time. By Miocene time, the Pennales comprised about 2000 known species and had already assumed the role that they play today as primary freshwater producers both in the planktonic realm and on lake and river bottoms.

Life on the Land

Climatic changes exerted a profound influence over Neogene terrestrials biotas, and the geographic and evolutionary modifications of biotas represented in the fossil record help us reconstruct these changes. As in Late Cretaceous and Paleogene times, flowering plant fossils represent our best gauge of climatic shifts.

Flowering plants

Climatic deterioration and an explosion of herbs. In the world of plants the Neogene Period might be described as the Age of Herbs. *Herbs*, or herbaceous plants, as we have seen are small, nonwoody plants that die back to the ground after releasing their seeds. (Defined in this way, herbs include many more plants than the few that we use to season our food.) The recent success of herbs is primarily the result of the worldwide climatic deterioration that took place during Oligocene and Miocene times when cooler, drier conditions caused forests to shrink and opened up new environments to plants such as herbs and grasses, which prefer open habitats and can withstand low rainfall. Today, there are some 10,000 species of grasses alone.

The *Compositae*, an important family of herbs that includes such seemingly diverse members as daisies, asters, sunflowers, and lettuces, appeared near the beginning of the Neogene Period only 20 to 25 million years ago, and yet today this family contains some 13,000 species, including the plants that ecologists refer to as "weeds". As any gardener knows, weeds are exceptionally good invaders of bare ground. They may not compete successfully against other plants to retain the space that they invade, but they soon disperse their seeds to other bare areas created by such destructive agents as fires, floods, or droughts, and spring up anew.

Neogene climatic changes were in part the result of local tectonic events that caused areas in eastern Africa, western North America, and South America to receive less rainfall than they had during the Paleocene Epoch. The regional changes were also partly due to global climatic trends. The presence in the Southeast Pacific of ice-rafted, coarse sediments of Early Miocene age shows that by this time Antarctic glaciers had begun to flow to the sea. Cores of deep-sea sediment also reveal that the belt of siliceous diatomaceous ooze that encircled Antarctica simultaneously expanded northward at the expense of carbonate ooze, which tends to accumulate where climates are warmer. Because of this polar cooling, the Early Miocene Age marked the beginning of strong latitudinal gradients in the distribution of oceanic plankton. Since that time, assemblages of species at high latitudes have differed greatly from those at low latitudes.

Analyses of leaf margins from terrestrial floras of North America reveal only modest fluctuations during Neogene time, a trend that contrasts sharply with the dramatic drop in temperatures that marked the end of the Eocene Epoch. There was, however, a slight cooling trend from early in the Oligocene Epoch until late in the Pliocene Epoch, when the Ice Age began. At this time, climates not only became slightly cooler but also grew drier and more seasonal. As a result, dry grasslands expanded into areas that had once been open woodlands and dense forests.

Modernization of terrestrial vertebrates

Because we are naturally interested in the origins of large mammals, we often ignore the great success of smaller creatures—or we assume that because these creatures are small, they are also primitive and have changed relatively little in the more recent geologic periods. In fact, the Neogene Period might well be labeled the Age of Frogs, the Age of Rats and Mice, the Age of Snakes, or the Age of Songbirds, since all four of these groups have undergone tremendous adaptive radiations over the past few million years.

Many species of rats and mice dig burrows in dry terrane and eat the seeds of grasses and herbs. To some extent, the success of these small rodents during Neogene time may be related to that of both the grasses and the Compositae and also, more fundamentally, to the drying and cooling of climates that favoured these plants. Perhaps we can attribute the success of modern frogs and toads, whose species number about 2000, to their remarkable ability to catch insects through the quick protrusion of their long tongues. In any case, the snakes have obviously flourished largely because of the proliferation of frogs and rodents, since few other predators can pursue small rodents down their burrows without digging. Before the start of the Neogene Period, there were few snakes except members of the primitive boa constrictor group. Today, however, the more advanced snakes of the family Colubridae include about 1400 species, many of which are poisonous.

Also poorly represented before Neogene time were the *passerine* birds, or songbirds and their relatives, which are highly conspicuous today. Presumably, these birds have also benefited from the diversification of seed-bearing species of herbs, but like frogs, they owe much of their success to the fact that they are well equipped to capture flying insects. It is probable that many types of flying insects were not heavily preyed upon until groups of passerine birds took to the air.

Of course, groups of large animals also developed their modern characteristics during of the Neogene Period. Among the herbivores, for example, the horse and rhinoceros families dwindled after mid-Miocene time in a continuation of the general decline of the odd-toed ungulates. Meanwhile, the even-toed, or cloven-hoofed, ungulates expanded, especially through the adaptive radiation of both the *deer* family and the family called the *Bovidae*, which includes cattle, antelopes, sheep, and goats. The *giraffe* family and the *pig* family also radiated during the Miocene Epoch, but the number of species in these families has since declined. Similarly, many types of elephants, including those characterized by long trunks, experienced great success during the Miocene and Pliocene intervals but later declined. Today, there are only two elephant species: the large-eared African elephant and the smaller, more docile Indian elephant that is commonly trained to perform in circuses.

The carnivorous mammals also assumed their modern character in the course of the Neogene Period; this group included the dog and cat families, both of which had appeared during Paleogene time. The *bear* and *hyena* families were also important Miocene additions to the carnivore group.

Many of the Neogene mammal groups expanded successfully because of the spread of open country in the form of savannahs and open woodlands. Several radiating herbivore groups, such as the antelopes and cattle, included many species that were well adapted for long-distance running over open terrane and that grazed on harsh grasses with the aid of high-crowned, continuously growing teeth. Many grasses contain tiny fragments of silica that offer resistance to grazing, and only continuously growing teeth

Fig. 9.3. A reconstruction of the so-called Hipparion fauna of Asia. This diverse fauna occupied open country in Asia about 10 million years ago, in Late Miocene time. Hipparion is the galloping horse (center). The elephant on the lieft, with downward-directed tusks, is Dinotherium.

can tolerate the resulting wear. Also on the increase were the groups of rodents that are adapted for burrowing in prairies and the elephants, which require open country simply to move around. As we might expect, the diversification of herbivores in savannahs and woodlands in turn fostered the success of groups of carnivores that were well adapted for attacking herbivores in open country—groups such as hyenas, lions, cheetahs, and longlegged dogs.

Ultimately, however, the greatest change in the terrestrial ecosystem was wrought by primates, simply because the ecologically disruptive humans belong to this group. As a group, primates tend to favour forests over savannahs; in fact, most live in trees. As we have seen, monkeys were presented by Oligocene time; the oldest group includes the so-called *Old World Monkeys*, which now live in Africa and Eurasia. Before the end of the Oligocene interval, however, a distinctive group of monkeys reached South America. These *New World monkeys*, which differed from their Old World counterparts in that most possessed prehensile (or grasping) tails, probably had a separate evolutionary origin. In any event, monkeys on both sides of the Atlantic underwent adaptive radiations during Neogene time.

Apes, which evolved in the Old World, flourished for a time but have since declined in number of species. We will discuss apes and apelike animals when we examine the origins of humans, which belong to the same superfamily, the Hominoidea . The most recent phases of human evolution have taken place within the climatic context of the recent Ice Age (Pleistocene Epoch); therefore, before we discuss the Hominoidea, it is appropriate that we examine the major global events of this fascinating interval of geologic time and of the Pliocene Epoch that preceded it.

Late Neogene Climatic Change

The cooling and drying of climates that characterized Oligocene and Miocene time caused grasses and weedy plants to expand their coverage of the land, and Early Pliocene time saw modest climatic changes as well. In contrast, Late Pliocene and Pleistocene time were marked by strong, rapid climatic fluctuations in the Northern Hemisphere—changes that characterized the modern Ice Age.

Pliocene equability

Stratigraphic unconformities in various parts of the world indicate that at the very end of the Miocene Epoch, between 6 and 5 million years ago, global sea level fell by perhaps as much as 50 meters. This so-called *Messinian Event*, as we will see, isolated the Mediterranean Sea from other oceans, causing the Mediterranean to dry up temporarily. It is generally believed that the Messinian Event can be attributed to removal of water from the ocean by the sudden expansion of glaciers in Antarctica. There is, in fact, evidence that climates became cooler in the Souther Hemisphere at this time. Siliceous, rather than calcareous, sediments were suddenly deposited over larger areas of southern oceans than before. The success of diatoms, which produced these sediments, indicates that upwelling had increased, apparently because of steepening temperature gradients. This cooling was not worldwide, however; terrestrial floras and marine microfossils near the Pacific coast of

North America show a slight warming trend across the Miocene-Pliocene boundary.

As the Pliocene Epoch got underway, about 5 million years ago, sea level rose again, and the seas continued to stand well above their present level between about 4.5 and 3.5 million years ago, leaving marine deposits inland of coastlines in areas like California, eastern North America, and countries bordering the North Sea and the Mediterranean. Fossil faunas and floras also reveal that global climates during this time were more equable than they are today; pollen analyses, for example, indicate that southeastern England was subtropical, or nearly so, and that northern Iceland enjoyed a temperate climate. Especially in the Hemisphere, however, this warm interval came to a sudden close with the start of the modern Ice Age slightly more than 3 million years ago. Almost certainly, the glacial interval that ensued has not yet ended.

Pleistocene continental glaciation

The evidence supporting the existence of a recent ice age is now so conclusive that it is difficult for us to understand why so many competent naturalists during the last century were skeptical that continental glaciation had indeed occurred. Early in the century, naturalists generally invoked the rushing waters of the biblical flood to explain the appearance throughout much of Europe of large boulders far from their sources. This idea gave way to a second hypothesis: that the boulders had been rafted by icebergs floating on floodwaters. Not until the 1830s, in large part because of the efforts of the great geologist Louis Agassiz, was it widely recognized that the so-called *erratic boulders* and other coarse sedimentary debris had been transported great distances and deposited by glaciers of continental proportions. We now know that the recent Ice Age has consisted of many intervals of glacial expansion that have been separated by warmer interglacial intervals, including the one in which we almost certainly live today. There is no reason to believe that the Ice Age has ended.

The marks of recent glaciation tell of ice sheet development primarily in the Northern Hemisphere. They take many forms. Moraines and other glacial deposits are scattered over large areas of Eurasia and North America. Some, like the moraines that form much of Cape Cod, Massachusetts, extend into the marine realm. Terminal moraines often flank depressions in the earth that were

Fig. 9.4. Reconstruction of glaciers as they existed during a typical interval of the Pleistocene Epoch. Large continental glaciers were centered in North America, Greenland, and Scandinavia.

left after glaciers retreated, and many of these depressions became the beds of modern lakes, including the Great Lakes of North America. Hudson Bay is not bounded by a terminal moraine, but it is an arm of the ocean that spread into a region where the earth's crust was depressed by the thickest continental glaciers in North America and has not yet rebounded fully. Figure 18-8 shows the amount of crustal uplift that has taken place in Scandinavia, which was similarly burdened with a large ice cap until the crust began to rebound about 10,000 years ago.

Some elevated regions, such as Mount Monadnock, New Hampshire, provide clues to the maximum thickness of the Pleistocene glaciers that flowed past them. Like some mountains in Antarctica today, Mount Monadnock once stood partially above surrounding ice sheets. The maximum elevation reached by the neighbouring glaciers can be determined from the position of the line that separates the lower part of the mountain, which was smoothed by glacial flows, from the still-rugged upper part, which the glacier didn't reach.

Even in areas far removed from continental glaciation, smaller glaciers—which are known as Alpine, or mountain, glaciers—modified the landscape; these glaciers flowed from high elevations, where climates are cold, through valleys between mountains to lower elevations. Their most spectacular Pleistocene productions are the U-shaped valleys sculpted by glaciers flowing through valleys that were originally more nearly V-shaped in cross section. Even in the tropical Hawaiian Islands, small moraines attest to the growth and movement of glaciers at high elevations during glacial intervals.

The alternations of Pleistocene glacial and interglacial intervals have caused climatic belts and the floras and faunas that occupy them to shift over distances measured in hundreds of kilometers. Thus, fossils of mammals such as the muskrat, which today does not range south of Georgia, reveal that climates in Florida were cool when glaciers pushed southward into the northern United States. Other fossil occurrences, such as those of hippopotamuses in Britain, show that during at least some interglacial intervals, climates were warmer than they are today.

One of the most useful fossil indicators of Pleistocene climates is the pollen of terrestrial plants. Pollen assemblages reveal climatic change by indicating the shifting of floras to the north or south shows the southward movement of floras in Europe by about 20° latitude during the most recent glacial interval there. Beetles are also extremely valuable indicators of Pleistocene climates, since some are adapted to cool climates and others to warm climates. Beetle species can be readily identified from their genitalia, which survive well in fine-grained sediments, and virtualy all beetle species known from the Pleistocene are alive today, so their ecologic requirement are well understood.

During glacial intervals, some mountain tops that stood above ice sheets served as refuges for plant and animal species that could not live on the ice. Some species that were stranded during the most recent glacial advance could not migrate across warmer lowlands when the glaciers melted back and thus today remain isolated from other members of their species, which generally live farther north. In his book *On the Origin of Species*, Charles Darwin discussed the isolation, during the Ice Age, of these still-marooned populations, which thus served as early evidence of continental glaciation.

Large refuges between major centers of ice accumulation also existed during glacial intervals. The most famous of these refuges was an area known as *Beringia*, which was so designated because it included the region of the Bering Strait. During glacial episodes, regression of the seas turned the Bering Strait into a land corridor between Asia and North America by means of which not only lower mammals but also the first humans entered the New World. Ironically, Beringia, which was hospitable to terrestrial mammals during the height of glaciation, included portions of Siberia and Alaska—areas that we now view as inhospitable to most species but that happened to remain unglaciated when other parts of the Northern Hemisphere were covered with ice.

Today, there remain only two ice caps of the sort that many times during the Pleistocene Epoch expanded to cover broad areas. One of these modern ice caps covers much of Greenland, while the other covers nearly all of Antarctica. Today, about threequarters of the world's freshwater is locked up in glacial ice, and most of this belongs to the Antarctic ice cap. It may seem impressive that glaciers now contain about 25 million cubic kilometers of ice, but it has been estimated that during glacial advances of the Pleistocene Epoch, the volume of ice was nearly three times as great, with larger ice sheets measuring an average of about 2 kilometers (~ 1.2 miles) in thickness. Shelves of ice projected into the sea, and these shelves, together with the icebergs and pack ice that broke loose from them, spread over half the world's oceans.

One important effect of each major expansion of ice sheets during the Pleistocene Epoch was a partial draining of the oceans, which took place as great quantities of water were locked up on the land. During major glacial expansions, most of the surfaces that now form continental shelves stood above sea level. Rivers cut rapidly downward through the soft sediments of many continental shelves to form valleys that exist today as submarine canyons, following further excavations by submarine turbidity currents. During some glacial episodes, sea level dropped slightly more than 100 meters (~ 330 feet) below its present position.

The chronology of glaciation

The microplankton fossil record of the North Atlantic suggests that rapid cooling occurred slightly more than 3 million years ago, apparently in response to the initial formation of continental ice sheets. The beginning of continental glaciation at this time is confirmed by the first appearance of coarse, ice-rafted sedimentary debris in the North Atlantic and also of glacial deposits in Iceland above well-dated lava flows. It was somewhat later that the first glaciers spread southward into the central United States, and the beginning of the Pleistocene Epoch is now placed at 1.8 million years before the present. The sequence of glacial tills in the United States and Europe suggests that there were four major glacial intervals during the Pleistocene Epoch; these intervals are formally recognized as glacial ages, and the intervening intervals are known as interglacial ages. In accordance with normal stratigraphic practice, each age corresponds to a stage in the geologic column.

Table 9.1. Glacial and interglacial stages of the Pleistocene Epoch.

North America	*Europe*
Wisconsin	Riss-Würm
(Sangamon interglacial)	(interglacial)
IIlinoisan	Mindel
(Yarmouth interglacial)	(interglacial)
Kansan	Günz
(Aftonian interglacial)	(interglacial)
Nebraskan	Danube

In recent years, it has become evident not only that the major glacial ages were characterized by smaller pulses of glaciation but also that minor glacial advances occurred during interglacial ages as well. These smaller fluctuations are reflected in sediments obtained from deep-sea cores. The crucial data are the oxygen-isotope ratios of foraminifer skeletons preserved in deep-sea sediments. Oxygen, as we have seen, is present in sea water as the stable isotopes ^{16}O and ^{18}O, and these isotopes are incorporated into the growing skeletons of foraminifers in the same proportions in which they occur in the surrounding sea water.

As it turns out, the relative proportions of these two isotopes in fossil foraminifer skeletons fluctuate greatly when traced downward through cores of deep-sea sediments. When these fluctuations were first discovered, some scientists erroneously assumed that they reflected variations in water temperature during the Pleistocene Epoch. Soon, however, it was found that similar fluctuations were exhibited not only by planktonic foraminifers but also by foraminifers that lived on the sea floor, an area that has remained at near-freezing temperatures since the origin of the psychrosphere. As we have previously noted we now know that

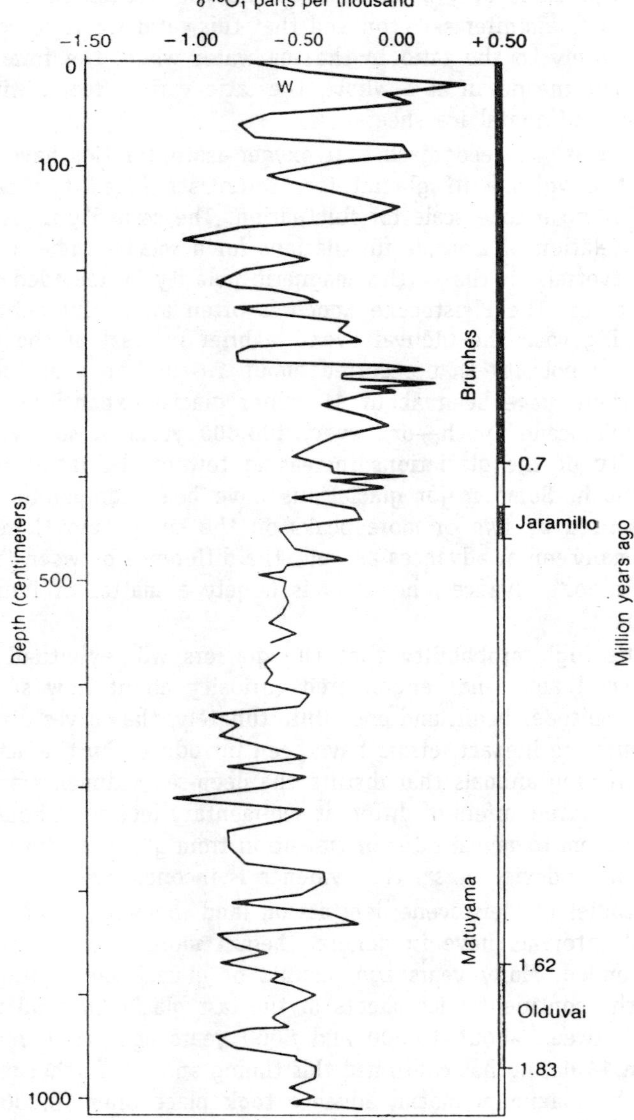

Fig. 9.5. Oxygen-isotope fluctuations in planktonic foraminifers within a deep-sea core. The index of oxygen-isotope composition plotted here records decreases in temperature toward the right and increases toward the left. On the right is a paleomagnetic time scale for the core. W=the Wisconsin, or Riss-Wurm, Stage.

temperature has only a minor effect on the oxygen-isotope ratio within a foraminifer skeleton and that this ratio tends to conform more closely to the ratio in the sea water where the foraminifer lived. For the ocean as a whole, the ratio varies greatly with the size of continental ice sheets.

Once it was recognized that oxygen-isotope ratios have varied with the volume of glacial ice, scientists faced the task of establishing a time scale for fluctuation. The scale Figure is based on correlation of isotopic fluctuations for a small number of cores with reversals in the earth's magnetic polarity as recorded in the same cores. The Pleistocene Epoch is often arbitrarily defined as beginning with the Olduvai Event, a brief reversal of the earth's magnetic polarity that occurred about 1.8 million years ago. In fact, there have been about 18 minor glacial expansions during the Pleistocene Epoch—one every 100,000 years or so—with the intensity of the glaciations increasing toward the latter part of the epoch. Some major glaciations have been compound events represented by two or more peaks on the curve, and there have been many minor advances as well. The difference between "major" and "minor" advances, however, is largely a matter of individual judgment.

The high probability that the glaciers will eventually flow southward again has engendered curiosity about how suddenly glacial episodes begin and end. Unfortunately, the curves displayed in Figure are inexact: errors have been introduced by the activities of burrowing animals that disturb the deep-sea sediments and mix up fossil foraminifers of different sedimentary levels. Although the curves seem to reveal rates of transition from glacial to interglacial intervals and vice versa, the evidence is inconclusive.

Studies of Pleistocene deposits on land, however, indicate that glacial intervals have in general, begun more slowly than they have ended. Many years ago, counts of glacial varves suggested that the continental ice sheets of the last glacial interval melted back between about 15,000 and 8000 years ago. More recently, carbon 14 dating has confirmed this timing and has further revealed that the maximum glacial advance took place only 18,000 years ago, at the time the southern most moraines formed. Thus, the last major glacial interval, which began more than 100,000 years ago, ended not only recently but also suddenly, on a geologic scale of time.

Cyclical Pleistocene deposits in Czechoslovakia and Austria suggest that this pattern is typical. Each cycle includes deposits of one major interglacial and glacial interval. The lowest interglacial unit in each of these cycles is a forest soil that contains pollen from a humid interglacial climate slightly warmer than the present one. Above this are soils representing progressively cooler conditions that supported prairies and conifer forests. Sometimes soils representing even cooler but nonglacial Arctic climates follow. The entire interglacial sequence represents a long, gradual episode of cooling. Above each interglacial sequence are loess deposits formed during a glacial interval. *Loess* is windblown silt that accumulated in front of the ice sheets, in the frigid Pleistocene deserts that developed there. Most of this silt was derived from glacial meltwaters.

The Pleistocene deposits of Czechoslovakia and Austria were not themselves glaciated, which is fortunate in that glaciation would have damaged their excellent stratigraphic record. These areas were, however, close enough to the periodically advancing glaciers to record their presence through the accumulation of loess. In contrast to the depositional evidence for a normally slow onset of glacial conditions, an abrupt transition from loess to forest soil typically indicates the sudden ending of a glacial interval.

The nature of glacial and interglacial intervals

Because of their relative recency, the last glacial interval and the interglacial interval that preceded it have left the best record of Pleistocene climatic fluctuations. A major international project called CLIMAP was recently formed to reconstructed the oceangraphic conditions and climates that characterized this glacial interval. Many of CLIMPA's conclusions have been based on studies of the geographic distributions of fossil coccoliths and planktonic foraminifers that have been obtained from deep-sea cores but represent species that remain alive today. Knowledge of the environmental requirements of these planktonic species has permitted paleontologists to reconstruct the distributions of water masses associated with particular temperatures, and these temperatures, combined with information on the distribution of terrestrial floras and faunas, have made it possible to map the distribution of Pleistocene vegetation and environments on land.

During the Wisconsin glacial interval, north-south temperature gradients steepened in the Northern Hemisphere both in shallow seas and on land. Winter temperatures fell only slightly in most tropical areas but plummeted at latitudes north 30° in the Northern Hemisphere. Other patterns of climatic change, however, were more complex. Some areas that lay only a few degrees south of continental glaciers, for example, became much wetter than they are today. Among these was the Great Basin in the American West, which drew rainfall from glaciated terranes to the north and accumulated numerous lakes in areas that are now arid. Subtropical deserts, such as portions of the Sahara near the Mediterranean, also became wetter in their northern regions. In contrast, environments of low rainfall—steppes, semideserts, savannahs, and dry grasslands—stretched in a broad belt across Eurasia south of glaciers. At the same time, the tropical rain forests of South America and Africa, which lay relatively far from the lowered oceans, shrank back in areas where drying conditions came to prevail, eventually giving way to vegetation that required less moisture.

In Europe, the Alps and the Pyrenees developed glaciers of their own under the cold climatic conditions, and the north-south trend of these cold barriers blocked the southward migration of many species that were moving ahead of continental glaciers advancing from the north. Many forms of life, such as magnolia trees, were caught in this glacial "trap" and disappeared from Europe but still managed to survive in North America, where southward migration to Florida and Central America remained unimpeded.

Antarctica, which was positioned over the South Pole, accumulated a continental glacier long before the beginning of the Pleistocene Epoch, but the isolation of this island continent following the breakup of Gondwanaland prevented its glaciers from spreading over a vast continental area in the manner of the late Paleozoic glaciers of Gondwanaland. Nonetheless, snow and ice accumulated over the southern tip of Africa, an eastern segment of Australia, and a large part of southern South America. A large lake occupied central Australia, and much of this continent was wetter and more heavily vegetated than today.

Three large glacial centers developed in the Northern Hemisphere—one in North America, one in Greenland, and one in Scandinavia. The northern Atlantic Ocean, being adjacent to these three glacial centers, was more profoundly affected by the Pleistocene glaciers than any of the world's other major oceans except the Arctic, but land areas adjacent to the Atlantic also suffered marked climatic change. During glacial intervals, pack ice similar to that which now occupies bays adjacent to northern Canada must have choked large areas of the North Atlantic. Farther south, along the east coast of North America, glaciers flowed southward to New Jersey, and tundra occupied what is now Washington, D.C. Today, the Gulf Stream swings far northward, maintaining a relatively warm climate in the British Isles and northwestern continental Europe at about 55° north latitude, but planktonic foraminifers in deepsea cores have revealed that the Gulf Stream has periodically shifted from a northern interglacial position, like that of the present, to a glacial position that carried its warm waters more directly eastward, toward Spain. As a consequence of this pattern, northern Europe remained frigid during glacial intervals.

Another geographic change compounded this cooling trend in the northern Atlantic. Because major wind systems are driven by convection produced by temperature contrasts, the steepening of north-south temperature gradients during glacial intervals strengthened the trade winds of the Northern Hemisphere. As Figure 18-18 illustrates, this caused the winds to blow more strongly along their diagonal paths toward the equator—as a result of which the westward-flowing equatorial currents must have been pushed southward. Today, these currents are deflected northward by the large hump of South America at Brazil, feeding warm water to the Northern and Southern Hemispheres, with the portion of the Atlantic north of the equator capturing more than its share of the warm equatorial currents. It is assumed that when strong northeasterly trade winds pushed the equatorial currents southward during recent glacial intervals, less of the warm water of these currents was captured by the hump of South America, causing the northern Atlantic gyre to remain cooler than it would have been otherwise.

The causes of glaciation

What factors have brought about Pleistocene glaciation? This is a double question. On the one hand, we must know why the Ice Age began in the first place. On the other, we must determine the reason, once the Ice Age had begun, that the glaciers waxed and waned at frequent intervals. In considering these questions, it is important to understand that both the expansion and the contraction of glaciers are unstable processes; once either of these processes has begun, it usually accelerates automatically. This happens because ice reflects a much higher percentage of warming sunlight than does land or water: its albedo is higher. Thus, if a small amount of climatic cooling generates glacial ice, this ice reflects a higher percentage of sunlight than was reflected before the glacier developed, and this leads to further cooling and glacial expansion, which leads to even less absorption of sunlight, and so on. In the same manner, climatic warming melts ice, resulting in exposure of light-absorbing land, which produces even more warming. These conditions tell us that the question of why glaciers expand and contract is to a large extent a question of why a new trend (expansion or contraction of ice) gets started. A small change in climatic conditions can ultimately have an enormous effect.

The first question—What might have initiated the Ice Age?—Is one that geologists have not been able to resolve with any degree of certainty. The formation of ice earlier in geologic history can be partly attributed to the movement of continents across poles; whereas polar seas are often warmed by the exchange of water with warmer regions, a large landmass centered over a pole tends to become quite cold. Apparently it is no coincidence that the major ice caps of both the Ordovician Period and late Paleozoic time developed at times when large continents were passing over the South Pole. The Pleistocene Ice Age, in contrast, was focused in the Northern Hemisphere, with the North Pole itself centered in the Arctic Ocean. It is true that the North pole was partly isolated from warmer oceans by neighboring landmasses, but this condition existed for millions of years before ice sheets formed and does not in itself provide the answer.

It has been suggested that the sudden elevation of mountain ranges might have triggered the Ice Age by creating high-altitude mountain glaciers which expanded by reflective cooling and eventually lowered global temperatures. It has also been hypothesized that a reduction in the sun's energy output might have led to the Ice Age. These theories are neither supported nor refuted by existing evidence. Perhaps the most reasonable idea yet put forward to explain the onset of glaciation is that when the Isthmus of Panama formed about 3.5 million years ago, the Gulf Stream was strengthened by the northward deflection of the equatorial current. Moisture that the warm Gulf Stream waters thus supplied to northern regions led to an increase in snowfall here and to a buildup of ice caps. We will return to this idea when we review Neogene events in the Atlantic Ocean.

More readily evaluated is the notion that changes in the relationship of the earth to the sun may cause glacial fluctuations. Supporters of this theory point to the observation that the earth's orbit periodically changes shape as a result of the movements of other planets, which exert a gravitational pull on the earth. The most pronounced orbital changes of the earth follow a 92,500-year cycle, which approximates the periodicity of the oxygen-isotope cycle recorded in deep-sea cores.)

Astronomical arguments have been carried even further in attempts to explain-smaller scale climatic cycles, such as changes in sea level. Thorium dating of coral reefs that now stand about 6 meters (~20 feet) above sea level has revealed that about 125,000 years ago the world's oceans stood much higher than they do today. Sea level subsequently dropped but rose quickly about 105,000 years ago, and it then dropped and rose once more about 82,000 years ago. The second and third high stands were lower than the first, but it is not known precisely how far sea level fell in between the high stands. Terraces formed by reef growth are found at many localities, including New Guinea, Barbados (an island in the West Indies), and the Florida keys. They appear to record minor oscillations that occurred during a single larger oscillation, which is the last major interglacial interval, known as the Sangamon or Eem.

It has been suggested that the minor oscillations recorded in the fossil-reef terraces represent an astronomical cycle known as the *precession of the equinoxes*. Procession is a slight wobble in the axis of the earth's rotation caused by the gravitational pull exerted by the sun and the moon on this planet. The wobble causes each hemisphere of the earth to experience its longest and shortest days at a different position in its elliptical orbit each year. Winter in the Northern Hemisphere is likely to be coldest when, on the longest day of the year (June 21), the earth happens to be positioned at the point in its elliptical orbit that lies farthest from the sun. The earth is in this position once every 22,000 years, the time required for completion of one precession cycle. Those who favour the theory of astronomical control of glacial cyclicity point to the fact that the earth's 125,000-, 105,000- and 82,000- year-old high stands of sea level were separated by intervals of approximately 22,000 years. They argue that precession of the equinoxes imposes small-scale cycles on larger (92,500-year) cycles that reflect changes in the shape of the earth's orbit. Skeptics question the certainty of this temporal correspondence maintaining that the effects of the astronomical cycles have not been proven strong enough to influence continental glaciation. Instead, some scientists favour the notion that fluctuations in the intensity of solar radiation have governed Pleistocene climatic fluctuations.

Climatic cycles, which are reflected in glacial movements, have also occurred on smaller scales. Between about 1500 and 1850 A.D., for example, the world experienced a minor episode of refrigeration known as the *Little Ice Age,* which reached its peak about 1700. During this period, areas like Scandinavia and New England suffered bitter winters and short summers that caused numerous major crop failures. George Washington and his troops at Valley Forge were actually fortunate, in 1777 and 1778, to have experienced a winter that, for the times, was relatively mild. Studies of glacial moraines show that over the last 10,000 years (since the last major glacial interval), the world has experienced three other cold intervals comparable to the Little Ice Age. The warm interval preceding the first of these peaked about 7000 years ago, at which time global temperatures were warmer than they have been at any time since.

On a still smaller scale, numerous cycles are evident in climatic records of the hundred years or so. Beginning around the turn of

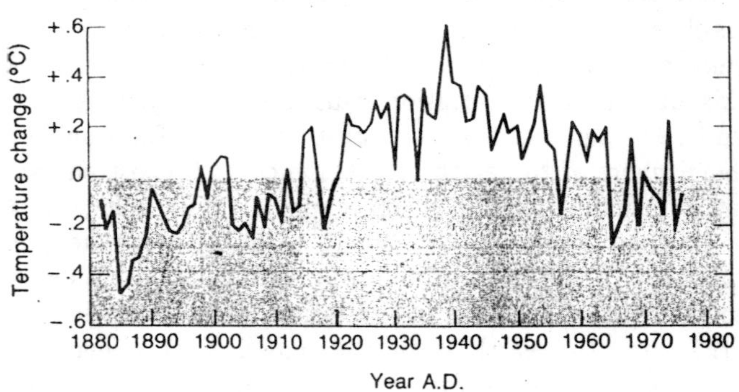

Fig. 9.6. Climatic changes over the past 100 years. The graph represents changes in average annual temperature from year to year. We cannot be certain whether the cooling trend of the past few years will soon end or whether it will continue and produce another Little Ice Age.

the century, they average annual temperature at the earth's surface showed a rising trend, reaching a maximum rate to increase in 1938. Since that time, the rate of change has followed a zigzag course downward, and the average annual temperature has dropped many times. Does this indicate that we are entering another Little Ice Age? Or will the trend reverse itself so that we will climb further out of the cold period that reached a climax about 1700? Unfortunately, attempts to answer these questions represent little more than conjecture.

One factor that concerns many scientists is the possibility that the burning of wood and fossil fuels for energy, a process that adds carbon dioxide to the atmosphere, may eventually cause the earth's climate to warm up considerably as a result of the so-called greenhouse effect. If the temperature of the earth were to rise by only a few degrees, the partial melting of glaciers would elevate the oceans to such an extent that the major port cities of the world would be flooded. Unfortunately, we cannot yet assess the risk that we are incurring by burning carbon compounds to obtain energy.

We have been discussing how humans may be influencing the climate without first having discussed where our species came from to begin with. Having reviewed global events of Neogene time, we can now step back and review what is known about human origins.

Human Evolution

In addition to the single species that now constitutes the human family, the superfamily Hominoidea currently consists of just four species of the ape family—the common chimpanzee, the pigmy chimpanzee, the gorilla, and the orangutan—together with six species of the gibbon family. The human family, Hominidae, did not evolve from the modern ape family, Pongidae; instead, the two families have followed independent lines of evolution. Although it is possible that Hominidae and Pongidae evolved independently from a single family of primitive apes, the case is not clear. Thus, we will first review the evolution of the Pongidae and then outline that of the Hominidae.

Early apes in Africa and Asia

Although an extensive Plio-Pleistocene fossil record has been uncovered for the Hominidae, very few known fossil remains of any age represent the Pongidae. Furthermore, the fossil record of the superfamily Hominoidea in latest Miocene and earliest Pliocene time (8 to 4 million years ago) is almost entirely barren. Sediments representing this interval in Africa, where apes and human may have evolved, are rare and poorly studied.

Farther back in the Miocene Series, numerous fossils represent two families of apelike species that have both been considered potential ancestors of modern apes and humans. These early families are the *Dryopithecidae* and the *Ramapithecidae*. The record of the group, the Dryoithecidae, begins in Africa at about the base of the Miocene Series, extending upward to sediments that are of mid-Miocene age (about 14 million years old). In Eurasia, dryopithecids first apper in sediments about 14 million year old, and their last occurrence is in sediments 8 or 9 million year old. The fact that dryopithecids did not spread from Africa to Eurasia until about 14 million year ago is not surprising in that the African plate did not collide with Asia until about 20 million years ago. As a result of this collision, not only hominoids but also many other previously isolated groups of African mammals, including elephants and giraffes, spread into Eurasia. Until recently, dryopithecids were thought to have resembled and given rise to modern apes, but closer examination has revealed that the dryopithecids were actually not so similar to apes; their tooth proportions resembles those of Old World (African and Eurasian) monkeys, and their limb structure suggests a life of four-legged motion in the trees, perhaps with occasional forays onto the ground.

In Asia, ramapithecid fossils have not been found in sediments older than about 14 million years, but a jaw fragment from Africa extends their range back to about 17 million years. The ramapithecids were generally assumed to have evolved from the dryopithecids, from which they differ in their reduced canine teeth (or eyeteeth) and in their large, relatively flat cheek teeth, which in advanced forms had thick enamel. Unfortunately, few ramapithecid limb or pelvic bones are known, so we have little information about modes of locomotion within the group. Fragmentary limb bones suggest that these early apes ranged in weight from 20 kilograms (~44 ponds) to 70 kilograms (~ 154 pounds). *Gigantopithecus*, which is probably the largest of the ramapithecids, moved about on the ground like a gorilla, but the smaller forms may have been arboreal.

The ramapithecids became so diverse that by Late Miocene time the Old World was much more heavily populated by apes than

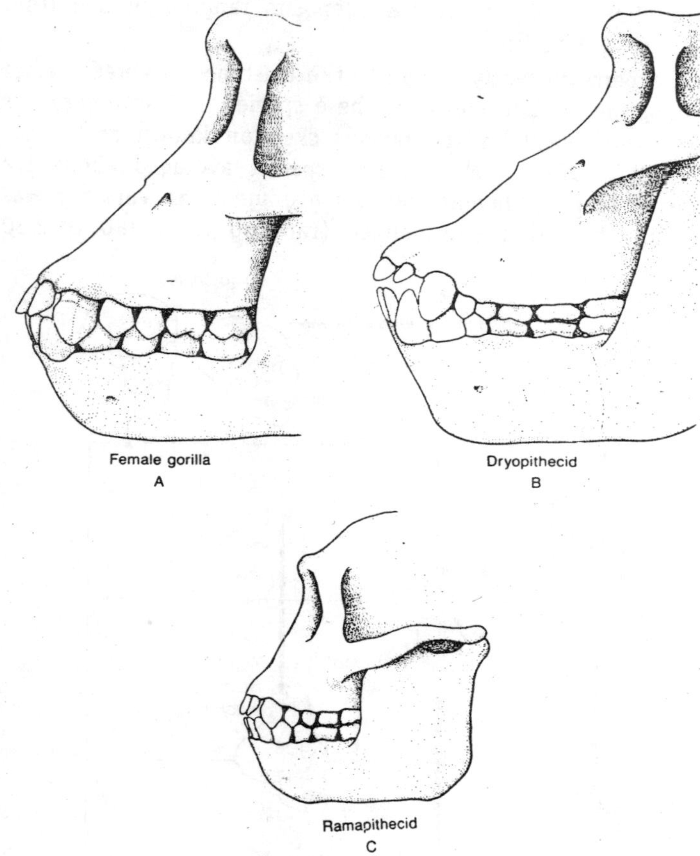

Fig. 9.7. Skulls of apelike animals. The skull of the dryopithecid (Sivpithecus) in the center in many ways resembles that of a modern ape. The ramapithecid skull, represented by Ramapithecus on the right has reduced canine eyetheeth and broad molars, and the muzzle is less elongate than in the other forms.

Africa is today. These animals virtually disappeared before the end of Miocene time, however.

Species of Australopithecus: Ape people of Africa

The fossil record of the Hominidae is so poor for the latter part of the Miocene Epoch that little is known of the history of the family during this interval. By comparison, the Pliocene fossil record is a treasure trove, yielding numerous remains of the genus *Australopithecus*, the oldest known genus of the human family, which has been found in African deposits ranging in age from about 1.3 to 4.0 million years.

Most experts currently recognize the existence of three species of *Australopithecus*. The oldest of these species is *Australopithecus afarensis*, which includes the famous skeleton known as "Lucy." East African fossils reveal that this species averaged about 1.2 meters (~ 4 feet) in height, but its average brain capacity was larger than that of a chimpanzee (ranging from 380 to 450

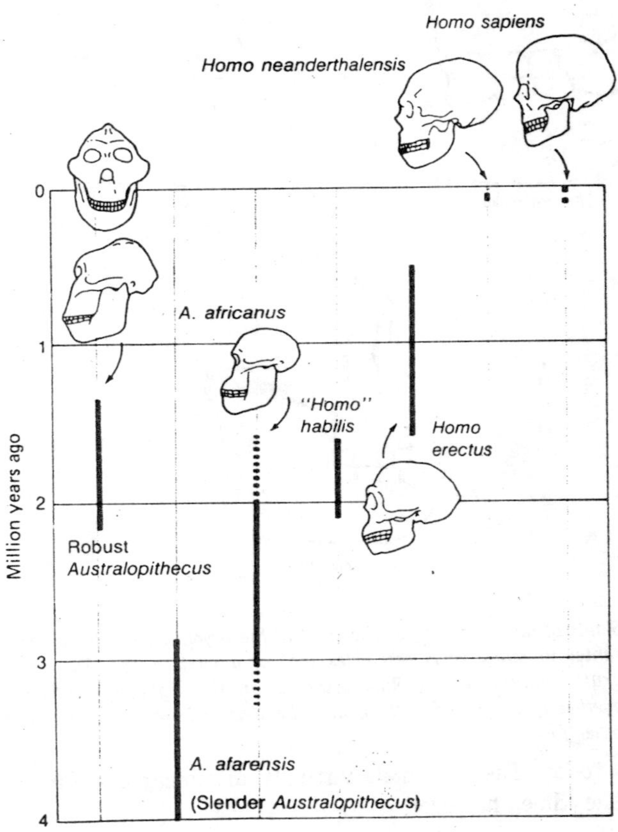

Fig. 9.8. Stratigraphic ranges of species of the Hominidae (human family) as currently recognized from fossil data. Two slender species and one robust species of Australopithecus are represented on the left. Species of the genus Homo are represented on the right. Some workers place the species labeled "Homo" habilis in the genus Australopithecus rather than in the genus Homo. Some workers classify Neanderthal as a subspecies of our own species, Homo sapiens.

centimeters, compared to 300 to 400 for a chimp). By comparison, the modern human brain averages about 1330 cubic centimeters. Our larger body requires a somewhat larger brain simply for control of bodily functions. Even in relation to body size, however, the human brain is much larger than that of *Australopithecus afarensis*, which for simplicity we will call "Afarensis."

On the other hand, the pelvis of Afarensis is remarkably like that of humans. Unlike the elongate pelvis of apes, the pelvis of Afarensis was adapted to support an upright body. This upright posture is also evidenced by a spectacular set of tracks left by either Afarensis or a close relative in soft volcanic ash in Tanzania slightly more than 3 million years ago.

Although its pelvis was similar to that of a human, Afarensis had a skull that retained many apelike features, including massive

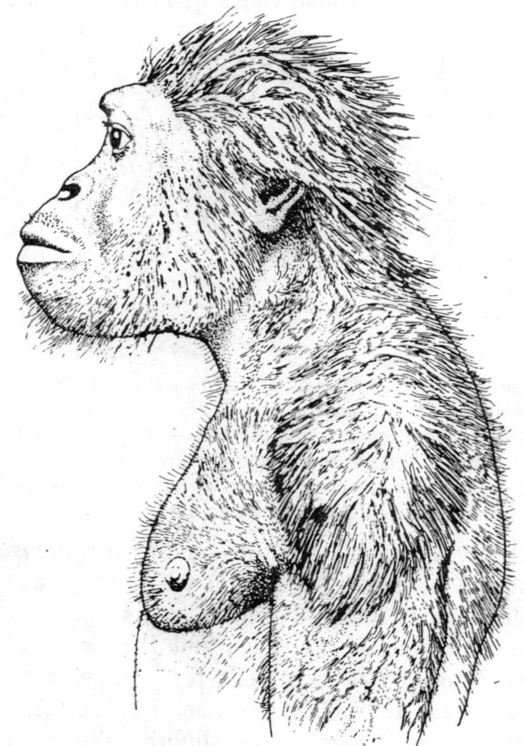

Fig. 9.9. Reconstruction of the oldest known species of the human family, Australopithecus afarensis. The low fore head, heavy brow, and projecting lower face of this species are all evident here.

brow ridges, a low forehead and projecting, muzzlelike mouth. A reconstruction of the animal in side view illustrates these features especially well.

The teeth of Afarensis are in many respects intermediate between those of apes and humans. Compared to humans, apes have large incisors (or front teeth) in relation to the size of their molars. Apes also have large projecting canines (eyeteeth) and a space between their incisors and canines in the upper jaw to accommodate the upward-projecting canine of the lower jaw. Humans do not have this space; Afarensis does, but it is reduced in size, in keeping with the size of the canine teeth of Afarensis, which are larger than a human's but smaller than an ape's. The incisons of Afarensis are smaller for the size of the jaw than those of an ape, but the molars are larger, and so the relative proportions of the teeth more closely resemble those of modern humans.

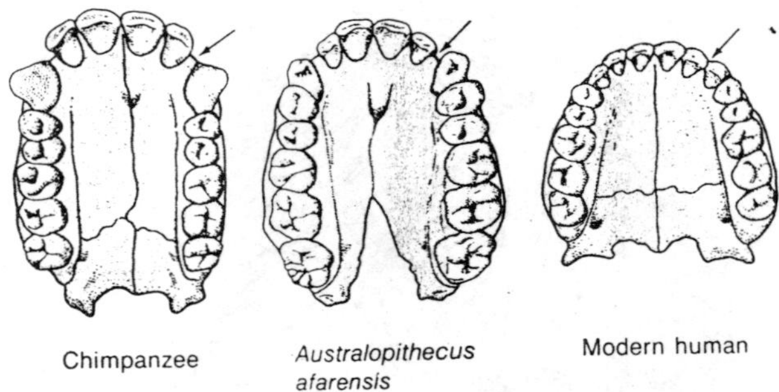

Fig. 9.10. Comparison of the upper teeth of Australopithecus afarensis with those of a chimpanzee and a modern human. The chimpanzee has larger canine (eyeteeth) than humans, with a gap between each canine and the adjacent incisor (arrow). The incisors of the ape are also large in relation to the molars. In these important features, Australopithecus afarensis is intermediate between apes and humans.

Fossils clearly assignable to *Australopithecus afarensis* range in age from about 4 to about 2.8 million years, thus spanning more than 1 million years. About 3 million years ago, Afarensis was joined by *Australopithecus africanus*, and the two species appear to have shared to African continent for a time, with Africanus surviving until about 1.6 million years ago in southern and eastern Africa. *Australopithecus africanus* was slightly taller than Afarensis, averaging perhaps 1.4 meters (~4½ feet). Well-preserved Africanus skulls exhibit the heavy brow and the long, sloping facial region below the eyes that characterize all member species of the genus *Australopithecus*.

The third species of *Australopithecus* was *Australopithecus robustus*, whose geologic "lifetime" overlapped with that of Africanus, spanning an interval from about 2.3 to 1.3 million years ago. Robustus lived up to its name in many ways—its head and teeth were heavily constructed; the crown of its skull was elevated into a bong crest similar to that of a gorilla, thereby providing for the attachment of strong jaw muscles; and its wide, flangelike cheekbones and massive lower jaws provide further evidence of powerful chewing muscles. Despite its presumably formidable appearance, the robust *Australopithcus* was not necessarily a fierce creature. Nor was it large in its overall proportions; it is thought to have averaged not much more than 1.5 meters (~5 feet) in height set in its huge jaws were exceptionally broad, flat cheek teeth. The power of the Robustus jaw was apparently directed toward grinding up coarse plant food.

It is widely argued that the specialization of *Australopithecus robustus* for chewing coarse food makes it an unlikely ancestor of modern humans. It is almost universally agreed, however, that one or both of the more slender australopithecine species have a place in our evolutionary history.

The human genus makes its appearance

We do know that our own genus, *Homo* was alive in Africa about 1.6 million years ago in the form of *Homo erectus*. In contrast to *Australopithecus*, which is unknown from localities outside Africa, *Homo erectus* was a widely traveled species. Known as *Pithecanthropus* before its similarities to modern humans were fully acknowledged, *Homo erectus* lived not only in Africa and Europe but also in China, where it has been referred to as "Peking man" and in java, where it has been called "Java man." Fossil skulls of *Homo erectus* represent a long interval of time, extending from about 1.6 million years ago to perhaps 400, 000 years ago.

Homo erectus resembled modern humans and differed from Australopithecus in several ways, but the recent discovery in Africa of the 1.6 million year old skeleton of a twelve-year-old boy indicates that *Homo erectus* rivaled us in size. The cranial capacity of *Homo erectus* skulls that are currently available for study ranges from about 800 to 1300 cubic centimeters. The maximum size approximates the average for modern humans (about 1300 cubic centimeters), but the minimum size is much lower. There is some evidence, however, that the brain size of *Homo erectus* increased somewhat over the course of the species' existence. The brain case of *Homo erectus* was also relatively longer and lower than ours; furthermore, its forehead was low and interrupted by brow ridges that were even bigger than those of the two slender species of *Australopithecus*. These features reflect derivation from *Australopithecus,* as do the projecting mouth and heavy lower jaw—but the mouth and jaw as well as the cheek bones were less highly developed than those of *Australopithecus*. The teeth of *Homo erectus* were smaller than those of Australopithecus and though larger than the teeth of modern human, nevertheless bear a close resemblance to them. *Homo erectus* was also a toolmaker. Its stone culture, known as *Acheulian,* included the production of magnificent hand axes.

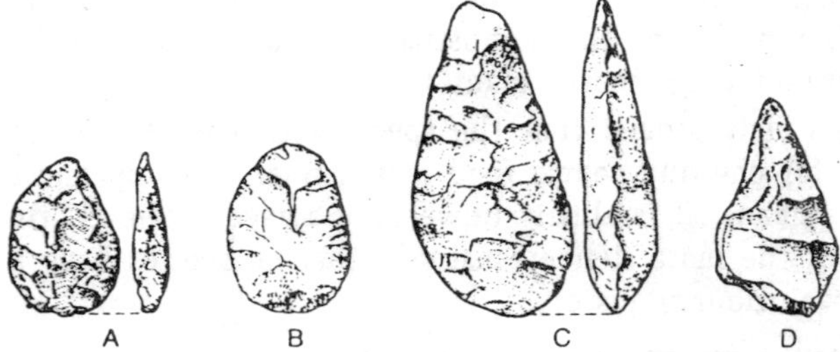

Fig. 9.11. Tools of the widespread Acheulian culture attributed to Homo erectus. A—Twisted oval tool from Saint-Acheul near Amiens, France. B—Oval hand ax from south of Wadi Sidr, Israel. C—Large hand ax from Orgesailie, Kenya. D—Hand ax from Hoxne, Suffolk, England.

One problem in tracing the evolution of our genus concerns the classification of a species that made its earliest known appearance below *Homo erectus,* in strata about 2 million years old. This problematic species has traditionally been designated *"Homo" habilis*. (The generic name is in quotation marks because some workers consider the skull of this species to be of the *Australopithecus* type). The brain capacity of Habilis (650 to 800 cubic centimeters) was larger than that of other *Australopithecus* species but smaller than that of most specimens of *Homo erectus*. Part of the reason that Habilis has been placed in the genus *Homo* is that stone tools of the so-called Oldowan type have been found in associated sediments of Olduvai Gorge, the locality that gave the stone culture its name and it was once believed that only true

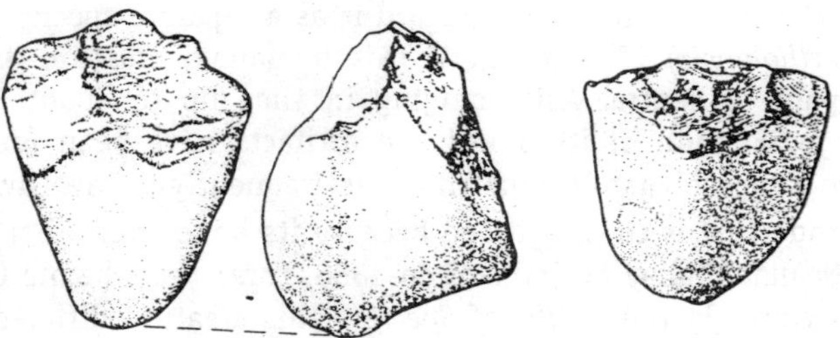

Fig. 9.12. Simple stone implements from Olduvai Gorge, Tanzania, representing the Oldowan culture that existed about 2.5 million years ago. These are "core" implements from which flakes were struck.

humans, or members of the genus *Homo*, could fabricate tools. This belief has no real foundation and is damaged by recent discoveries of Oldowan tools in sediments about 2.5 million years old-about half a million years older than the oldest known Habilis skull. The Oldowan tools comprise sharp flakes of stone and many-sided "core" implements that represent pieces of stone left after the removal of the flakes. The flakes may have served for slicing and the core implements that represent pieces of stone left after the removal of the flakes. The flakes may have served for slicing and the core implements for pounding and chopping. Members of the genus *Australopithecus* remain strong suspects in our search for the producers of these implements, Which are the oldest of all known tools.

Was Neanderthal one of us?

Just as there are gaps in our knowledge of the evolutionary pathway that led to *Homo erectus*, so are there gaps in the evolutionary pathway connecting this species to our own, *Homo sapiens*. Unfortunately, sediments in Africa representing the interval above the youngest specimens of *Homo erectus* have yielded a poor suite of fossil remains. As a result, we are not certain exactly what evolutionary pathway led from *Homo erectus* to our species, *Homo sapiens*. The interval of poor fossil remains extend from about 400, 000 to 100, 000 years ago, and most of the partial skulls found within this interval have robust brow ridges and low foreheads reminiscent of *Homo erectus*.

In sediments about 100,000 years old, the hominid fossil record improves with the appearance of well-preserved fossils of the creature known as Neanderthal. Some scientists regard Neanderthal as a variety or subspecies of our species (*Homo sapiens neanderthalensis*), while other regard it as a separate species (*homo neanederthalensis*). The record if this humanoid creature extends from Spain to central Asia, ranging in time up to about 35,000 years ago. Enough of its bones and artifacts have been found in caves to suggest that Neanderthal was frequently a cave dweller.

Neanderthal is so designated because its bones were first found in the Neander Valley of Germany in 1856, three years before Charles Darwin wrote *On the Origin of Species*. This creature differed from modern humans in a number of skeletal features. Neanderthal resembled *Homo erectus* in that its skull was long and low with prominent brow ridges, a projecting mouth, and a receding chin. On the other hand, its brain was quite large—slightly larger, on average, than that of modern humans. The large brain is by no means indicative of superior intellect, however; Neanderthal's body was somewhat more massive than ours, though slightly shorter, and may have required some extra brain cells simply for the purpose of motor control.

The reputation of Neanderthal has been damaged by generalizations that have been unfairly gleaned from the remains of a single individual found at Lachapelle-aux-Saints, France, which attracted much attention early in the twentieth century. The stooped posture of this individual was taken as evidence that Neanderthals as a group were primitive ceratures, lacking fully erect posture. The fact is that the French specimen was poorly reconstructed. Neanderthal normally stood as erect as we do, although his bones were more dense, and he probably moved more slowly.

Neanderthal also differed from modern humans in the presence of a gap between its cheek teeth and the back of its jaw and in certain distinctive features of the shoulder blade, pelvis, and hand. The finger bones of Neanderthal have stronger attachment areas for tendons than those found in modern humans, suggesting that this creature possessed a grip stronger than our own. Those who favour the idea the Neanderthal was a species separate from *Homo sapiens* cite these features along with the *Homo sapiens* cite these features along with the *Homo erectus*—like shape of its skull as evidence. Those who prefer to include Neanderthal in our species emphasize similarites in teeth and brain size.

The distribution of the two species in time and space suggests that Neanderthal and *Homo sapiens* may, in fact have been

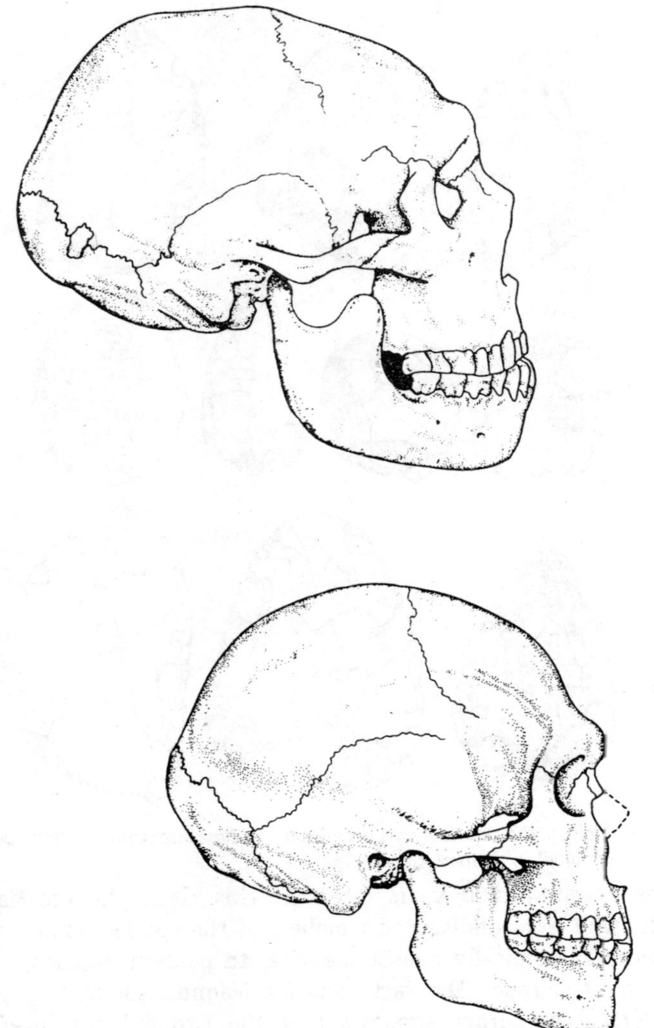

Fig. 9.13. A Neanderthal skull (above) and a skull of the modern human (below). The Neanderthal brain case is larger and lower in the forehead region. The brow is also heavier, as is the lower jaw. A characteristic of Neanderthals is the gap between the rear teeth and the jawbone.

reproductively isolated. The interval during which Neanderthals lived represented the most recent glacial stage, the Riss-Wurm stage, but Neanderthals disappeared before the glaciers melted back. Radiocarbon dating tells us that Neanderthal vanished from eastern

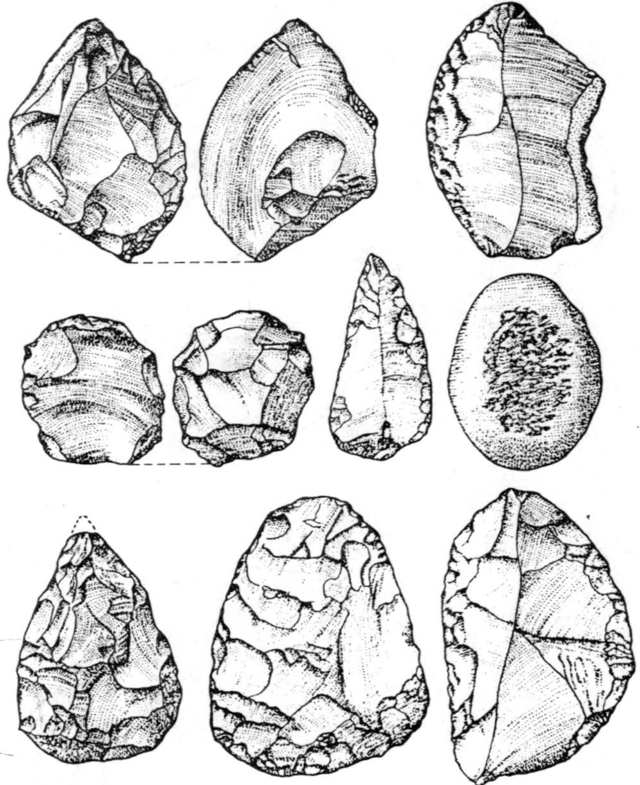

Fig. 9.14. Tools of the Cro-Magnon people. Among these are scrapers (top row) and an anvil or hammerstone.

Europe about 40,000 years later. At this time, the Cro-Magnon people, who were undisputed members of the species *Homo sapiens* and were anatomically almost identical to modern humans, spread throughout Europe. The fact that Cro-Magnon suddenly replaced Neanderthal in Europe suggests that the two did not interbreed successfully and thus may well have been distinct species.

Where did the biologically modern population of *Homo sapiens* known as Cro-Magnon originate? One possibility is that they evolved in Europe from a population of Neanderthals. Another is that they migrated north from Africa about 40,000 years ago. There are, in fact, incomplete African skulls dated at approximately 100,000 years of age that seem to have a modern appearance. Perhaps descendants of *Homo erectus* in Africa were the predecessors of our species, while Neanderthal originated from *Homo erectus* descendants in Europe.

In any event, Neanderthals had developed a distinctive stone culture known as *Mousterian* in which flakes of stone were fashioned into knives, scrapers, and projcetile heads—far more sophisticated tools than the Acheulian implements of *Homo erectus*. The Cro-Magnon culture that followed at first incorporated Mousterian elements but then emerged as a more sophisticated culture known as *Late Neolithic* in which a variety of specialized tools were invented.

Even more innovative was the artwork of the Cro-Magnon people, which seems to reflect a new use of the powers of imagination. Their magnificent cave paintings, primarily of animals, can still be admired in France and Spain, while other artifacts shown that their artistic efforts included modeling in clay, carving friezes, decorating bones, and fabricating jewelry from teeth and shells.

Thus we are left with several intriguing questions. Did modern humans, in the form of Cro-Magon, evolve from a population of Neanderthals? There is no question that the two groups occupied Europe simultaneously for a brief time between about 40,000 and 35,000 years ago. Did early Cro-Mangnon people retain Neanderthal culture briefly because they evolved from Neanderthals, or did they as similate the culture when they came into contact with Neanderthals following a separate origin? Were the two groups members of the same species, or did they represent different species? Did *Homo sapiens* exterminate Neanderthal through warfare in Europe between 40,000 and 35,000 years ago? And finally, does the advanced art and technology of Cro-Magnon, together with the phenomenal cultural expansion that followed Cro-Magnon's ascendency reflect the fact that modern humans have a substantially more powerful intellect than that of Neanderthals? Although we know that our brain case differs in shape from the long, low structure of the Neanderthal, we do not know whether our brains actually function in a different manner.

One intriguing fact is that, whatever his mental powers may have been, Neanderthal did have religion. Burial sites reveal that Neanderthals sometimes prepared their dead for a future life by interring them with flint tools and cooked meat. In the Zagros Mountains of Iraq, a Neanderthal man who died following a skull injury was buried in bed of boughs and flowers that can be indentified from the pollen they left beneath the skeleton!

Human expansion and the extinction of large mammals

Anthropologist continue to debate not only how the various races of *Homo sapiens* spread throughout the world but also whether all of these races evolved from Cro-Magnon or simply share with cro-Magnon features characterizing all older populations of *Homo sapiens*. One particularly noteworthy question regarding the geographic spread of *Homo sapiens* is, When did our species first reach the Americas? This issue is intertwined with another: the cause of an extinction of many large terrestrial mammals at the end of the last glacial advance about 10,000 years ago. As a result of this extinction event, the world today is impoverished with regard to large mammals; only the faunas of the African savannahs and open woodlands give us a hint of what the rich Pliocene terrestrial ecosystems must have been like, not only in Africa but also in Eurasia and North America. In fact, the modern African fauna, which is rich by today's standards, is impoverished in comparison with the diverse faunas that characteriszed the Pliocene and pleistocene epochs. Climatic deterioration led to a reduction of faunal diversity in Eurasia during the Pleistocene Epoch, but even the fossil record of that epoch has yielded a fauna of more than 50 genera of Bovidae (cattle, antelope, sheep, goats, and related hoofed animals).

Within the rich Plio-Pleistocene faunas were many species of immense proportions. Because increased body size reduced heat loss (big animals have smaller surface areas for their body volume than do small animals), the large size of some Pleistocene species has been considered to have been an adaptation to cold climates. Among the huge Pleistocene species were an elephant-sized bison of the American plains whose horns spread more than 2 meters (~7 feet), or three times the span of the living species; a North American beaver (*Castoroides*) that was nearly as large as a black bear; horses of the modern genus Equus, which evolved to the size of our artifically bred Clydesdale draft horses; and mamoths (members of the elephant genus Mammuthus) that stood about 4.5 meter (~15 feet) at the shoulder, about 30 percent taller than an average African elephant!

Near the end of the Pleistocene Epoch, about 11,000 years ago, at the time the glaciers were receding from Eurasia and North America, numerous species of large mammals disappeared throughout the world. Some have argued that these extinctions occurred at the hands of humans armed with newly developed weapons for which large conspicuous mammals were easy marks. Because they were represented by few individuals to begin with, such species were eventually hunted out of existence. This the ory suggests that a wave of humans swarmed into North America about 11,000 years ago, by way of the Bering land bridge, and were such successful big-game hunters that their populations expanded and spread rapidly—perhaps by about 16 kilometers (~10 miles) per year—in their efforts to find new game. Radiocarbon analyses of wood associated with stone weapons reveals that humans at this time employed sophisticated projectile systems in the New World; they were known to have launched flint tips of many types at the ends of lances and may also have propelled darts with throwing sticks. The famous Clovis and Folsom projectile points represent this stage of weaponry development.

Regional Events

Because many shifts in climate and in biogeograophic distribution have been reconstructed in detail for that Neogene period, geoglogists have been able to relate many regional geologic events to these phenomena. This relationship represetnts a recurring theme in the review of Neogene regional events that follows.

We will begin our regional tour of the Neogene Period by reviewing the history of the western United States, which is highlighted not only by the elevation of imposing mountains that form part of our scenery today—the Cascade Range, the Sierra Nevada, and the Rocky Mountains—but also by climatic changes that resulted from the uplifting of these mountains. Next, we will examine events in and around the Atlantic Ocean; among our topics here will be the origin of the modern mountainous topography of the Appalachians, the uplift of the Isthmus of Panama, the origin of the Caribbean sea, and the cooling of the Atlantic with the onset of the Ice Age. We will then review the Neogene history of the African continent and its famous rift valleys, which have contributed valuable fossil remains of human ancestors. Following this, we will learn how the northward movement of Africa against Eurasia closed the venerable Tethys Sea and, in the process, formed the Mediterranean. At one point, this remnant of the Tethys suddenly dried up, but it was soon refilled with sea water. Finally, we will observe how to northward movement of the Australian plate has brought biotas of the Southern Hemisphere into contact with animals and plants of Eurasian ancestry.

Development of the American West

The pre-Neogene history of mountain building in the Cordilleran region was described in earlier chapters. By late Paleozoic time, uplifts resulting from the final mountain-building episode of the western interior, the Laramide orogeny, had been largely subdued by erosion thereby setting the stage for the Neogene events that produced the Rocky Mountains. In the broad region west of the Rockies, the Neogene period was a time of widespread tectonic and igneous activity, which built most of the mountains standing there today.

Provinces of the American West

Lying between the great plains of the United States and the Pacific Ocean are several distinctive physiographic provinces that have taken shape largely in Neogene time, primarily as a result of uplift and igneous activity. Let us briefly review the present cheracteristics of these provinces before considering how they have come into being.

The lofty, rugged peaks of the *Rocky Mountains*, some of which stand more than 4.5 kilometers (~14,000 feet) above sea level, could only be of geologically recent origin. We have seen that the widespread subsummit surface of the Rockies was all the remained of the Laramide uplifts by the end of the Eocene Epoch about 40 million years ago. One question we must answer, then is how the Rocky Mountain region became mountainous once again.

Centered adjacent to the Rockies in the "four corners" area where Colorado, Utah, New Mexico, and Arizona meet is the oval-shaped *Colorado Plateau*, much of which stands about 1.5 kilometres (~ 1 mile) above sea level. The Phanerozoic sedimentary units here are not intensively deformed. Some, however are gently folded in a steplike pattern, and others, especially to the west, are offset by block faults. Cutting through the plateau is the spectacuar Grand Canyon of the Colorado River, but about 10 million years ago there was neither a Colorado Plateau nor a Grand Canyon. The origin of these features forms another part of our story.

West of the Rockies and the Colorado Plateau, within the belt of Mesozoic orogeny, lies the *Basin and Range Province*. This is an area of north-south trending block-fault valleys and intervening ridges—features of Neogene origin. A large area of this province forms the *Great Basin*, an arid region of interior drainage. Volcanism has been associated with some faulting episodes, and sediment eroded from the ranges blankets the valleys to depths ranging from a few hundred meters to about 3 kilometers (~ 2 miles). The thickness of the earth's crust in the Basin and Range Province ranges from about 20 to 30 kilometers compared to thicknesses of 35 to 50 kilometers in the Colorado Plateau. The thinning and block faulting in the Basin and Range Province point to extension of the crust by at least 65 percent and perhaps by as much as 100 percent.

Farther north, centered in Oregon, is a broad area covered by volcanic rocks of the *Columbia River* and *Snake River plateaus*. Today the climate here is cool semiarid; only about one-quarter of the plateau area is cloaked in forest and woodland, while sagebrush and drier conditions characterize about half of the terrane. In Onligocene time, however, lavas had not yet blanketed the region and, as revealed by fossil plant remains, a large forest of redwood trees grew there.

Along the western margin of the Columbia Plateau stand the lofty peaks of the *Cascade Range*. These are cone-shaped volcanoes that represent the volcanic are associated with subduction of the Pacific plate along the western margin of the continent. Volcanism began here in Oligocene time and continues to the present, as manifested by the recent eruptions of Mount St. Helens.

The Cascade volcanic belt passes southward into the *Sierra Nevada Range*, a mountain-sized fault-block of granitic rocks. The plutons forming the Sierra Nevada were emplaced in east-central California during Mesozoic time, before igneous activity at this latitude shifted inland. As we will see, however, the present topography of the Sierra Nevada is of Neogene origin. This mountain range is unusual in that throughout its length of some 600 kilometers (~350 miles), it is not breached by a single river. This is why it represented such a formidable obstacle to early pioneers attempting to reach the Pacific.

The Sierra Nevada Range stands between the Basin and Range Province to the east and the *Great Valley* of California to the west. The Great Valley is an elongate basin containing large volumes of Mesozoic sediment (the Great Valley Sequence) eroded from the plutons of the Sierra Nevada region long before the modern Sierra Nevada formed by block faulting. Above these are Cenozoic deposits, some of which accumulated during marine invasions of the Great Valley and others during times of nonmarine sedimentation.

West of the great Valley are the California *Coast Ranges*, which consist of slices of crust that include crystalline rocks representing Mesozoic orogenic activity, Franciscan rokcs of deep-water origin, and Tertiary rocks. To the south, the *Transverse* and *Peninsular ranges* are formed of similarly faulted and deformed rocks, but these ranges lie inland of the main belt of Franciscan rocks in the region of intensive Mesozoic igneous activity. Striking features of all of these mountainous terranes are the *great faults*—including the Garlock Fault—that divide the crust into sliver-shaped blocks. The longest and most famous of these faults is the San Andreas, which extends for about 1600 kilometers (1000 miles). Until the great San Francisco earthquake of 1906, it was not widely recognized that movement along the San Andreas persisted. The earthquake of 1906, was produced by a sudden movement of up to 5 meters (~16 feet) along the fault. Study of geologic features cut by the San Andreas Fault shows that its total movement during the past 15 million years has amounted to about 315 kilometers (190 miles). Continued movement at this rate for the next 30 million years or so would bring Los Angeles northward to the latitude of San Francisco—through which the fault passes. As we will see, the faulting and uplifiting of the Coastal Ranges of California are probably causally related not only to the Neogene uplift of the Sierra Nevada but also to the origins of the Basin and Range topography to the east.

The *Olympic Mountains* of Washington have quite a different history. These relatively low mountains, which lie to the west of the Cascade Volcanics, consist of oceanic sediments and volcanics, that were deformed primarily during Eocene time in association with subduction along the continental margin.

Development of the American West: The Miocene Epoch

We will begin our story of the development of the modern Cordilleran provinces described above with a summary of major events of the Miocene Epoch. Then we will examine the Pliocene events that followed, and we will conclude by analyzing the tectonic mechanisms that may account for events of both epochs.

Geologic features of the far west in Miocene time. Subduction continued beneath the continental margin in the northwestern United States, and the resulting volcanic arc produced peaks in the Cascade Range, where volcanism continues today. To the south, in California, the mid-Miocene interval was a time of faulting and mountain building; elements of the modern Coast Ranges, and other nearby mountains were raised and the seas were driven westward. Meanwhile, as in Paleozoic time, the Great Valley remained a large embayment, and during Miocene time it received great thicknesses of siliciclastic sediments, most of which were shed from the region of the modern Sierra Nevada. Although the block faulting that eventually produced the Sierra Nevada did not begin until Pliocene time, volcanoes of the southern end of the Cascade Island Arc were shedding sediments westward, from the position where the Sierra Nevada later stood.

Before the Sierra Nevada rose appreciably, the Basin and Range Province began forming to the east. During Paleogene time, light-coloured rhyolitic (felsic) ashfall deposits were occasionally spread across the province, but near the beginning of the Miocene Epoch basaltic volcanism predominated—and it was at this time or slightly earlier that the Basin and Range topography began to form. Since the beginning of block faulting, most lavas of the region have welled up along faults rather than reaching the surface through cylindrical vents. To the north of the Basin and Range Province, basalt also spread from fissures in much greater volume. Most of the great Columbia Plateau formed in this way between about 16 and 13 million years ago; here, individual basalt flows range in thickness from 30 to 150 meters (~100 to 500 feet), and in places, the total accumulation reaches about 5 kilometers (~3 miles).

One of the most important Miocene events in the Cordilleran region was a broad regional uplift that affected the Basin and Range Province, the Colorado Plateau, and the Rocky Mountains. Today, even the basins of the Basin and Range Province nearly all stand at least 1.3 kilometers (~ ¾ mile) above sea level. Thus, it is remarkable that mid-Cenozoic fossil floras of this region are characterized by species that could only have lived at low altitudes. Geologists have reconstructed even more precise histories of uplift for the Colorado Plateau and Rocky Mountains by studying the time at which rivers have cut through well-dated volcanic rocks. Many of the rivers of these regions existed before uplift began in the Miocene Epoch, and these cut rapidly downward as the land rose, producing deep gorges. A large part of the Grand Canyon, for example was incised during the rapid elevation of the Colorado Plateau between about 10 and 8 million years ago.

Uplift in the Rockies began slightly earlier, in Early Miocene time, and terrane that now forms the Southern Rockies has since risen between 1.5 and 3.0 kilometers (~1 to 2 miles).

Development of the American West: Plio Pleistocene time

In both the Colorado Plateau and the Rockies, the Miocene pulse of uplifting was followed by a lull and then by an episode of renewed elevation; in fact, large-scale uplifting was the dominant process in the Cordilleran region during Pliocene and Pleistocene time. We will discuss some of the details of this uplifting after noting several other key plio-Pleistocene event.

During the Pliocene and Pleistocene epochs, igneous activity continued in the volcanic provinces of Oregon, Washington, and Idaho (Figure 18-50). Many of the scenic volcanic peaks of the Cascades, including Mount St. Helens, have formed within the past 2 million years or so. Beginning in Late Miocene time and continuing sporadically to the present, the flowing of basalt from fissures has produced the Snake River Plain, which amounts to an eastward extension of the Columbia Plateau.

In California, faulting and deformation continued during the pleiocene and Pleistocene epochs. Since the beginning of the Pliocene Epoch about 5 million years ago, the sliver of coastal California that includes Los Angeles has moved northward on the order of 100 kilometers (~60 miles). The Great Valley has, of course, remained a lowland to the present day, but during Pliocene and Pleistocene time it became transformed from a marine basin into a terrestrial one. Early in the Pliocene Epoch, seas flooded the basin from both the north and the south. The sedimentary sequence of the basin reveals that several transgressions and regressions occurred during Pliocene time, but as the epoch progressed, uplift associated with movement along the San Andreas Fault eliminated the southern connection. Eventually, nonmarine deposition prevailed throughout the Great Valley, which is now one of the world's richest agricultural areas as well as the site of large reservoirs of petroleum.

Meanwhile areas to the east of the Great Valley underwent uplift. The Sierra Nevada had experienced considerable tilting during Miocene time, but fossil plants of Miocene age preserved on the crest of the Sierra Nevada were still types that could not have lived as much as a kilometer above sea level. Thus, it was not until Pliocene time that the Sierra Nevada became elevated to its present height, which exceeds 4.3 kilometers (~14,000 feet). The consequences of this uplift were enormous for the Basin and Range Province to the east, where uplift continued from Miocene time on a smaller scale. Sitting in the rain shadow of the Sierra Nevada, this area—which in Miocene time had been covered by evergreen forests—came to be carpeted by savannah vegetation. The trend toward increasing aridity, which was compounded by the global trend toward drier climates, continued into very late Neogene time, when the Great Basin became a desert.

The Colorado Plateau and Rocky Mountains, where uplift had slowed in Late Miocene time, experienced rapid elevation once again. Streams that had been established millions of years earlier cut rapidly downward as the uplift proceeded. The Rockies attained most of their modern elevation during Pliocene time, and the result was the origin of deep canyons that now carry streams through tall mountain ranges. One of the most scenic of these canyons is Royal Gorge near Canon City, Colorado. The Colorado Plateau also rose again during Pliocene and Pleistocene time, and the Colorado River responded by cutting swiftly downward. In fact, it appears that much of the Grand Canyon of the Colorado formed during just the past 2 or 3 million years!

The renewed elevation of the Rocky Mountains left the Great Plains to the east in a partial rain shadow. Sediments derived from the rejuvenated Rockies spread eastward, forming the Pliocene Ogallala Formation. Caliche nodules are abundant in many parts of the Ogallala, indicating the presence of seasonally arid climates. The Ogallala is a thin, largely sandy unit that lies buried under the Great Plains from Wyoming to Texas, and serves as a major source of ground water. Unfortunately, this is ancient water that is not being renewed as rapidly as it is drawn from the earth. As a result, severe water shortages may one day strike many areas of the central United States.

It is important to recognize that the rain-shadow effects of both Sierra Nevada and the Rockies were superimposed on the larger global trend toward drier, cooler climates that has characterized the post-Eocene world. Then, with the onset of the Pleistocene Epoch, frigid conditions brought glaciation to mountainous regions of the western United States just as they foster glaciation in Alaskan mountains today. The Sierra Nevada, for example, was heavily glaciated, as were portions of the Rocky Mountains. Today, broad U-shaped valleys in both mountain systems testify to the scouring activity of Pleistocene glaciers.

Possible mechanisms of uplift and igneous activity in the American West

What has led to the many tectonic and igneous events to Neogene time in the American West? It seems likely that the secondary uplift of the Colorado Plateau and the Rock Mountains, which took place long after the Laramide orogeny, may to a large extent represent simple isostatic adjustment. When uplifts in these areas were largely leveled during Eocene time, they left behind felsic roots of low density. With the weight of the mountains removed, these roots were apparently out of isostatic equilibrium, and hence they began to rise up toward positions of equilibrium during the Neogene period.

The elevation of the Basin and Range, with its block faulting and relatively thin crust, requires a different explanation. Basin and Range events, the spreading of the Columbia River Basalt, and the extensive faulting and folding along the California coast all began in Miocene time and seem to be related in some way to plate-tectonic movements, including those that simultaneoulsy occurred along the Pacific Coast. One idea is that *East Pacific Rise*, a large oceanic rift that passes into the Gulf of California, breaks up as it passes inland through thick continental crust and branches out to become the many normal faults of the Basin and Range Province that have caused thinning and elevation of the crust. The high heat flow that is associated with such a spreading zone is also alleged to have caused the Neogene elevation of the Basin and Range Province.

Objection have been raised against the rifting hypothesis for Basin and Range development based on the belief that spreading should have ceased when the East Pacific rise came into contact with the subduction zone at the western boundary of the North American plate; instead, it is argued, movements must have been propagated along one or more transform faults such as the San Andreas, passing along the continental margin. This argument leads to an alternative hypothesis of Great Basin development. The crux of this hypothesis is that crustal shearing adjacent to a strike-slip fault like the San Andreas will automatically cause extensional faulting similar to that of the Great Basin. This second hypothesis is deficient in one regard: it fails to account for the broad elevation of the Basin and Range Province during Neogene time.

Still another hypothesis is that heat from the subducted Pacific plate elevates the crust in the Basin and Range regions, placing it under tension that results in block faulting and in thinning of the crust. At present, we have no actual evidence that this mechanism has operated.

We are more certain, however, about the general Pattern of tectonism along the Pacific coast. North America became attached to the Pacific plate near the beginning of the Miocene Epoch. Movements along the San Andreas and other faults that have formed since that time account for the complex slivering and deformation in the Coast Ranges and neighbouring areas.

The Atlantic Ocean and its Environs

Although the margins of the Atlantic Ocean were relatively quiescent during the Neogene Period, they did experience mild vertical tectonic movements and these movements together with more profound changes in sea level, had major effects on water depths and shoreline positions. We will review events in these regions first for Miocene and then for plio-Pleistocene time.

Miocene event

Global sea level has never stood as high during the Neogene Period as it did during much of Cretaceous or Paleogene time. For this reason, along the Atlantic Ocean, Neogene marine sediments stand above sea level in only a few low-lying areas. Among the most impressive of the Miocene deposits found here are those of the Chesapeake Group which frorm cliffs along the chesapeake Bay in Maryland. The Chesapeake Group accumulated during a worldwide high stand of sea level between about 16 and 14 million years ago, and they were deposited in the Salisbury Embayment, one of several downwarps of the American continentl margin. Inhabiting the waters of the Salisbury Embayment was a rich fauna that included many large vertebrates, especially whales, dolphins, and sharks. Most of the fossils of baleen whales represent juvenile animals which suggests that the embayment may have been a calving ground. Perhaps sharks were numerous because the young whales were especially vulnerable prey. Landmammal bones are also found here and there in the Chesapeake Group, indicating that the waters of the embayment were shallow. Pollen from nearby land plants settled in the Salisbury Embayment, leaving a fossil record that shows a warm, temperate flora near the base of the Chesapeake Group slowly giving way upward in the sedimentary sequence to a slightly cooler but still temperate flora.

The Chesapeake Group and other buried deposits of earlier age to the south consist primarily of siliciclastic sediments shed from the Appalachians to the west. Erosional features associated with the Appalachians reveal that these ancient mountains have a complex history. Like the modern Rockies, the existing topographic mountains that we call the Appalachians are the product of secondary uplift. The Appalachian orogenic belt was largely leveled by erosion by the end of the Mesozoic Era, as a result of which the *Schooley Peneplain*, or "almost plain" formed the surface of the land on top of the deformed Appalachian rocks. Three intervals of uplift and erosion followed. None of the intervals of erosion leveled the region, but each left erosional surfaces that can be identified in the present complex topography. The pulses of secondary uplift in the Appalachians presumably reflect isostatic response to the persistence of low-density roots of the original mountain system. Unfortunately, geologist have found it impossible to correlate well-dated deposits of the Atlantic Coastal Plain of North America with pulses of uplift and erosion in the Appalachians. These pulses remain poorly dated.

On the other side of the Atlantic, Miocene seas also encroached only modestly beyond the present shorelines. The North Sea did extend indland beyond the present coastline, and the Rhine River mouth lay in the area now occupied by the city of Cologne, where a large swamp formed in the area where the Rhine met the sea. During Early and Middle Miocene time, logs of swampdwelling trees accumulated to produce lignite, an immature brown coal that is now of considerable economic value.

In late Miocene time sea level declined to some extent, and fewer marine deposits formed on either side of the Atlantic inland of the present coastlines.

Plio-Pleistocene events

After the Messinian Event at the end of the Miocene Epoch, the Antarctic ice cap melted back, and sea level rose throughout the world. This trend, which reached a peak about 4 million years ago, resulted in marine deposition in areas that are now exposed above sea leval-on both sides of the North Sea, for example, as well as on the other side of the Atlantic from Virginia to Florida and the Caribbean. The Caribbean region deserves special mention here because it underwent major changes late in the Neogene Period. Although the Caribbean Sea is now an embayment of the

Atlantic Ocean, it was once connected to the pacific. During the Cretaceous Period, the floor of the Caribbean, which consists of oceanic rocks, was a small segment of the Pacific plate that was pushing toward the Atlantic, but during the Cenozoic Era the Caribbean sea floor has lain along the north coast of South America while the Atlantic plate has been subducted beneath it. The Caribbean plate became a discrete entity late in Cenozoic time, when a new subduction zone came to connect the subduction zone bordering North America with the one bordering South America. The Greater Antilles—Cuba, Puerto Rico, Jamaica, and Hispaniola—represent an ancient mountain belt that is actually the southern end of the North American Cordillera. The Lesser Antilles represent an island arc west of the subduction zone, together with islands formed by deformation associated with subduction. The Yucatan Peninsula is a broad carbonate platform that lies to the west of the Caribbean, and, as we have seen, the Bahamas an ancient carbonate platform lying to the north.

During the Pliocene Epoch the waters of the Caribbean became separated from those of the Pacific when tectonic activity formed the norrow *Isthmus of Panama*. Deep-sea cores reveal that different panktonic foraminifer species lived on opposite sides of the isthmus about 3.5 million years ago, suggesting that the isthmus had formed by this time.

The most profound effect of the newly formed isthmus lay in its role as a land bridge that allowed mammals to migrate between North and South America. Previously, in Neogene time, the terrestrial faunas of North and South America had remained largely separate from each other—although a few species had passed from one continent to the other early in the Neogene Period, perhaps by swimming or floating on logs. South America had been a great island continent and like Australia, was populated by many marsupial mammals. The marsupials of Australia and South America share common ancestors that populated these continents as well as Antarctica when all three were part of a single landmass late in the Mesozoic Era; in fact Eocene mammal faunas resembling those of South America occur in Antarctica. By Pliocene time, however, when the isthmus developed, South American marsupials differed greatly from Australian marsupials, and the South American fauna included several groups of placental mammals whose ancestors had reached the continent from the north at the beginning of Cenozoic time or even earlier. Among the South American marsupials present when the land bridge formed were members of the opossum family, and among the placentals were sloths and armadillos that dwarf their relatives in the modern world.

More North American species invaded South America than vice versa. Among those that reached south America were members of the camel, pig, deer, horse, elephant, tapir, rhino, rat, skunk, squirrel, rabbit, bear, dog, raccon, and cat families. Migrating in the oppostie direction were monkeys, anteaters, armadillos, porcupines, opossums, and other, less familiar animals.

The formation of the isthmus of Panama was associated with a Plio-Pleistocene pulse of orogeny that elevated the Andean mountain ranges to the south by 2 to 4 kilometers (~1.5 to 2.5 miles). Large areas of South America thus came to lie in the rain shadow of the Andes. Dry conditions intensified during Pleistocene glacial episodes, at which time parts of the great Amazonian rain forest were fragmented into small pockets separated by grassland barriers or by poorly wooded areas. Even today, certain groups of rain-forest–dwelling birds and lizards remain clustered in the areas where they were concentrated during the Pleistocene Epoch.

The formation of the isthmus of Panama also had profound oceanographic effects. Recall the suggestion that it may even indirectly have triggered the Ice Age in the Northern Hemisphere by deflecting warm waters into the Gulf Stream, which then increased humidity and snowfall at high latitudes. The Caribbean Sea was also affected. Today the Caribbean is isolated from other tropical areas and consequently harbors a unique fauna of reef-building corals. Early in the Cretaceous Period, however, the coral fauna here closely resembled that of the Tethyan Seaway in what is now the Mediterranean region, since the spreading of the Atlantic had not yet separated the two regions by a vast expanse of ocean. Then in Late Cretaceous time, the width of the Atlantic became so great that very few coral larvae were able to cross it, and the fauna of the Caribbean began to diverge from that of the Tethys Sea. Finally, the uplift of the Isthmus of Panama separated the Caribbean coral fauna from that of the Pacific. Since Miocene time, both the restricted area of the isolated Caribbean and a progressive climatic deterioration have caused the coral fauna here to dwindle. Especially during glacial episodes of the Pleistocene Epoch, the Caribbean has been near the lower temperature limit for coral-reef

Fig. 9.15. Animals that took part in the great faunal interchange between North and South America when the Isthmus of Panama was elevated, connecting the continents.

growth. What remains today is a small coral fauna consisting of about 60 species, whereas nearly 10 times as many species inhabit the vast biogeographic province formed by the western Pacific and Indian oceans.

In fact, marine life in and around the Atlantic Ocean suffered greatly during the expansion of glaciers and nearby land areas slightly more than 3 million years ago. During the Pliocene high stand of sea level just before the glacial interval began, enormous molluscan faunas occupied the borders of the Atlantic under relatively warm conditions. Seas that invaded Virginia supported a subtropical marine fauna, and even northern Iceland was bordered by a temperate fauna. In mid-Pliocene time, the global high stand of sea level, together with temperate climates, permitted a wave of Pacific mollusks to spread over the polar region into the Atlantic. One of the immigrants was *Mya arenaria*, the "steamer clam" that is widely consumed in eastern North America.

Pliocene strata in Iceland have been dated by the application of radiometric and paleomagnatic methods to interbedded lavas associated with the Mid-Atlantic Rift, which passes through Iceland. Consequently, geologists can estimate with reasonable accuracy when the richly fossiliferous Pliocene beds in Iceland suddenly gave way to tillites—a transition marking the expansion of ice sheets across the island.

The refrigeration that accompanied continental glaciation reached far southward. We have seen, for example, that during glacial episodes of the Pleistocene Epoch, the lowering of sea level and the accumulation of pack ice reduced the size of the clockwise gyre of ocean currents north of the equator and, at the same time, pushed the Gulf stream diagonally across the Atlantic toward Spain. Recall, too, that warm equatorial waters, which the hump of South America had deflected into the Caribbean, now passed southward under the influence of stronger trade winds. Studies of planktonic foraminifers in deep-sea cores suggest that February sea-surface temperatures in the central Caribbean during the most recent glacial advance (the Wisconsin or Riss-Wurm Stage) were about 4°C (9°F) cooler than they are today. Long before this time, during the early phases of glacial cooling, thousands of species of marine mollusks disappeared from the Atlantic Ocean and neighbouring seas in the Northern Hemisphere. Along the Atlantic coast of North America, for example, about 70 percent of these species disappeared, and the Caribbean fauna was similarly decimated. Faunas of the North Sea and the Mediterranean experienced less severe but still noteworthy pulses of extinction. Tropical species suffered especially severe losses.

Africa, Southern Europe, Asia, and Australia

We have already noted that the landmass of Africa has remained unusually stable for hundreds of millions of years, in part because most of its area lay well within Gondwanaland for a long interval of time. Currently, however, the strength of the African craton is being tested by continental rifting, and it is not certain whether this great continent will survive or be torn asunder. We will now examine this situation and also the geographic consequences of the movements of two lithospheric plates: the northward movement of the African plate, which is associated with the uplift of the Alps and other mountain chains, and the north ward movement of the Australian plate, which produced the Himalayas.

Africa: Rifting and climatic change

Near the beginning of Miocene time, two large domes, the Ethiopian Dome and the Kenyan Dome, began to rise up in eastern Africa—a process that was followed by volcanism and crustal rifting. The Ethiopian Dome in the north was the site of the large three-pronged rift that split the Arabian Peninsula from the rest of Africa, allowing the Red Sea and Gulf of Aden to open in Early Miocene time. The southern prong was crossed by the western rift, which eventually propagated all the way to the southwestern margin of Africa. Since early in the Miocene Epoch, the ever-deepening valleys of this rift system have received large quantities of sediment and have also harbored some large lakes.

Neogene tectonism in Africa has had important biotic consequences that were enhanced by trends toward more arid climatic conditions—as were similar changes in western North America and in South America. With respect to the moist winds from the South Atlantic, much of eastern Africa now lies in the rain shadow of highland of the Kenyan Dome east of the western rift. East of the eastern rift tall volcanic mountains like scenic Mount Kilimanjaro trap moisture from trade winds moving inland from the Indian Ocean.

The emergence of wind barriers, together with the global trend toward cooler, more arid conditions, caused grassy terrane to replace most forests in eastern Africa—a change analogous to that in western North America. Fossil mammal faunas reveal that early in the Miocene Epoch, eastern Africa was cloaked in equatorial rain forest. By mid-Miocene time, however there was some open terrane in this area, as evidenced by the presence of ostriches and of herbivorous mammals that possessed continuously growing molar teeth adapted to grinding up hard grasses. By Late Miocene time, the savannah fauna resembled that of the modern Serengeti Plains in that there was a preponderance of large ground-dwelling mammals, including many species of gazelles and antelopes.

Another geologic development of great biotic consequence was the collision of the Arabian section of Africa with Eurasia, between 20 to 18 million years ago. This contact allowed for biotic interchange between Africa and Europe for example, elephants for the first time escaped from their African birthplace and, in the course of the Miocene Epoch, spread throughout Eurasia and across the Bering land bridge to the Americas. As we have seen, the attachment of Africa Eurasia also afforded apes their first opportunity to emigrate northward from Africa.

Rifting and volcanism have continued to the present day in Africa without fully separating eastern Africa from the remainder of the continent. Like rift valleys of other regions, those of Africa have subsided rapidly and have received large volumes of sediment interlayered with volcanic rocks. It is because of the almost continuous sedimentation here that the stream-cut gorges subsequently cut down through the layers of deposits in African rift basins have proven to be fruitful collection sites for hominid remains. Olduvi Gorge, the most famous of these sites, is incised into the Serengeti Plain at the margin of the eastern rift. From Olduvai have come not only fossils representing the human family but also their very early stone implements. The relatively complete depositional sequence at Olduvai for the past 2 million years has made the stratigraphic section here a standard against which other Africa sequences are often compared. The presence of volcanic rocks that can be dated by radiometric and fission-track techniques and that can be correlated paleomagnetically further enhances the value of the Olduvai section as a standard for worldwide comparison.

Closure of the Tethyan Seaway

The collision of the African plate and southern India with Eurasia during the Cenozoic Era destroyed what remained of the Tethyan Seaway. Today only vestiges of the seaway remain in the form of the isolated Mediterranean, Black, Caspian, and Aral seas.

We have seen that between 18 and 14 million years ago, Africa and Eurasia collided, creating a land corridor across which terrestrial animals began to migrate. This event marked the end of the eastward connection between the western Tethys and the waters of the Indo-Pacific region. By Early Miocene time, the eastern Tethys was divided into a northern and a southern arm; thus, during the collision in which the Dinaride and Hellenide ranges of Yugoslavia and Greece as well as the Taurus Mountains of southern Turkey rose up as barriers, the two arms became separate seas. Consequently, the Mediterranean Sea was born and also, to the north of it, an isolated body of water known as the *Paratethys*. The Paratethys received freshwater from rivers in Eurasia and survived as a brackish inland sea from the middle to the end of Miocene time.

At the end of the Miocene Epoch, both the Mediterranean and the Paratethys underwent spectacular changes. The first strong hint that geologists had of these changes was the discovery in 1961 of pillar-shaped structures in seismic profiles of the Mediterranean sea floor. These structures looked very much like the salt domes of Jurassic age in the Gulf of Mexico, but if the strange features were indeed salt domes, the salt could only have formed by evaporation of the Mediterranean. In 1970, the presence of evaporites here was confirmed by drilling that brought up anhydrite in cores representing the latest Miocene Epoch. Also present here was gravel that represented shallow marine or nonmarine conditions. The idea that the Mediterranean had somehow turned into a shallow evaporite basin was confirmed by the discovery of halite (rock salt) near the center of the eastern Mediterranean. Because it is highly soluble and does not precipitate until after the other salts in a solution have precipitated, halite is usually found in the center of an evaporite basin.

Further evidence that the Mediterranean dried up at the end of Miocene time was the discovery of deep valleys filled with Pliocene sediments lying beneath the present beds of rivers like the Rhone in France, the Po in Italy, and the Nile in Egypt. Rivers like the Rhone and the Nile were already flowing into the Mediterranean earlier in the miocene Epoch, and when the waters of the sea fell, the rivers cut deep canyons to the basin floor. In attempting to find solid footing for the Aswan Dam, Soviet geologists discovered the canyon buried beneath the present Nile delta and judged it to rival the modern Grand Canyon of Arizona in size!

Clearly, what happened at the end of the Miocene Epoch was that the single narrow connection between the Mediterranean Sea and the Atlantic Ocean closed—probably as a result of a lowering of sea level in the Atlantic. Rates of evaporation similar to those of the Mediterranean region today would dry up an isolated sea as deep as the Mediterranean in a mere thousand years. Not only evaporites but also fossils of freshwater (or brackish-water) ostracods occur in the latest Miocene record of the Mediterrranean sea floor, suggesting that some of the rivers that flowed into the evacuated Mediterranean basin not only cut rapidly downward to reach the basin floor but also cut northward, capturing freshwater from the Paratethys and carrying it to the Mediterranean basin to form scattered lakes. Soon the Paratethys disintegrated into several smaller bodies of water, including the ancestral Black, Caspian, and Aral seas.

All of this happened between about 6 million years ago, when the eastern passage to the Atlantic closed, and 5 million years ago, when the Mediterranean basin refilled with deep water. Five-million-year-old deep-water micro fossils in sediments on top of evaporites attest to the refilling. The brief nonmarine interlude corresponded to most of the Messinian Stage (the final stage of the Miocene Series in Italy, where the disappearance of the Mediterranean obviously had special significance), and it is for this reason that the temporary demise of the sea is often referred to as the Messinian Event.

Apparently the Mediterranean-Atlantic connection was re-established when the natural dam at Gibraltar was suddenly breached. It has therefore been suggested that the first Atlantic waters must have been carried into the deep basin by a waterfall that would have dwarfed Niagara Falls.

As for the remnants of the Paratethys, many simply dried up. Fossils show that the Black Sea, which survived, remained isolated from marine waters as a freshwater lake until very recently, when erosion finally connected it with the Mediterranean by way of the narrow Bosporus. Thus the Black Sea became a brackish basin once again.

Farther east, the Tethyan Seaway was interrupted during Miocene time by the attachment of the Indian Peninsula to the Eurasian Plate, where molasse that shed southward from the newly forming Himalayas produced a lengthy and relatively complete fossil record for mammals. The famous Siwalik beds of Pakistan and India, for example, provide a nearly continuous record for the interval from 11 to 1 million years ago, documenting the composition of the rich faunas that occupied the spreading savannahs of Late Miocene and Pliocene age. The great rivers of eastern Asia, which flow the Himalayas to the sea, alsc formed in Miocene time during the uplift of the Himalayan regicn, Because of the high relief and abundant rainfall of the region, the Indus and Ganges of India and the several large rivers of Indochina contribute huge volumes of sediments to sediment to the ocean each year.

Australia's northward journey

After Australia broke away from Antarctica, Antarctica remained positioned at the South Pole, but Australia moved northward along with new Zealand and New Guinea, both of which are part of the same lithospheric plate—the plate that moved against Asia to form the Himalayas. Paleomagnetic data enable us to track the progress of the Australian plate on its journey.

One of the consequences Australia's move to lower latitudes is that its climate became warmer and drier. Today, much of Australia lies in the latitudinal belt extending from 20 to 30° south latitude. Deserts tend to form within this belt because it is where equatorial air descends after rising, cooling, and losing much of its moisture. Therefore, deserts now occupy a large area in the central and western parts of Australia, while easterly trade winds bring rain to the east coast of the continent. Pollen analyses show that quite different climatic conditions prevailed during the Miocene Epoch, when Australia lay largely withi the belt of westerly winds that were laden with moisture. Rain forests clocked the estern and south eastern parts of the continent, and central Australia bordered rivers and streams, although some open grassland was also present.

The migration of Australia toward the equator has also permitted the Great Barrier Reef to flourish. It was only late in Oligocene time that the northeastern shore of Queensland, which is now bordered by the reef, came to lie fully within the latitudinal zone that accommodates luxuriant reef growth.

One of the most interesting effects of the northward drift of Australia was that it brought into contact two entirely distinct terrestrial biotas: the biota of Australia and that of Asia The zone of contact between the two biotas, located in the Malay Archipelago, was first recognized by Alfred Russel Wallace, an English naturalist who, working independently from Charles Darwin, also conceived of the idea of evolution by natural selection. In fact, it was in 1859—the very year that Darwin published *On the Origin of Species*—that Wallace, who was traveling in the Malay Archipelago, published a paper on the biogeographic line of demarcation, which then became known as *Wallace's line*.

Wallace originally placed his line north of the island of Sulawesi, but later in life he decided that the Sulawesi fauna was dominated by Asian elements, and, accordingly, he shifted the line to position south of the island. This decision is accepted by most modern biogeographers, since only one marsupial mammal from the Australian region has reached Sulawesi. Still, of the diverse Asian fauna of mammals to the northwest, only a few species of shrews, two monkeys, a deer, a pig, and a porcupine have reached Sulawesi. Migration in both directions has been difficult because of the presence of water barriers. Plants migrate more easily than mammals, however, and Australian and Asian floras intermingle over a broader Zone.

Until the advent of plate-tectonic theory, there was no explanation for the narrow zone of overlap between Asian and Australian biotas in the vicinity of Sulawesi. If landmasses were immobile, why had there not been greater intermingling in the long course of Cenozoic time? Geologic evidence now shows that the Australian plate, which was bounded on the north by New Guinea, collided with the Asian plate to which Sulawesi belongs in Middle Miocene time, only about 15 million years ago. Furthermore, it was only in Late Miocene and Pliocene time that the islands between New Guinea and Sulawesi emerged to form stepping stones for the limited interchange of species.

INDEX